D1251547

KNOWLEDGE AND ERROR

VIENNA CIRCLE COLLECTION

VOLUME 3

EDITOR: BRIAN MCGUINNESS

ERNST MACH

ERNST MACH

KNOWLEDGE AND ERROR

Sketches on the Psychology of Enquiry

With an Introduction by

ERWIN N. HIEBERT

D. REIDEL PUBLISHING COMPANY

DORDRECHT-HOLLAND / BOSTON-U.S.A.

Library of Congress Cataloging in Publication Data
Mach, Ernst, 1838–1916.
Knowledge and error.
(Vienna circle collection ; v. 3)
Translation of the 5th ed. of Erkenntnis und Irrtum.
Bibliography: p.
Includes index.
1. Knowledge, Theory of. 2. Science—Philosophy. 3. Science—Methodology. 4. Thought and thinking.
I. Title. II. Series.
BD163.M173 121 73–75641
ISBN 90–277–0281–0
ISBN 90–277–0282–9 pbk.

ERKENNTNIS UND IRRTUM

First published by Johann Ambrosius Barth, Leipzig, 1905

This translation from the 5th edition, 1926

Translation from the German by Thomas J. McCormack (Chapters xxi and xxii) and Paul Foulkes (all other material)

Published by D. Reidel Publishing Company,
P.O. Box 17, Dordrecht, Holland

Sold and distributed in the U.S.A., Canada, and Mexico
by D. Reidel Publishing Company, Inc.
Lincoln Building, 160 Old Derby Street, Hingham,
Mass. 02043, U.S.A.

Printed in The Netherlands by D. Reidel, Dordrecht

To the memory of

DAVID HUME, RICHARD AVENARIUS

and

WILHELM SCHUPPE

CONTENTS

ERNST MACH'S *KNOWLEDGE AND ERROR*

Introduction by Erwin N. Hiebert

The clearest and most comprehensive statement of Mach's mature scientific epistemology is given in the compilation of twenty-five essays that he published in 1905 under the title: *Erkenntnis und Irrtum. Skizzen zur Psychologie der Forschung*. These essays were drawn, as Mach tells us, from his winter semester lecture course delivered in 1895/6 in Vienna and entitled *Psychologie und Logik der Forschung*. Translated, at least in part, into French (1908), Russian (1909) and Turkish (1935), it seems puzzling, to say the least, to see that the crowning work of Mach's philosophical deliberations – *Knowledge and Error* – should have waited seventy years to appear in English.[1] This delay is all the more surprising in view of the attention given to Mach's works by English scholars since at least the 1890's. The incentive for Mach's wide and sustained readership in the English-speaking world is, of course, closely related to the efforts of Paul Carus, editor of the Open Court Publ. Co. of LaSalle, Illinois. Mach's *Die Principien der physikalischen Optik* was dedicated to Carus posthumously in 1921. By 1926 all of Mach's important monographs had been brought out in English by Open Court with the exception of *Wärmelehre* and *Erkenntnis und Irrtum*. The first American edition of Mach's *Mechanics,* for example, was published in 1890, and the sixth, with a new introduction, in 1960. With the publication of the present volume, and the *Wärmelehre* (now in process of being translated), all of Mach's major treatises, finally, will be available in English.

It may not be inappropriate to mention that this would have pleased Mach enormously. He was an outspoken Anglophile and attached himself to the empirical tradition of Berkeley, Hume and Mill. He championed the approach to physics of Faraday and Maxwell. Among the staunchest of allies for his epistemological views, Mach included economist-logician W. S. Jevons, the mathematician-physicist-philosophers W. K. Clifford, P. E. B. Jourdain, and Karl Pearson. To Pearson he dedicated the third edition of *Die Analyse der Empfindungen* (1902). On the Continent Mach's favorite philosopher-scientists were Kirchhoff, Helmholtz, Hertz, Duhem and Poincaré.

Erkenntnis und Irrtum.

Skizzen
zur Psychologie der Forschung.

Von

E. MACH

Emer. Professor an der Universität Wien.

LEIPZIG
Verlag von Johann Ambrosius Barth
1905.

On a number of occasions Mach expressed the sentiment, especially in his correspondence, that America was the land of intellectual freedom and opportunity, the coming frontier for a new radical empiricism that would help to wash metaphysics out of philosophy. In 1901 he sponsored the German edition of *Concepts and Theories of Modern Physics* (1881) by J. B. Stallo, Cincinnati lawyer and philosopher.[2] Mach warmly endorsed Stallo's book because his scientific aims so closely approximated his own, and because Stallo rejected the latent metaphysical elements and concealed ontological assumptions of the mechanical-atomistic interpretation of the world. The second edition of *Wärmelehre* was dedicated to Stallo in 1900.

The fourth edition of *Populär-wissenschaftliche Vorlesungen* (1910), containing seven new essays, was dedicated to Harvard Professor of physiology, philosophy, and psychology, William James. Mach had a strong intellectual affinity for James' pragmatism because, like himself, he recognized that James had come to radically empirical views from science. Both men took pure pre-conceptualized experience, from which the mental and physical predicates of experience are composed, to be neutral rather than real, unreal, objective or subjective.

Mach was so taken with James that he thanked him for giving him his first understanding of Hegel. James was so impressed with his visit with Mach in Prague in 1882 that he wrote his wife:

As for Prague, *veni, vidi, vici.* I went there with much trepidation to do my social-scientific duty.... I heard Hering give a very poor physiology lecture, and Mach a beautiful physical one.... Mach came to my hotel and I spent four hours walking and supping with him at his club, an unforgettable conversation. I don't think anyone ever gave me so strong an impression of pure intellectual genius. He apparently has read everything and thought about everything, and has an absolute simplicity of manner and winningness of smile...[3]

In a letter to Carl Stumpf at the same time, James wrote:

[In Berlin] Helmholtz... gave me the very worst lecture I ever heard in my life except one (that one was by our most distinguished American mathematician). The lecture I heard in Prague from Mach was on the same elementary subject as Helmholtz's, and one of the most artistic lectures I ever heard.... *Überhaupt* I must say that the hospitality of Prague towards wandering philosophers much surpasses that of Berlin and Leipzig.[4]

Stumpf thought Mach's "positivistic theory of knowledge... impossible and unfruitful".[5]

Bibliothèque de Philosophie scientifique

ERNST MACH

PROFESSEUR A L'UNIVERSITÉ DE VIENNE

LA

Connaissance et l'Erreur

TRADUIT SUR LA DERNIÈRE ÉDITION ALLEMANDE

PAR

le Dr MARCEL DUFOUR

PROFESSEUR AGRÉGÉ
A LA FACULTÉ DE MÉDECINE DE NANCY

PARIS

ERNEST FLAMMARION, ÉDITEUR

26, RUE RACINE, 26

1908

Droits de traduction et de reproduction réservés pour tous les pays,
y compris la Suède et la Norvège.

Gand, juin 1911
George Sarton.

The circumstances that led to Mach's *Knowledge and Error* are significant for an understanding of the dual historical and philosophical focus and implications of the work. Its completion, as a synthesis of epistemological and methodological inquiries, was coupled with Mach's decision to begin a new chapter in his professional career when the opportunity presented itself. In 1895 Mach was invited to accept the third chair in philosophy at the University of Vienna. He was then fifty-seven years of age and had been for twenty-eight years teaching experimental physics at the Charles University in Prague. Mach replied that he would welcome the philosophy professorship on condition that he would be able to deliver lectures on psychology.[6] The formal academic title that Mach settled for within the philosophy faculty was: Professor of the History and Theory of the Inductive Sciences (*Geschichte und Theorie der induktiven Wissenschaften*).

The new post in Vienna, in fact, provided Mach with just the optimum environment and intellectual stimulant for clarifying his philosophical position. The chief task at hand, he had always felt, was not so much to develop a new philosophy as to discard an old one. Whether or not philosophers *per se* would pay attention to his attempts he did not much care. His investigations were directed toward practising scientists. They were focused upon the critical analysis and interrelationship of ideas internal to the physical sciences, physiology and psychology. Although the term 'philosophy of science' was then not current, that is what Mach's intellectual peregrinations were all about. He was, and wanted to be, the scientist's philosopher and not the philosopher's philosopher. In 1886 he wrote in his *Analysis of Sensations:*

> I make no pretension to the title of philosopher. I only seek to adopt in physics a point of view that need not be changed immediately on glancing over into the domain of another science; for, ultimately, all must form one whole.[7]

In *Knowledge and Error* Mach makes it very clear that whereas great philosophers like Plato, Aristotle, Descartes and Leibniz opened up new avenues of scientific investigation, there were others, not labelled philosophers, like Galileo, Newton and Darwin, who powerfully advanced philosophical thinking. What the philosopher tries to begin with, he suggested, appears to the scientist as the distant goal of his work. By the turn of the century, Mach, the physicist, had acquired the reputation of

Э. Махъ.

ПОЗНАНІЕ
и
ЗАБЛУЖДЕНІЕ.

ОЧЕРКИ ПО ПСИХОЛОГІИ ИЗСЛѢДОВАНІЯ.

Разрѣшенный авторомъ переводъ со второго, вновь просмотрѣннаго нѣмецкаго изданія

Г. Котляра

подъ редакціей профессора

Н. Ланге.

ИЗДАНІЕ
С. СКИРМУНТА.

being so outspoken and caustic a scientific revisionist that it had become impossible to ignore him as a philosopher. Still, we note from his remarks in the Preface to *Knowledge and Error* that he was adamant in his rejection of any claim to being a philosopher; that he denied the existence of a Machist philosophy. These assertions are not to be taken at face value. When Mach refers to himself as a mere weekend sportsman *(Sonntagsjäger)* we may accept this self-appraisal as no more than a characterization of his attitude and scholarly activity for the first three decades of his academic life. There is ample evidence in *Knowledge and Error* to show that Mach realized that his own intellectual evolution had moved far beyond mere auxiliary reflections about certain aspects of science and its history. Mach had, in fact, adopted a methodology, if not a metaphysics of his own, although he would have denied that.

In his third chapter on *Philosophy of Mind, circa* 1898, C. S. Peirce (who lost no love for metaphysics and spoke mostly of bad metaphysics) wrote:

> Mach belongs to that school of *soi disant* experiential philosophers whose aim it is to emancipate themselves from all metaphysics and go straight to the facts. This attempt would be highly laudable, – were it possible to carry it out. But experience shows that the experientialists are just as metaphysical as any other philosophers, with this difference, however, that their pre-conceived ideas not being recognized by them as such, are much more insidious and much more apt to fly in the face of all the facts of observation.[8]

Metaphysics for Mach? Never! Transcendentalism? Impossible! Philosophy? Perhaps! Toward the end of his life Mach became more tolerant about philosophy and even expressed appreciation for the fact that

> a whole host of philosophers – positivists, critical empiricists, adherents of the philosophy of immanence, and certain isolated scientists as well – have all, without any knowledge of one another's work, entered upon paths which, in spite of all their individual differences, converge almost towards one point.[9]

That point of convergence, Mach implied, was not far removed from his own anti-metaphysical and critical positivism and empiricism. He had arrived, so to speak, on the threshold of philosophy by simultaneously pursuing physics, physiology, and psychology, while critically searching out and historically clarifying the underlying foundations and meaning of scientific puzzlements that commanded his attention as a young physicist.[10]

There are other indications that Mach tempered his views on philosophy in his later years. For example, in his preface to Stallo's work, Mach remarked that Stallo proceeded, as a philosopher would, to analyse scientific conceptions by advancing from very general philosophical considerations to an examination of the discovered laws of physics. By contrast, Mach characterized his own approach as that of a physicist who dissects specific concepts historically in order to gain general philosophical perspectives. The implication, clearly, was that Mach did not dismiss Stallo's approach as an unfruitful one. He wrote: "During the mid-sixties when I began critical work, it would have been very encouraging and beneficial to have known about the related endeavors of a comrade like Stallo".[11]

There can be no doubt but that Mach wanted to be recognized as a physicist. He was trained as a physicist and spent all but the last few years of his life teaching physics. He was, however, no orthodox pedagogue. His life-style as a teacher and scholar, can be characterized by what Einstein called an incorruptible scepticism and independence.[12] The deep-seated epistemological puzzles that Mach released from the hidden hornets' nest of physics were by-products of an impenitent curiosity and critical posture towards the historical tradition that physics had inherited. To put it squarely, Mach was initially drawn toward the history of science because of what it might teach him about the internal structure of physics.

As a *Privatdozent* in Vienna in 1863, Mach had been convinced that his scientific inquisitiveness and his effectiveness as a teacher might be sharpened by the study of the history of science. He argued that students should not be expected to adopt, as self-evident, propositions that had cost several thousand years of thought. In 1872, in his first major historical work – *Die Geschichte und die Wurzel des Satzes von der Erhaltung der Arbeit* – he conjectured that there is "only one way to (scientific) enlightenment: Historical studies!"[13] The investigation of nature, he believed, should be backed up with a special classical education that "consists in the knowledge of the historical development of... science".[14] It was not the logical analysis of science but the history of science that would encourage the scientist to tackle problems without engendering an aversion to them. There were two paths that the scientist might follow in order to become reconciled with reality: "Either one grows accustomed to the puzzles and they trouble one no more, or one learns to understand them with the help of history and to consider them calmly from that point of view".[15] What Mach discovered from his study of history was that, with the passage of time, the historically acquired had come to be philosophically affirmed. Accordingly, he stressed the need for scientists to acknowledge and combat the subtle ways in which scientific conceptions take on a status of philosophical necessity rather than one of historical contingency.

Mach's various escapades into the history of science as a tool to illuminate physics led him straightaway into the exploration of problems connected with the psychology and philosophy of sensation and perception, i.e. directly into the yawning jaws of the psychology and philosophy of science. He had recognized that the gulf that separates physical from physiological and psychological investigation was to be found not in a difference of subject matter but in different modes of investigation in the three domains. For example, to refer to that which is physical in an investigation, he felt, was but to designate one method of cognitive organization of which there are many. It was not the facts but the points of view that would distinguish the disciplines. He perceived that the process of cognition was anchored to the adaptation of thoughts to facts and thoughts to othert houghts within a psycho-physiological framework.

Mach concluded from his historical studies that the conceptual creations of science, always tentative and at best incomplete, take on a configuration at any time that reveals the attendant historical circumstances and the convergence of interest and attention of those scientific investigators at work on the problems – now physicists, now physiologists, now psychologists. Implied in this analysis was the notion that the form given to a scientific construct depends in large part on the whims of history and on the environmentally conditioned process of cognitive organization employed by scientists.

The hundred or so scientific papers that Mach published from the early 1860's on, while in Graz and Prague, cover just as great a span of topics in conventional physics as in the unconventional application of physics to problems in physiology and psychology. As is evident from *Knowledge and Error,* the next logical step for Mach was to unpack and examine such related issues as memory, association, reflex, instinct, intuition, perception, the will, the ego, phantasy, hypothesis and concept formulation, analogical reasoning, thought experimentation, the psychic aspects of induction and deduction, and the growth of individual and societal outlooks on the sciences. Physics, physiology, psychology? No! This was philosophy. The more Mach struggled to integrate such complex and interrelated matters, and the more he wrote to clarify his position, the more his views crystallized and took on dogmatic form. In doing so he had moved towards working out arguments that no longer were focused primarily on the historical explanation of scientific enigmas. Mach, the puzzled physicist, who had turned historian, had been transfigured into a philosopher, much as he eschewed that title.

Mach recognized this, but insisted that he was still primarily a physicist, albeit one who was unwilling to be identified with any school of thought or to restrict his work to the conventional domains of physics. Reminiscing about his stand toward relativity, late in 1915, Mach recalled that what had interested him most was to step outside the limitations and proscriptions that great thinkers had drawn around themselves.

> For that reason I was actively occupied with questions about physiological sensation and psychology at the same time.... In my old age I can no more accept relativity theory than the existence of atoms and other such dogmas. Nothing was more alien to me than to found a school of thought. To the contrary, my irresistible inclination to wander away from the high-road was opposed to that and presupposed a far-reaching understanding for dissidents without however becoming their followers. For nothing is more wholesome than the doubt awakened by opposition.[16]

In light of such remarks it remains to be seen how Mach's contemporary colleagues in physics evaluated his work. What did they think of his physics, and what of his devious flirtations with the philosophy of physics?

In his day Mach invariably was recognized as an enterprising, imaginative and ingenious experimentalist. The insistent driving force and success that represents Mach's efforts to master physiology and psychology – in part by means of the techniques of physics – provides evidence for the authenticity of his search for novel problems and perspectives beyond the traditional physics of his day. He managed to accomplish this without surrendering his hold on professional physics and experimental investigation.

Mach's reputation as a theoretician, however, often was challenged. Most of his close contemporaries discovered him to be hypercritical, negative (intrinsic to positivism), and more destructive than creative of new and useful ideas. Mach was too myopically oriented to the compendious manipulation of the given data of the phenomenal world to allow himself to explore grand and unifying conceptual frameworks for science. In fact, everything that Mach wrote is saturated with a naive fascination and intrusive meddlesomeness about his immediate and empirically discoverable environment. This attitude was in keeping with his mistrust of over-arching theoretical syntheses held together, as he thought they were, by clandestinely imported metaphysical suppositions. Instead he fostered a radical pluralism of conceptual ideas because he believed that the deep understanding of natural phenomena could be furthered by examining the given data in many different, although ultimately complementary, ways.

Albert Einstein wrote in 1916 in his obituary for Mach:

> The unmediated pleasure of seeing and understanding, Spinoza's *amor dei intellectualis*, was so strongly predominant in him that to a ripe old age he peered into the world with the inquisitive eyes of a carefree child taking delight in the understanding of relationships.

Einstein felt that even when Mach's scientific investigations were not founded on new principles, his work at all times displayed "extraordinary experimental talent."[17]

Apparently Mach's philosophy did get in the way of his science, but his scientific intuition and skill were extolled nevertheless. Wilhelm Ostwald wrote in 1927:

> So clear and calculated a thinker as Ernst Mach was regarded as a visionary [*Phantast*], and it was not conceivable that a man who understood how to produce such good experimental work would want to practise nonsense [*Allotria*] which was so suspicious philosophically.[18]

Arnold Sommerfeld referred to Mach as a brilliant experimentalist but a peculiar theoretician, who, in seeking to embrace the 'physiological' and 'psychical' in his physics, had to relegate the 'physical' to a less pretentious level than physicists were accustomed to expect from a colleague.[19] This too is what philosophers had accused Mach of doing, viz. confusing physics and psychology in a manner that leads to a Humian skepticism, thus vitiating his whole theoretical analysis of science.

Whenever Mach encountered theoretical statements that in any way struck him as unwarranted – and that happened rather frequently – he became a detective, the Sherlock Holmes of science. In *The Adventure of the Priory School,* Watson says to Holmes: "Holmes, this is impossible". Holmes to Watson: "Admirable! A most illuminating remark. It *is* impossible as I state it, and therefore I must in some respect have stated it wrong".[20] For Mach, "to state it wrong" as, for example, Newton had it wrong, became the point of departure and the whetstone of ingenuity for critical insight and analytical dissection.

Mach, it seems, was downright uncomfortable, intellectually, about general theoretical syntheses. The more general, the more uncomfortable he became. He was at ease with facts *(Tatsachen),* thoughts *(Gedanken),* the adaptation of thoughts to facts, that is, observation *(Beobachtung),* and the adaptation of thoughts to other thoughts, that is, theories *(Theorien).* More than that was unjustifiable. We might suggest that Mach was accustomed to treating facts and theories in the way that Sherlock Holmes did when he remarked in *A Scandal in Bohemia* (1891): "I have no data yet. It is a capital mistake to theorize before one has data. Insensibly one begins to twist facts to suit theories, instead of theories to suit facts".[21]

When Mach joined the philosophical faculty in Vienna in October of 1895 he lost no time in pouring all of his accumulated erudition into an inaugural lecture course that would lay out, in psychophysiological perspective, his mature thought on the origins, structure, process, and conceptual roots of the natural sciences. The product of these efforts was an integrated presentation of views scattered throughout earlier treatises on energy conservation, mechanics, heat theory, optics, the physiological study of animal behavior, and the analysis of the elements of sensation. The year after Mach had delivered these inaugural lectures in Vienna, in 1897, he suffered a stroke that paralysed the right side of his body. He retired from the Vienna professorship in 1901 and devoted the rest of his life to bringing out *Erkenntnis und Irrtum* (1905), *Space and Geometry* (1906), *Kultur und Mechanik* (1915), and *Die Principien der physikalischen Optik* (posthumous 1923).

Ludwig Boltzmann (1844–1906) of the University of Leipzig succeeded Mach in the chair in Vienna in 1902, but with the new title of Professor of theoretical physics and natural philosophy. Boltzmann, who taught in Vienna from 1873 to 1876 and 1894 to 1900, in his inaugural lecture on the principles of mechanics in October of 1902 mentioned neither Mach nor the history of science. He remarked:

It is customary in inaugural lectures to begin with a hymn of praise to one's predecessors. Today I can save myself this sometimes troublesome task, for even if Napoleon the First never managed to be his own great-grandfather, I on the other hand am currently my own predecessor. I can therefore embark straightaway on the treatment of my real topic.[22]

Boltzmann's 'real topic' was to bring home to Vienna the message that the logical extension of mechanics was still of great significance for the growth of all the other domains of physical inquiry; that the difficulties encountered in mechanics were no cause for abandoning mechanics, but only signals of caution against "overshooting the mark". Obviously Mach's criticism of mechanics had overshot the mark.

How did Mach, who was constantly making derisive comments about philosophers, come to dedicate his *Erkenntnis und Irrtum* to a genuine *Fachphilosoph* like Wilhelm Schuppe (1836–1913) of the University of Greifswald? In the letter requesting Schuppe's permission to do so, Mach wrote:

I come to you with the question whether you would accept the dedication of this treatise? I do not fail to recognize that this borders on making a wild demand on a professional philosopher. But perhaps no one will burden you with the responsibility for the offenses committed by a weekend sportsman (*Sonntagsjäger*). Because of the impetuosity of some of my followers, under whom I already have had to suffer much, I will, at any rate, declare again that I am not a philosopher at all but only a scientist (*Naturforscher*) who seeks in his own way a tenable point of view.[23]

Schuppe, a realist, and the best known representative of 'immanent philosophy', held that the world is not transcendent but is immanent in consciousness. He opposed realists, materialists, positivists and idealists who in any way supported hypotheses rooted in transcendent postulates. The idea of immanent consciousness, the ego, he saw as the point of departure for the development of an epistemology. In his *Erkenntnistheoretische Logik* of 1878, which Mach discovered for himself in 1902,[24] Schuppe maintained that, in spite of differences between individuals (because they have different sense experiences), the real world is the content and object of consciousness *(Bewusstsein überhaupt)* that is common to mankind. It is not that individual consciousnesses are merely similar, they are the same in essence, being trans-subjective. Schuppe emphasized, as Fichte had, that conscious content, consisting of sense data, is partly *a posteriori.* The categories of understanding that are valid for all phenomena, however, are contributed *a priori* by the common consciousness. Thus, like Mach and the nativists, and unlike Kant, Schuppe did not derive the given data, from 'things-in-themselves', and as *a priori* forms of sense, but saw that the data contain a spatial and temporal determinacy regarded as *a posteriori* elements of the given. Thus neither Schuppe nor Mach would allow for any *a priori* forms except categories of the understanding – an epistemology that leads to the conclusion that what man knows is independent of individual consciousnesses, and partakes, instead, of the perceptible content of common consciousness.

In various philosophically unbuttoned comments, strewn throughout his works, Mach had taught a phenomenalistic and sensationalistic conception of experience that points toward a radical scepticist theory of cognition. Beginning with the assertion that mental statements do not exist, it had been close to hand for Mach to conclude – although this is a *non sequitur* – that no persisting actual entities whatsoever exist. Thus both minds and material objects had been reduced to nothing but elaborate complexes of the elementary sensations of experience. Such claims obviously go back to Berkeley and Hume and were laid against the view that the ultimate natures of mental and physical events lie behind or beyond phenomena, and thus beyond the reach of experience. Conceptual scientific schemes were interpreted as mere short-hand expressions invented as calculation formulae to which no empirical reality corresponds.

Mach had been led to assert that particular sensations alone are real, while universals and mathematical relationships are mere inventions. He must have realized, somewhere along the way, that such a view had never been able to explain why mathematical inventions had proved to be such useful tools for penetrating into the unknown aspects of nature, and why so many new discoveries had been made by pure mathematical means. It had been said that Mach's philosophy seemingly required no theories, and accordingly yielded no insights. He had attempted to found a realistic philosophy on the analysis of sensations.

In Russia Mach's works were translated, studied, transformed, and eventually seen as a threat to Marxist dialectical materialism. In his *Materialism and Empirio-criticism,* under the section on 'Parties in Philosophy and Philosophical Blockheads', Lenin was to bring to light in 1909 what had been brewing a decade earlier. He wrote:

True! The 'realists', etc., including the 'positivists', the Machists, etc., are all a wretched mush; they are a contemptible *middle party* in philosophy, who confuse the materialist and idealist trends on every question. The attempt to escape from these two basic trends in philosophy is nothing but 'conciliatory quackery'.[25]

Small wonder that Mach, who believed himself to be innocent of all philosophical '-isms', was greeted with a barrage of labels too fierce to mention: Berkeleyian subjective idealism, Kantian objective idealism, Humian skepticism, metaphysical experientialism, neutral monism, anti-materialistic positivism, inconsistent solipsism, immanentism, and the fictions of pragmatism. Was Mach being framed by his own philosophical pronouncements?

The way out of the thicket of mixed-up, despicable, and sterile '-isms' for Mach was toward the light of a more realistic and optimistic epistemological theory of knowledge. It was opportune to identify with the philosophy closest to positivism, sensationalist realism and critical empiricism – the philosophy of immanence – and still remain far removed from the infectious threat of transcendentalism and the metaphysics of mechanistic materialism. It is by acknowledging and identifying the changes associated with Mach's own intellectual expansion vis-à-vis philosophy of science over three decades that the reading of *Knowledge and Error* takes on meaning and melds with ideas expounded in his earlier works. As already mentioned, Mach in this work goes over most of the same ground that he had worked over before, but the tone, and emphasis, and conscious search for integrative elements is different.

Two main themes – epistemological psychology *(Erkenntnispsychologie)* and the methodology of the natural sciences *(naturwissenschaftliche Methodologie)* – form the backbone of Mach's treatise. More convincingly than his other works, this volume demonstrates Mach's firm belief that one should not place undue emphasis upon the particular science from which the philosophy of science is approached. Philosophy, for him, was made up of views he thought sufficient for scientific purposes. It was an interpretation of method in the sciences rather than a philosophy. The adopted title *Erkenntnis und Irrtum* was borrowed from the seventh of Mach's essays in the volume where he discusses how knowledge and error, derived from the same psychic source, come to be distinguished only on the basis of experienced outcome of specific circumstances.[26]

Thus error, when clearly recognized, serves as a corrective to advance cognition just as much as does positive knowledge. Error results from circumstantial distortion. Facts have no absolute status and are what they are only by virtue of their observed relational circumstances.

When Mach sent William James his new book, James responded: "Your *Erkenntnis & Irrtum* fills me with joy – and when I'm able to get to it I shall devour it greedily".[27] James did, and he understood Mach's terms precisely:

> Error in Mach's eyes has the exclusively practical meaning of a concept that leads to disappointment in expectation. *Täuschung* [deception] leads to *Enttäuschung* [disappointment], *Wahrheit* [truth] to *Bestätigung* [confirmation]. But all these terms are immanent in experience.[28]

The purpose of the 'concepts' *(Begriffe)*, as Mach saw it, was to help find one's way about amidst this intricate entanglement of the facts. Only James felt that Mach had gone too far, had been 'too absolutistic', in referring to the 'fiction of isolated things';[29] and Mach had not gone far enough, had not been 'emphatic enough', about the importance of conceptual formulations as practical motives for scientific research.[30] A year earlier, in 1904, James had seen pragmatism strengthening its base in Europe:

> Thus has arisen the pragmatism of Pearson in England, of Mach in Austria, and of the somewhat more reluctant Poincaré in France, all of whom say that our sciences are but *Denkmittel* – 'true' in no other sense than that of yielding a conceptual shorthand, economical for our descriptions.[31]

The first edition of *Erkenntnis und Irrtum* was exhausted in less than a year. In the preface to the second edition of 1906 Mach mentioned how pleased he was to have discovered that Pierre Duhem, another physicist, working independently, had reached conclusions remarkably similar to his own, viz. in *La Théorie physique, son objet et sa structure,* of 1906. Duhem wrote to Mach, appreciatively, saying that to emphasize the agreement between their views was for him the best indication that he was on the right track.[32]

Mach's colleague in Vienna, Ludwig Boltzmann, on returning from a trip to California acknowledged with thanks receipt of a copy of *Erkenntnis und Irrtum* and added:

> I have many greetings to bring you from America, especially from Professor Löb [Jacques Loeb of the Department of Physiology, in Berkeley], who is attached to you with affectionate veneration and devotion. More about that when we meet.[33]

With these introductory remarks about some few chosen aspects of Mach's life and work, and with the primary emphasis on the circumstances leading up to *Knowledge and Error,* the reader should be in a favorable position to examine and evaluate the merits and shortcomings of Mach's mature philosophical ideas. I have not thought it appropriate in this place to summarize or even list the main arguments that Mach seeks to elucidate as he moves within his three-level world to relate the physical, the physiological, and the psychical. The work largely speaks for itself, and, at best, the important task, as in all attempts to understand and get behind the thought of an author, is to read and query the primary document itself. The perceptive reader will have no difficulty in reading into the book both more than and less than is printed on the page. In fact, it is not at all surprising to learn that Mach's contemporaries – scientists, historians, and philosophers alike – could not resist taking up positions for or against Mach and his unconventional epistemology. Attitudes of indifference or neutrality toward Mach were rare during the years that he flourished as chief philosopher of science within the Austro-Hungarian empire. In any case, even where Mach's views now may seem superficial, narrow, erroneous, cantankerous or misdirected, the scope and implication of his views, and the influence that his work exerted on others, are such as to merit study for the light that they shed on the state and evolution of the art of historical and philosophical discussions in the sciences around the turn of the century.

Mach exerted an influence – if not always in a positive way – on persons such as Einstein, Planck, Schrödinger, Russell, Moore, Peirce, and James. In turn these thinkers provoked new revolutionary trends in America, Austria and England, viz. pragmatism, logical positivism, and analytical philosophy. The appointment of an experimental physicist from Prague to the first professorship of the history of science in Vienna indirectly contributed to setting the stage for significant interactions that came to be explored and cultivated in our time between the sister disciplines of the history of science and the philosophy of science.

Mach, alias Blaise Pascal: "Se moquer de la philosophie, c'est vraiment philosopher".[34]

Mach, alias Christian Morgenstern: "Und er kommt zu dem Ergebnis: Nur ein Traum war das Erlebnis. Weil so schliesst er messerscharf, nicht sein kann, was nicht sein darf".[25]

Mach, alias William James:

Believe truth! Shun error! – these, we see, are two materially different laws; and by choosing between them we may end by colouring differently our whole intellectual life. We may regard the chase for truth as paramount, and the avoidance of error as secondary; or we may, on the other hand, treat the avoidance of error as more imperative, and let truth take its chance.... Our errors are surely not such awfully solemn things. In a world where we are so certain to incur them in spite of all our caution, a certain lightness of heart seems healthier than this excessive nervousness on their behalf. At any rate, it seems the fittest thing for the empiricist philosopher.[36]

ERWIN N. HIEBERT

Cambridge, Mass.
November 1975

ACKNOWLEDGEMENTS

The title pages of *Erkenntnis und Irrtum* (Leipzig 1905 edition, William James' copy) and *La Connaissance et l'Erreur* (Paris 1908 edition, George Sarton's copy) are reproduced from volumes owned by the Harvard College Library. The title page of *Poznanye y Zabluzhdenye* (Moscow 1909 edition, Ernst Mach's copy) is reproduced from a volume owned by the Ernst-Mach-Institut, Freiburg i.Br. The permission to include them in this edition is acknowledged with gratitude.

E.N.H.

NOTES

[1] Four of the twenty-five essays in *Knowledge and Error* had appeared earlier, two of them in English: 'Über Gedankenexperimente', *Zeitschrift für den physikalischen und chemischen Unterricht* **10** (1897), 1–5; 'Die Ähnlichkeit und die Analogie als Leitmotiv der Forschung', *Annalen der Naturphilosophie* **1** (1902), 5–14; 'On the Psychology and Natural Development of Geometry', *The Monist* **12** (1902), 481–515; and 'Space and Geometry from the Point of View of Physical Inquiry', *The Monist* **14** (1903), 1–32.

[2] J. B. Stallo, *Die Begriffe und Theorien der modernen Physik*, Leipzig, 1901. Translated from the English by Hans Kleinpeter, with a *Vorwort* by Mach.

[3] Letter dated: Aussig, Bohemia, Nov. 2, 1882, as quoted in *The Letters of William James*, ed. by his son Henry James, Boston, 1920, Vol. 1, pp. 211–212.

[4] James to Stumpf, dated: Paris, Nov. 26, 1882, as quoted in Ralph Barton Perry, *The Thought and Character of William James*, Vol. 2, Boston, 1935, pp. 60–61.

[5] *Ibid*, p. 202, Stumpf to James dated: Berlin, May 8, 1907.

[6] Letters of the classical philologist Theodor Gomperz of the University of Vienna, to Mach, dated: Wien, 2 Dez. 1894, 16.12.94, and 22.2.95, in Ernst-Mach-Institut, Freiburg i. Br. See also Josef Mayerhöfer, 'Ernst Machs Berufung an die Wiener Universität 1895', *Clio Medica* **2** (1967), 47–55.

[7] Ernst Mach, *Beiträge zur Analyze der Empfindungen*, Jena, 1886 p. 21; English edition, *The Analysis of Sensations*, New York, 1959, p. 30.
[8] Arthur W. Burks (ed.), *Collected Papers of Charles Sanders Peirce*, Vol. VII, Cambridge, Mass., 1966, Section 485, p. 292.
[9] Ernst Mach, *The Analysis of Sensations*, 1959, p. xli (Preface to the 4th edition of 1903). See also Mach's comment in the Preface to the 7th (1912) edition of *The Science of Mechanics:* "I could hardly avoid touching upon philosophical, historical, and epistemological questions".
[10] See Erwin Hiebert, 'Mach's Philosophical Use of the History of Science', in Roger H. Stuewer (ed.), *Historical and Philosophical Perspectives of Science*, Minneapolis, 1970, pp. 184–203.
[11] Stallo, *Die Begriffe*, 1901, pp. xii–xiii.
[12] P. A. Schilpp (ed.), *Albert Einstein, Philosopher-Scientist*, Evanston, Illinois, 1949, Einstein's 'Autobiographical Notes', p. 21.
[13] Ernst Mach, *Die Geschichte und die Wurzel des Satzes von der Erhaltung der Arbeit*, 2nd unaltered ed., Leipzig, 1909, p. 2. English edition, *History and Root of the Principle of Conservation of Energy*, Chicago, 1911, p. 16.
[14] *Ibid., Geschichte*, p. 3; *History and Root*, p. 18.
[15] *Ibid., Geschichte*, p. 1; *History and Root*, pp. 15–16.
[16] 'Vorwort zur neunten Auflage', Mach, *Die Mechanik*, Darmstadt, 1963, pp. xviii–xx.
[17] Albert Einstein, 'Ernst Mach', *Physikalische Zeitschrift* **17** (1916), 101–104.
[18] Wilhelm Ostwald, *Lebenslinien, Eine Selbstbiographie*, Vol. 2, Berlin, 1927, p. 171.
[19] Arnold Sommerfeld, 'Nekrolog auf Ernst Mach', *Jahrbuch der bayerischen Akademie der Wissenschaften* (1917), 58–67.
[20] Arthur Conan Doyle, *The Annotated Sherlock Holmes*, ed. by W. S. Baring-Gould, New York, 1967, Vol. 2, p. 620.
[21] *Ibid.*, Vol. 1, pp. 348–349.
[22] Ludwig Boltzmann, *Theoretical Physics and Philosophical Problems*. Selected writings edited by Brian McGuinness, Dordrecht 1974. See pp. 146–152.
[23] Letter Mach to Schuppe dated: Wien, 7, II, 1905, in *Erkenntnis* **6** (1936), 79.
[24] Letter Mach to Schuppe dated: Wien, den 29. III, 1902, *Ibid.* p. 74.
[25] V. I. Lenin, *Collected Works*, Moscow, 1962. Vol. 14, p. 340.
[26] Marcel Dufour, the translator of the French edition, was dissatisfied to render *Erkenntnis und Irrtum* as *La Connaissance et l'Erreur* because he felt that the antithesis of *Erreur* was *Vérité;* and so suggested *Vérité et Erreur* or *Qu'est ce que la science?* See letters Dufour to Mach dated: Nancy, 13 Mars 1907, and 19 Mars 1907 in Mach-Institut-Freiburg i. Br.
[27] James to Mach, dated: Cambridge, (Mass.), Aug. 9 [19] 05 in Mach-Institut-Freiburg i.Br.
[28] Marginalia at the end of Chapter 7 of James' copy of the 1905 edition of *Erkenntnis und Irrtum* in the Houghton Library of Harvard University. We may note here that James carefully read and annotated most of Mach's major treatises.
[29] *Ibid.*, Marginalia to Section 10 of the first chapter: "? too absolutistic".
[30] *Ibid.*, Marginalia to Section 12 of the eighth chapter: "W. J. But not emphatic enough".
[31] William James, *Collected Essays and Reviews*, ed. by Ralph Barton Perry, New York, 1920, pp. 449–450.
[32] Letter Duhem to Mach dated: Bordeaux, le 4 oct. 1906: "Je me suis empressé d'y jeter les yeux et j'ai été à la fois très heureux et très confus de la préface, si aimable à

mon égard, que vous y avez mise. Je n'aurais pu souhaiter, en faveur de mon modeste ouvrage, suffrage plus flatteur et plus autorisé que le vôtre; et l'accord de mes méditations avec les vôtres est, pour moi, l'une des meilleurs marques que je me trouve dans la bonne voie." Mach-Institut-Freiburg i.Br.

[33] Letter Boltzmann to Mach dated: Wien, 25/8, 1905, in Mach-Institut-Freiburg i.Br.

[34] Blaise Pascal, *Pensées* (I, 4), text of Jacques Haumont, Paris, 1972, p. 5.

[35] Christian Morgenstern, *Alle Galgenlieder*, Berlin, 1932, 'Die unmögliche Tatsche', p. 164.

[36] William James, *The Will to Believe*, New York, 1897, Sections VII & VIII, pp. 18–19.

AUTHOR'S PREFACE TO THE FIRST EDITION

Without in the least being a philosopher, or even wanting to be called one, the scientist has a strong need to fathom the processes by means of which he obtains and extends his knowledge. The most obvious way of doing this is to examine carefully the growth of knowledge in one's own and the more easily accessible neighbouring fields, and above all to detect the specific motives that guide the enquirer. For to the scientist, who has been close to these problems, having often experienced the tension that precedes solution and the relief that comes afterwards, these motives ought to be more visible than to others. Since in almost every new major solution of a problem he will continue to see new features, he will find systematizing and schematizing more difficult and always apparently premature: he therefore likes to leave such aspects to philosophers who have more practice in this. The scientist can be satisfied if he recognizes the conscious mental activity of the enquirer as a variant of the instinctive activity of animal and man in nature and society – a variant methodically clarified, sharpened and refined.

The work of schematizing and ordering methodological knowledge, if adequately carried out at the proper stage of scientific development, must not be underrated[1]. But one must emphasize that practice in enquiry, if it can be acquired at all, will be furthered much more through specific living examples, rather than through pallid abstract formulae that in any case need concrete examples to become intelligible. Those whose guidance has been really useful to the disciples of scientific research were therefore in the main scientists, men like Copernicus, Gilbert, Kepler, Galileo, Huygens, Newton, and more recently J. F. W. Herschel, Faraday, Whewell, Maxwell, Jevons and others. Men of great merit like J. F. Fries and E. F. Apelt, who have greatly advanced certain areas of scientific methodology, have not entirely succeeded in freeing themselves from preconceived philosophical views. These philosophers, and even the scientist Whewell, through adherence to Kantian notions, have been forced to rather weird conceptions as regards very simple scientific questions, as we shall see in the sequel. Amongst the older German philosophers, F. E. Beneke seems to be the only one who has been able to free himself completely from such prejudices, giving credit unreservedly to British scientists.

In the winter of 1895/96 I gave a course on 'psychology and logic of enquiry', in which I tried to reduce the psychology of enquiry as far as possible to notions native to science. The present volume is essentially a free treatment of a selection from that material. I hope that this will inspire younger colleagues and especially physicists to further reflection, and direct their attention to some adjacent fields which they tend to neglect, but which afford any enquirer much clarification as to his own thinking.

The execution will doubtless be marred by various deficiencies. For although I have always been vitally interested in the fields bordering on my speciality and in philosophy, it is naturally as a weekend sportsman only that I could roam through some of these areas, especially the last one. If in the process I happily found my scientific standpoint approaching that of well-known philosophers such as Avenarius, Schuppe, Ziehen and others, and their younger colleagues such as Cornelius, Petzoldt and v. Schubert-Soldern, as well as of certain eminent scientists, it lies in the nature of contemporary philosophy that my views have drawn me well away from other important philosophers[2]. I have to say with Schuppe: for me, the region of the transcendent is sealed. If moreover I confess openly that its denizens cannot even arouse my curiosity, the wide gap between myself and many philosophers becomes manifest. Therefore I have explicitly stated already that I am not a philosopher but only a scientist. If nevertheless I am at times somewhat obtrusively counted amongst philosophers, the fault is not mine. But neither, clearly, do I wish to be a scientist who blindly entrusts himself to the guidance of one single philosopher, somewhat in the way that a physician in Molière might expect and demand from his patients.

The work I have attempted in the interest of scientific methodology and the psychology of knowledge proceeds as follows. First I have aimed not at introducing a new philosophy into science, but at removing an old and stale philosophy from science – an endeavour that even some scientists rather resent. For amongst the many philosophical statements that have been made over the years some have been recognized by philosophers themselves as errors, or have been set out so clearly that any unprejudiced person could easily recognize them as such. In science, where they met with less alert criticism, they have survived longer, just as a defenceless species might be spared on a remote island free from predators. Such dicta, which in science are not only useless but produce obnoxious and idle pseudoproblems, deserve nothing better than being discarded. If in so doing I have done something useful, the credit really belongs to the philosophers. Should they decline it, then the next generation may perhaps be fairer to them than they themselves wished to be. Besides, over a period of more than forty years I have had the opportunity as a naïve observer free from any system to see, both in laboratory and lecture hall, the ways in which knowledge advances. I have tried to set them down in various writings. However, what I have found there is not my exclusive property. Other attentive enquirers have often made the same or very similar findings. Had the scientists' attention been less absorbed with the urgent special problems of research, causing many a methodological discovery to relapse into oblivion, what I can contribute to the psychology of enquiry must long since have been securely in their possession. That is precisely why I think that my work will not be in vain. Perhaps even philosophers may one day recognize my enterprise as a philosophical clarification of scientific methodology and will meet me half-way. However, even if not, I still hope to have been useful to scientists.

Vienna, May 1905 E. MACH

NOTES

[1] A systematic account with which in all essentials I agree and which ingeniously excludes controversial psychological questions whose resolution is neither urgent nor essential for epistemology is given by Prof. Dr. H. Kleinpeter (*Die Erkenntnistheorie der Gegenwart*, Leipzig, J. A. Barth, 1905).

[2] In one chapter of each of *M* and *A*, I have answered those objections to my views that have become known to me. Here I need merely add a few comments on Hönigswald's *Zur Kritik der Machschen Philosophie* (Berlin 1905). Above all, there is no Machian philosophy, but at best a scientific methodology and cognitive psychology and both are provisional, imperfect attempts, like all scientific theories. I am not responsible for a philosophy that might be constructed from this with the help of alien additions. That my views cannot agree with Kant's results must from the start be evident not only to any Kantian but also to myself, given the different premisses which exclude even a common basis for discussion (cf. Kleinpeter, l.c., and the present volume). Still, is Kant's the only infallible philosophy that it should be entitled to warn individual

sciences not to try in their own fields in their own ways to achieve what more than a hundred years ago that philosophy promised to furnish to those sciences yet never did? Without therefore in the least doubting Hönigswald's honest intentions, I rather think that it would have been more fruitful for him and others to argue with 'empiriocriticism' or 'immanentism' with whom he surely has more points of contact. Once philosophers are at one amongst themselves, agreement with scientists will no longer be so difficult.

AUTHOR'S PREFACE TO THE SECOND EDITION

The text of the second edition is not essentially different from that of the first. Time did not permit, nor indeed was there occasion for, a radical revision. I must add that a number of critical observations came to my knowledge too late to be taken into account.

In cases where works of related content appeared simultaneously with the first edition of this book or shortly thereafter, I have added references to them, in the form of notes. A closer relationship between my own fundamental views and those of Jerusalem has been revealed by his book *Der kritische Idealismus und die reine Logik* ['Critical Idealism and Pure Logic'] (1905). No doubt it was the difference of our specialist standpoints that prevented us from realizing how near we were to one another. The cause of this nearness is to be sought, most probably, in biology (particularly evolutionary theory), which has been a stimulus to both of us. I have found a number of points of contract and a great deal of stimulation in Stöhr's highly original *Leitfaden der Logik in psychologisierender Darstellung* ['Guide to Logic from a Psychological Point of View'] (1905). Duhem's *La Théorie physique, son objet et sa structure* (1906) [E.T.: *The Aim and Structure of Physical Theory*, Princeton, 1954] has given me great pleasure. I had not hoped to find so soon such far-reaching agreement in any physicist. Duhem rejects any metaphysical interpretation of questions in physics. He sees it as the aim of that science to determine the facts in a conceptually economic way. The historical and genetic method of presenting physical theories seems to him the only correct one and pedagogically the most effective. These are views that I have championed over a good three decades. I value the agreement between us all the more because Duhem arrived at the same results quite independently. At the same time, it must be said that whereas I, in the present book at any rate, principally stress the kinship between common-sense thinking and that of science, Duhem throws particular light on the differences between the observation and thinking of common-sense and the critical observation and thinking of the physicist. For that reason I should like to recommend his book as a complement and illumination of my own. In what follows I shall often have occasion to refer to Duhem's pronouncements and only rarely and on subordinate points to note a difference of opinion.

Vienna, April 1906 E. MACH

EDITOR'S NOTE TO THE ENGLISH EDITION

After Mach's death, in 1916, there were three further editions of the present work, but it would require a microscopic eye to discern the differences from the second edition of 1906. The third edition (1917) incorporated according to Ludwig Mach, only his father's marginal corrections. One of these must have been the change in dedication. Previously it had been to Wilhelm Schuppe with heartfelt respect; but Schuppe had died in 1913 and it is, all things considered, a mark of esteem for him that, when the book had to be dedicated to his memory, the names of Hume and Avenarius were joined to his.

The fifth edition (1926), from which – though it is a technicality to say so – the present volume is translated, contains a preface speaking of the removal of obvious errors and of the project, for the moment abandoned, of an appendix with material related to that of the main book:

> Should there later be a question of expansions of the book in the form of appendixes (wrote Ludwig Mach), the author's express wish is (sic) that as far as possible different approaches should be represented and on no account should a number of individual views be excluded in favour of a forced interpretation, quite alien to the author, of his views, an interpretation recommended by one-sided illumination and by its agreement with a currently favoured approach, such an interpretation, in fact, as I was unfortunately unable to prevent in the case of the 8th edition of the *Mechanics*.

The fifth edition of *Erkenntnis und Irrtum* also contains a conspectus of the pagination, in various editions, of the passages Mach quotes from his own works. In the nature of things this can interest only readers of German and it is omitted in the present volume.

Mach's footnotes have been translated with the correction only of obvious errors. They are printed, according to the publisher's practice, at the end of each chapter. Mach's own works are referred to by the following abbreviations:

Analyse der Empfindungen = *A*
Die Mechanik in ihrer Entwicklung = *M*

Populärwissenschaftliche Vorlesungen = *P*
Prinzipien der Wärmelehre = *W*
The edition, where given, is shown by placing the number after the letter.

Fuller details of other works referred to by Mach, none being later in date than 1906, are given in a special bibliography at the end of the book. There, also, will be found a Bibliography of Ernst Mach taken by kind permission of the authors and editors from that prepared by Otto Blüh and Wolfgang F. Merzkirch for *Ernst Mach, Physicist and Philosopher* 1970, edited by Robert S. Cohen and Raymond J. Seeger, being volume VI of Boston Studies in the Philosophy of Science. A few entries have been added, chiefly to cover recent publications. I am indebted to my colleagues, Professor Robert S. Cohen and Professor Erwin N. Hiebert, for bibliographical assistance.

Dr. Foulkes writes

As regards translation, there are some difficulties. On the one hand, there are certain established usages in philosophy which Mach took over in translation; on the other hand, the terms so used carry different connotations and overtones in Mach's German than did the originals in Hume's English, or than do those English terms that are generally used to render notions from German philosophy. The best one can do is to try to convey the sense of each passage and the general drift of the argument. The difficulty indeed starts with the very title, for 'Erkenntnis' is not quite the same as 'knowledge'; the former carries the suggestion of coming to be acquainted with, of recognizing. Moreover it is not possible to give a fair version by rigidly sticking to the same term in translation for each occurrence of a given term in the original.

A very few square brackets will be found in the book: they enclose opinions of my own.

B. MCGUINNESS

CHAPTER I

PHILOSOPHICAL AND SCIENTIFIC THOUGHT

1. Lower animals living under simple, constant and favourable conditions adapt themselves to immediate circumstances through their innate reflexes. This usually suffices to maintain individual and species for a suitable period. An animal can withstand more intricate and less stable conditions only if it can adapt to a wider range of spatial and temporal surroundings. This requires a farsightedness in space and time which is met first by more perfect sense organs, and with mounting demands by a development in the life of the imagination. Indeed an organism that possesses memory has wider spatial and temporal surroundings in its mental field of vision than it could reach through its senses. It perceives, as it were, even those regions that adjoin the directly visible, seeing the approach of prey or foe before any sense organ announces them. What guarantees to primitive man a measure of advantage over his animal fellows is doubtless only the strength of his individual memory, which is gradually reinforced by the communicated memory of forebears and tribe. Likewise, what essentially marks progress in civilization is that noticeably wider regions of space and time are drawn within the scope of human attention. With the partial relief that a rising civilization affords, to begin with through division of labour, development of trades and so on, the individual's imaginative life is focused on a smaller range of facts and gains in strength, while that of society as a whole does not lose in scope. Gradually the activity of thinking thus invigorated may itself become a calling. Scientific thought arises out of popular thought, and so completes the continuous series of biological development that begins with the first simple manifestations of life.

2. The goal of the ordinary imagination is the conceptual completion and perfection of a partially observed fact. The hunter imagines the way of life of the prey he has just sighted, in order to choose his own behaviour accordingly. The farmer considers the proper soil, sowing and maturing of the fruit of plants that he intends to cultivate. This trait of mental com-

pletion of a fact from partial data is common to ordinary and scientific thought. Galileo, too, merely wants to represent to himself the trajectory as a whole, given the inital speed and direction of a projected stone. However, there is another feature that often very significantly distinguishes scientific from ordinary thought: the latter, at least in its beginnings, serves practical ends, and first of all the satisfaction of bodily needs. The more vigorous mental exercise of scientific thought fashions its own ends and seeks to satisfy itself by removing all intellectual uneasiness: having grown in the service of practical ends, it becomes its own master. Ordinary thought does not serve pure knowledge, and therefore suffers from various defects that at first survive in scientific thought, which is derived from it. Science only very gradually shakes itself free from these flaws. Any glance at the past will show that progress in scientific thought consists in constant correction of ordinary thought. As civilization grows, however, so scientific thought reacts on those modes of thought that serve only practical ends: ordinary thought becomes increasingly restricted and replaced by technical thought which is pervaded by science.

3. The representation in thought of facts or the adaption of thought to fact, enables the thinker mentally to complete partially observed facts, insofar as completion is determined by the observed part. Their determination consists in the mutual dependence of factual features, so that thought has to aim at these. Since ordinary thinking and even incipient scientific thought must make do with a rather crude adaption of thoughts to facts, the former do not quite agree amongst each other. Mutual adaptation of thoughts is therefore the further task to be solved in order to attain full intellectual satisfaction. This last endeavour, which involves logical clarification of thinking though reaching far beyond this goal, is the outstanding mark that distinguishes scientific from ordinary thought. The latter is enough so long as it roughly serves the realization of practical ends.

4. Scientific thought presents itself in two seemingly different forms: as philosophy and as specialist research. The philosopher seeks to orient himself as completely and comprehensively as possible in relation to the totality of facts, which necessarily involves him in building on material borrowed from the special sciences. The special scientist is at first concerned only with finding his way about a smaller area of facts. Since, how-

ever, facts are always somewhat arbitrarily and forcibly defined with a view to the momentary intellectual aim, these boundary lines are constantly shifting as scientific thought advances: in the end the scientist too comes to see that the results of all other special enquiries must be taken into account, for the sake of orientation in his own field. Clearly in this way special enquirers also collectively aim at a total picture through amalgamation of all special fields. Since this is at best imperfectly attainable, this effort leads to more or less covert borrowings from philosophical thought. The ultimate end of all research is thus the same. This shows itself also in the fact that the greatest philosophers, such as Plato, Aristotle, Descartes, Leibniz and others have opened up new ways of specialist enquiry, while scientists like Galileo, Newton, Darwin and others have greatly furthered philosophic thought without being called philosophers.

Yet it is true that what the philosopher regards as a possible starting point appears to the scientist as a distant goal of his work; but this difference of view need not and indeed does not prevent enquirers from learning from each other. Through its many attempts to summarize the most general features of large areas, philosophy has gained ample experience in this line, even learning gradually to recognize and avoid some of its own mistakes that the philosophically untrained scientist is almost bound to commit even today. However, philosophy has furnished science with some positive notions of value too, for example ideas of conservation. Philosophers in their turn take from special sciences foundations that are sounder than anything from ordinary thought. Science, to him, is an example of a careful, solid and successful structure, whose excessive onesidedness at the same time affords him useful lessons. Indeed, every philospher has his own private view of science, and every scientist his private philosophy. However, these private scientific views are usually somewhat outdated and it is extremely rare that a scientist can respect the occasional scientific pronouncement of philosophers; whereas most scientists today adhere to a materialist philosophy of 150 years' standing, whose inadequacy has long since been recognized not only by professional philosophers but by any layman not too cut off from philosophic thought. Few philosophers today take part in the work of science, and only exceptionally do scientists address their own intellectual attention to philosophical questions: yet such efforts are essential for mutual understanding, since mere reading is here useless to either side.

Surveying the age-old paths that philosophers and scientists have trodden, we find that they are often well cleared. At some points however, they seem to be blocked by quite natural and instinctive philosophical and scientific prejudices that have remained as waste from old experiments and unsuccessful work. It would be advisable at times to clear these heaps of waste, or to sidestep them.

5. Not only humanity but each individual on reaching full consciousness finds within himself a view of the world to which he has not contributed deliberately. He accepts this as a gift of nature and civilization: everyone must begin here. No thinker can do more than start from this view, extend and correct it, use his forebears' experience and avoid their mistakes as best he may, in short: carefully to tread the same path again on his own. What, then, is this world view? I find myself surrounded by moveable bodies in space, some inanimate, others plants, animals and men. My body, likewise moveable in space, is for me a visible and touchable object of sense perception occupying a part of sensible space alongside and outside other bodies, just as they do. My body differs from those of other people in certain individual features but above all in that when objects touch my body peculiar feelings supervene that I do not observe when other bodies are touched. My body is not quite as accessible to my eyes as the bodies of others. I can see only a small part of my head, at least directly. In general my body appears to me under a perspective quite different from that of all others: towards them I cannot take up that optical point of view. Similarly for touch and the other senses[1]. For example, I hear my voice quite differently from that of others. Besides, I find memories, hopes, fears, drives, desires, a will and so on, of whose development I am as innocent as of the existence of the bodies in my surroundings. The foregoing considerations and the movement of the one definite body issuing from that will mark that body as mine. When I observe the behaviour of other human bodies, not only practical needs but also a close analogy force me, even against my will, to hold that memories, hopes, fears, motives, wishes, and will similar to those associated with my body are bound up with other human and animal bodies. The behaviour of other people further compels me to assume that my body and other objects exist as immediately for them as their bodies and other objects do for me; whereas my memories, desires and the like are for them the result of the same

sort of irresistible analogical inference as theirs for me. The totality of what is immediately given in space for all may be called the physical, whereas what is immediately given only to one while others must infer it by analogy may provisionally be called the mental. The totality that is given immediately only to one we shall call also his ego, in the restricted sense. We note Descartes' opposition of matter to mind, extension to thought. This is the natural basis of dualism, which may stand for a whole range of transitions from mere materialism to pure spiritualism, depending on how we value the physical and the mental, taking one as basic and the other derived. The dualist contrast may however become so acute as to exclude, contrary to any natural view, all contact between physical and mental; this gives rise to such monstrosities as 'occasionalism' or 'pre-established harmony' [2].

6. The findings in my spatial surroundings depend on one another. A magnetic needle is set in motions as soon as another magnet comes close enough. A body becomes warm near fire and cold when in contact with ice. A sheet of paper in a dark room becomes visible by the flame of a lamp. The behaviour of other people forces me to assume that in all this their findings resemble mine. A grasp of these mutual dependences within our findings and our experiences is of the greatest interest to us, both practically for the satisfaction of needs and theoretically for the mental completion of incomplete findings. In observing these dependences amongst bodies I can treat men and animals as though inanimate by abstracting from everything obtained by analogy. But I observe again that my body essentially influences this finding. A body can throw a shadow on a sheet of white paper; but if I have just been looking at a rather bright object, I shall see on the paper a spot similar to the shadow. By suitable position of the eyes I may see a body double, or two very similar bodies as three. If I have just turned round suddenly, I may see mechanically moving bodies as at rest and vice versa. If I close my eyes optical findings cease altogether. Analogous tactual or thermal findings and so on may be induced by corresponding bodily influences. If my neighbour conducts the same experiments on his body this does not alter my findings, although I learn from reports and have to assume by analogy that his findings are similarly modified.

The constituents of my spatial findings thus generally depend not only on one another but also on findings as to my body, and likewise for everybody. If one places excessive emphasis on the latter dependence while underrating the former, one may easily come to regard all findings as mere products of one's own body, that is as 'subjective'. However, the spatial boundaries of our own body are always present and we see that findings outside them depend on one another as well as on findings inside them. The investigation of external dependences is indeed much simpler and further advanced than that of cross-boundary ones. Still, we may expect these last to be of the same kind as the first, as we infer with noticeably growing certainty from research on the bodies of men and animals external to us. A developed physiology, increasingly based on physics, can unravel the subjective conditions of a finding. Naïve subjectivism, which construed variant findings of one person under variable conditions and of different persons as so many cases of appearance in contrast with a hypothetical constant reality, is no longer admissible. For what matters is only a full grasp of all conditions of a finding: that alone is of practical or theoretical interest.

7. As to the sum of my physical findings, these I can analyse into what are at present unanalysable elements: colours, sounds, pressures, temperatures, smells, spaces, times and so on. These elements[3] depend both on external and internal circumstances; when the latter are involved, and only then, we may call these elements sensations. Since another's sensations are no more directly given to me than mine to him, I am entitled to regard the elements of the mental as the same as those into which I have analysed the physical. Thus the mental and physical have common elements and are not in stark opposition as commonly supposed. This becomes even clearer if we can show that memories, ideas, feelings, will and concepts can be built up from traces left behind by sensations and are therefore comparable with them. If now I call the sum of my mental aspect, sensations included, my ego in the widest sense (in contrast with the restricted ego), then in this sense I could say that my ego contains the world (as sensation and idea). Still, we must not overlook that this conception does not exclude others equally legitimate. This solipsist position seems to abolish the world as independent, blurring the contrast between it and ego. The boundary nevertheless remains, only it no longer runs round the restricted ego but through the extended one, that is through 'conscious-

ness'. Indeed we could not have derived the solipsist position without observing the boundary and the analogy between my own and others' ego. Those who say that we cannot go beyond the ego therefore mean the extended ego, which already contains a recognition of the world and other minds. Nor is confining oneself to the 'theoretical' solipsism[4] of the enquirer any more acceptable: there are no isolated enquirers, each has practical ends of his own, can learn from others and works for their guidance too.

8. In making our physical findings we are subject to many errors and delusions. A straight rod dipped into water at an angle is seen bent, and the inexperienced might think that it will turn out to feel bent too. The virtual image in a concave mirror looks tangible. An object in glaring light is regarded as white and we are amazed to find that under moderate illumination we find it is black. The shape of a tree trunk in the dark reminds us of the figure of a man and we imagine him to be in front of us. All such 'delusions' rest on the fact that we do not know or fail to observe the conditions under which the finding is made, or that we suppose them to be other than they are. Besides, the imagination rounds off incomplete findings in the way that is most familiar to it, thus occasionally falsifying them. What in ordinary thought leads to the opposition between illusion and reality, between appearance and object, is the confusion between findings under the most various conditions with findings under very definite and specific conditions. Once this opposition has emerged, it tends to invade philosophy as well, and is not easily dislodged. The weird and unknowable 'thing-in-itself' behind appearences is the ordinary object's unmistakeable twin, having lost all other significance[5]. After misconstruing the boundary between the internal and external and thereby imposing the stamp of illusion on the ego's entire content, have we any further need for an unknowable something outside the confines that the ego can never transcend? Is it any more than a relapse into ordinary thought to see some solid core behind 'delusive' appearances?

When we consider elements like red, green, hot, cold and the rest, which are physical and mental in virtue of their dependence on both external and internal circumstances, and are in both respects immediately given and identical, the question as to illusion and reality loses its sense. Here we are simultaneously confronted by the elements of the real world and of the ego. The only possible further question of interest concerns their functional interdependence, in the mathematical sense. Such a connection might be called an object, though not an unknowable one: with every new observation or scientific theorem it becomes better known. If we look at the restricted ego without prejudice, it too turns out to be a functional connection between these elements, except that its form is here a little different from what we are used to in the physical field: consider the way ideas and their connections differ from physical elements. We need no unknown and unknowable something behind these processes, nor would it help in the least towards better understanding. Yet there is something all but unexplored standing behind the ego, namely our body; but every new observation in physiology and psychology makes the ego better known to us: introspection and experiment in psychology, brain anatomy and psychopathology, already the source of many valuable explanations, here work strongly in the direction of physics in the widest sense, combining with it into a more penetrating grasp of the world. We may expect all sensible questions gradually to approach being answerable[6].

9. In examining the mutual dependence of varying ideas one hopes to grasp mental processes and in particular one's own experiences and actions. One who still needs an observing and acting subject, has failed to see that he could have saved himself the whole trouble of the enquiry, for he has now gone full circle. It reminds us of the farmer who went to a factory to have the working of steam engines explained to him and then asked "Where are the horses that drive the machines?" It was Herbart's main merit to have examined the processes of ideas as such, yet even he spoiled his whole psychology by starting from the assumption that the soul is simple. Only lately have we begun to accept a psychology without soul.

10. In pushing the analysis of experience as far as currently untranscendable elements[7] our main advantage is that the two problems of the 'unfathomable' thing and the equally 'unexplorable' ego are presented in their simplest and clearest form, which is precisely what makes it easy to see them as sham problems. By elimination of what it is senseless to explore, what the special sciences can really explore emerges all the more clearly: the complex interdependence of the elements. While groups of such elements may be called things or bodies, it turns out that there are

strictly speaking no isolated objects: they are only fictions for a preliminary enquiry, in which we consider strong and obvious links but neglect weaker and less noticeable ones. The same distinction of degree give rise also to the opposition of world to ego: an isolated ego exists no more than an isolated object: both are provisional fictions of the same kind.

11. Our considerations offer little or nothing to the philosopher: they are not designed to solve one, or seven, or nine cosmic riddles; they merely lead to removing false problems that hinder scientific enquiry, while leaving the rest to positive research. We offer only a negative rule for scientific research which need not concern the philosopher, especially if he already possesses (or thinks he does) secure foundations for a world view. If then our account is to be judged primarily from a scientific standpoint, this cannot mean that philosophers are not to criticize it, modify it to suit their needs or even reject it altogether. However, for the scientist it is quite a secondary matter whether his ideas fit into some given philosophic system or not, so long as he can use them with profit as a starting point for research. For the scientist is not so fortunate as to possess unshakeable principles, he has become accustomed to regarding even his safest and best-founded views and principles as provisional and liable to modification through experience. Indeed, the greatest advances and discoveries have been possible only through this attitude.

12. To the scientist likewise, our account can show at best an ideal, whose gradual and approximate realization remains the task of future research. The finding out of the direct connections between the elements is so complex a task that it cannot be solved all at once but only step by step. It was much easier to ascertain a rough and ready outline of the way in which whole collections of elements or bodies depend on one another, and it was rather a matter of chance and practical need which elements seemed the more important, which were focused on and which remained unnoticed. The individual enquirer is in the midst of developing science and must start with his predecessors' incomplete findings which at best he can correct and perfect according to his ideal. In gratefully adopting for his own work the help and hints contained in these preliminaries, he often adds the errors of predecessors and contemporaries to his own. A

return to a quite naïve point of view, even if it were possible, would afford to one who had shed all the views of his contemporaries not only the advantage of freedom from prejudice, but also the drawback of confusion arising from the complexity of the task and the impossibility of even starting any enquiry. If therefore we seem here to be returning to a primitive standpoint, in order to conduct the enquiry along new and better paths, this is an artificial simplemindedness that does not give up the advantages gained through long periods of growing civilization, but on the contrary uses insights presupposing a fairly high level of thought as to physics, physiology and psychology. Only at such a level is a resolution into 'elements' conceivable. We are thus returning to the starting points of enquiry with the deeper and richer insight produced by previous enquiry. A certain stage of mental development must be reached before scientific considerations can start at all, but no science can use ordinary concepts in their vagueness: it must return to their beginning and origin in order to make the concepts more precise and pure. Should this be forbidden only to psychology and epistemology?

13. If we have to investigate a set of multiply interdependent elements there is only one method at our disposal: the method of variation. We simply have to observe the change of every element for changes in any other: it makes little difference whether these latter changes occur 'spontaneously' or are brought about through our 'will'. The dependences are ascertained by 'observation' and 'experiment'. Even if elements were linked only in pairs, being otherwise independent, the systematic study of these links would be troublesome enough: a simple mathematical argument shows that for elements interdependent in groups of three, four and so on, the task soon becomes practically inexhaustible. Any provisional neglect of less noticeable dependences, any anticipation of obvious connections, must therefore be felt as making the task significantly easier: both these simplifications were first discovered instinctively, under the influence of practical requirements, needs and mental constitution, and afterwards used with conscious skill and method by scientists. Without such moves, which might well count as blemishes, science could neither have arisen nor grown. Scientific research is somewhat like unravelling complicated tangles of strings, in which luck is almost as vital as skill and accurate observation. Research to the enquirer is just as

exciting as the pursuit of a rare beast over difficult terrain is to the hunter.

If one wants to investigate the interdependence of any elements, those whose influence is plain but felt as disturbing the enquiry had best be kept as constant as possible. That is the first and foremost way of making research easier. Since we know that every element depends both on external and internal elements we are led to begin by studying the concomitances of external elements while leaving internal ones (those of the observer himself) under conditions that remain constant as far as possible. By examining the interdependence of the luminosity of bodies, or their temperatures, or motion, under conditions that remain as far as possible the same for one or even different observers taking part, we make physical knowledge as free as possible from the influence of our own individual bodies. To complete this we have to study internal and cross-boundary physiological and psychological connections, a task significantly facilitated by the prior physical investigations. This division of labour too has arisen instinctively, and we need merely become conscious of its advantages to go on using it methodically. Scientific research is full of examples of analogous divisions in smaller fields of investigation.

14. Following these introductory remarks let us now look more closely at the leading themes of scientific enquiry. In this we lay no claims to completeness, and indeed would rather guard against premature philosophizing and systematizing. Let us take an attentive walk through the field of scientific enquiry and observe the detailed behaviour of the enquirer. By what means has our knowledge of nature actually grown in the past, and what are the prospects for further growth in the future? The enquirer's behaviour has developed instinctively in practical activity and popular thought, and has merely been transferred to the field of science where in the end it has been developed into a conscious method. To meet our requirements we shall not need to go beyond the empirically given. We shall be satisfied if we can reduce the features of the enquirer's behaviour to actually observable ones in our own physical and mental life (features that recur in practical life and in the action and thought of peoples); and if we can show that this behaviour really leads to practical and intellectual advantages. A natural basis for this purpose is a general consideration of our physical and mental life.

NOTES

[1] On good phonographs one recognizes the timbre of a friend's voice, but one's own sounds strange because the head resonance is missing.
[2] Euler, in the 83rd of his *Letters to a German Princess*, explains how ridiculous and counter to all ordinary experience it is not to assume a closer link between one's own body and psyche than between *any* body and *any* psyche. Cf. *M* 7, 1912, p. 431.
[3] Cf. *A* 4, 1903. I here wish to refer to the most interesting account of R. v. Sterneck, although I differ from him in many points ('Über die Elemente des Bewusstseins', *Ber. d. Wiener philosophischen Gesellschaft*, 1903).
[4] Cf. J. Petzoldt, 'Solipsismus auf praktischem Gebiet', *Vierteljahrsschrift f. wissensch. Philosophie* **XXV** 3, p. 339; Schuppe, 'Der Solipsismus', *Zeitschr. für immanente Philosophie* **III**, p. 327.
[5] Cf. Schuppe's excellent polemical passages against Ueberweg (Brasch, *Welt- und Lebensanschauung F. Ueberwegs*, Leipzig 1889).
[6] The account in sns. 5–8 has seemed to some readers to depart from that in *A*. However, this is not so. Without changing the essence of my account, I have adopted this form merely to make allowance for the reticence of scientists, above all as regards anything touching on psycho-monism. Besides, it is indifferent to me what name is given to my point of view.
[7] Decomposition into what I have here called *elements* is hardly conceivable at the quite naive level of primitive man who, like animals, probably takes the bodies in his surroundings as a whole without separating the contributions from the separate senses which are given to him only as a whole. Even less will he be able to separate colour from shape, or mixed colours into their components. All this is already the result of simple scientific experience and reflection. Decomposing noise into simple sensations of sound, touch into several partial sensations, light into sensations of basic colour and so on even belongs to more recent science. That here the limit of analysis is already reached and incapable of being pushed further by any physiological means we shall hardly believe. Our elements are thus only provisional, as those of alchemy were in the past and those of currently accepted chemistry are now. Although for our purpose of eliminating philosophical sham problems reduction to these elements seemed the best way, it does not follow that every scientific enquiry must begin with them. What is the simplest and most natural starting point for the psychologist need not at all be so for the physicist or chemist who faces quite different problems or different aspects of the same question. However, notice *one* thing. While there is no difficulty in building up every physical experience from sensation, that is mental elements, we can foresee no possibility of representing any mental experience in terms of the elements currently used in physics: i.e. from the masses and motions in the rigid form that alone is serviceable in that special branch of science. Although Dubois recognized this, he is nevertheless wrong in not even thinking of the inverse path and regarding reduction of the two fields to each other as in any case impossible. Remember that nothing can become the object of experience or science unless it can in some way enter consciousness. A clear recognition of this fact enables us to choose now the psychological, now the physical, approach as starting point, according to need and goal of the enquiry. One who thinks that because he has recognized his own ego as the medium of all knowledge he may no longer infer the ego of others by analogy thus likewise falls victim to a strange though widespread systematic superstition. For the same analogy serves the exploration of one's *own* ego.

It gives me pleasure that I can here refer to M. Verworn (*Naturwissenschaft und Weltanschauung* 1904), whose views are very close to mine. See especially the note on p. 45. Still, his term 'psycho-monism' seems to me less appropriate now than it would have been in a previous youthful phase of my thinking. H. Höffding (*Moderne Philosophen* 1905, p. 121) quotes the words of R. Avenarius: 'what I know is neither physical nor mental but only some third kind of thing'. This I should subscribe to at once if I did not suspect that by this third he means some *unknown* third, perhaps a thing-in-itself or some other metaphysical entity. For me the physical and the mental are essentially *identical*, immediately *familiar* and *given*, and different only as to the mode of viewing. This mode and therefore the distinction can supervene only with higher mental development and ampler experience. Prior to this the physical and the mental are indistinguishable. For me a scientific effort is lost if it does not keep firm hold on the immediately given, if, instead of investigating the connections between the characteristics of the given, it fishes somewhere in vacuo. If the connections are found one can still reflect on them in all sorts of ways, but that is not my business. My task is not philosophical, but merely methodological. Neither let it be thought that I wish to attack let alone abolish ordinary concepts developed instinctively on good empirical grounds, such as subject, object, sensation and so on. However these nebulosities, though adequate in practice, are unusable as regards method; there we must rather examine what functional dependences o 1the characteristics of the given have impelled us towards those concepts, as we have done here. No knowledge already acquired is to be thrown away, but to be preserved and used critically.

In our time there are once more natural scientists who are not completely absorbed in special research but are looking for more general guide lines. Höffding appropriately calls them 'philosophizing scientists', in order to distinguish them from philosophers proper. If we take two of them for a start, Ostwald and Haeckel, their outstanding importance in their own fields is certainly incontestable. As regards general orientation, these two are my valued companions in aim, even if I cannot agree with them in every aspect. In Ostwald I further honour a great and successful fighter against rigidity of method, in Haeckel an upright and incorruptible protagonist of enlightenment and freedom of thought. If in a word I were to formulate in which direction I differ most from these two men, it would be this: to me, mental observation is just as important and fundamental a source of knowledge as physical observation. Of the total research of the future it will no doubt be true to say what Hering (*Zur Lehre vom Lichtsinn* Vienna 1878, p. 106) said about physiology: that it will be like the construction of a tunnel from both ends at once, from the physical and the mental. Whatever else Hering's position, on this point I agree with him completely. The striving towards a bridge between and a uniform view of these apparently so different fields lies in the economic constitution of the human spirit. Nor would I doubt that if concepts are suitably transformed this goal is attainable from the physical and from the mental side, and will seem beyond reach only to one who from childhood on remains tied by rigid instinctive or conventional concepts.

If I am not mistaken, a striving after the same goal is also apparent in philosophical literature proper, which is more removed from my concerns. If one looks for example at G. Heymans's book *Einführung in die Metaphysik auf Grundlage der Erfahrung* (1905), most scientists will have as little objection to his clear and simple accounts as to his final position of 'critical psycho-monism'; except that strongly materialist thinkers might balk at the *name*. Still, one is curious why, if on Heymans's view the method of metaphysics is exactly the same as that of natural science, though extended to a wider field, he insists on calling it metaphysics, a term with so distasteful a flavour since Kant, and one that seems contradicted by the addition 'on the basis of experience'. Finally we must keep in mind that, since Newton, natural science has learnt to assess hypotheses, that is unknowns inserted between the given known, at their true and scant worth. It is not the provisional working hypothesis but the method of *analytic* examination that essentially advances science. If then it is very gratifying and encouraging that all of us are seeking in the same direction, nevertheless the remaining differences should warn each of us not to mistake the sought for as found, let alone as the sole doctrine of salvation.

CHAPTER II

A PSYCHO-PHYSIOLOGICAL CONSIDERATION

1. Experience grows through progressive adaptation of thought to fact. Mutual adaptation of thoughts produces the ordered, simplified and consistent system that we conceive as the ideal of science. My thoughts are directly accessible only to me, as my neighbour's only to him, belonging as they do to the mental field. Only when linked with physical features like gestures, facial expression, words, actions can I venture more or less certain inferences by analogy from my experience, which comprises physical and mental parts, to that of others. That same experience moreover teaches me to recognize my thoughts as depending on my surroundings, which includes my body and other people's behaviour. We cannot exhaustively consider the mental by 'introspection': this must go along with examining the physical.

2. How various the things I discover in myself, for example on the way to a lecture: my legs move, one step triggers off the next, without my making any special effort, even if I have to turn an obstacle. I walk past the municipal park and recognize the town hall which reminds me of gothic and Moorish buildings as well as of the mediaeval spirit that pervades its chambers. In the hope of conditions more fit to be called civilized I am just imagining the future when in crossing the road I am grazed by a speeding cyclist and made involuntarily to jump sideways. My phantasies of the future are replaced by silent grumbling against such reckless speedsters. The sight of the ramp of the university entrance finally reminds me again of my object, namely the task of the hour ahead, and speeds my pace.

3. Let us analyse this mental experience into its constituents. First we find those that are called *sensations* insofar as they depend on our bodies (the eyes being open, direction of ocular axis, normal condition and stimulation of the retina and so on), but are physical *qualities* insofar as

they depend on other physical features (presence of the sun, tangible bodies and so on): the green of the park, the greyness and shapes of the town hall, the resistance of the ground I tread, the grazing contact with the cyclist and so on. For the psychological analysis consider the term *sensation*. Towards sensations such as hot, cold, light, dark, vivid colours, the smell of ammonia or the scent of roses and the like, we are as a rule not indifferent. They are agreeable or disagreeable, that is to say our body reacts by means of more or less intense movements of approach or retreat, which in turn present themselves to introspection as complexes of sensations. In the beginning of our mental life distinct and vigorous memories remain of only those sensations to which strong reactions were attached. Indirectly, however, other sensations too may remain in our 'memory'. An otherwise indifferent look at the bottle that contained ammonia evokes the memory of the smell and so ceases to be indifferent. The whole prior experience of sensations, insofar as it is preserved in memory, tinges every new experience. The town hall I passed would be no more than a spatial arrangement of coloured patches, had I not previously seen many other buildings, walked their corridors and mounted their steps. Memories of the most varied sensations here become interwoven with the optical sensation to become a much richer complex, namely a perception, from which the mere immediate sensation is difficult to detach as such. If several people are offered the same optical field of vision, each has his attention excited in a specific direction; that is, their mental life is set into specific motion through strong individual memories. A middle-aged engineer, with his eighteen-year-old son and a little boy of five, goes for a walk through a street. Their eyes have taken in the same pictures: yet they notice different things: the engineer almost only trams, the young man especially pretty girls and the boy perhaps only the toys in shop windows. Innate and acquired organic circumstances intervene in this. Memory traces of earlier experiences that essentially help to determine the mental course of new sets of experiences, imperceptibly interweaving with them and assimilating them by extending the fabric, we may call *ideas*. These differ from sensations only by being less intense and more transitory and changeable, and by combining with one another through association. They are not a new kind of element in contrast with sensations, on the contrary they seem to be of the same nature[1].

4. At first glance, feelings, affects and moods like love, hate, anger, fear, depression, sadness, mirth and so on, seem to be new elements. On closer scrutiny, however, they are less analysed sensations linked with less definite, diffuse and vaguely circumscribed elements of internal space: they mark certain directed modes of bodily reactions known from experience, which, if strong enough, may actually break out into movements of aggression or flight. That this is of much less interest in general than to the individual and that even for him it is much more difficult to observe (for the body's elements are not so open to inspection as are generally accessible external objects and sense organs), makes such facts less well known and more difficult to describe and the terminology for them incomplete. Feelings may be linked with ideas as well as with external sensations. If a mode of reaction develops into a certain aggressive or defensive movement provoked by a set of sensations and aimed at a goal known beforehand, we speak of an *act of will.* If I speak of a walk to my *lecture*, if I am told of the visit of a *foreign scholar*, if a man is described as *just*, I may not be able to read the italicized terms as a definite complex of sensations, but through repeated and various use they have nevertheless gained the property of circumscribing and delimiting complexes within their scope to the extent that my behaviour and reaction to these complexes is thereby determined: words that cannot denote any such complexes would be unintelligible, without meaning. Even for words like red, green, rose, the idea that covers them spans a fairly wide range, in the above examples it becomes wider, and wider still with scientific concepts where at the same time the delimitation of, and way of reacting to, the relevant complexes becomes more sharply defined. There is a continuous transition from the most concrete ideas of ordinary thought to the most abstract scientific ones. This development, made possible through the use of language likewise begins quite instinctively, and issues in conscious methodical application only at the level of the definition of scientific concepts and vocabulary. The apparent gap between concrete ideas of sensation and concepts cannot deceive us as to the continuity between individual idea and concept or to the fact that sensations are the basic elements of all mental experience.

Thus there is no isolated feeling, willing and thinking. Sensations, being both physical and mental, form the basis of all mental experience. Indeed, sensations are always more or less active, triggering off the most varied bodily reactions, directly in lower animals and indirectly via the cortex in higher ones [2]. Mere introspection without constant regard to the body and so to the physical as a whole, of which the body is an inseparable part, cannot establish an adequate psychology. Let us therefore consider organic life as a whole, especially that of animals, attending now to the physical, now to the mental side, choosing examples in which this life takes on especially simple forms.

5. A butterfly flitting from flower to flower on splendid wings, a bee busily collecting honey and storing it, a bright and multi-coloured sand-hopper shrewdly escaping a pouncing hand, all these are familiar pictures of deliberate and thoughtful action. We feel ourselves akin to these small creatures. If however we observe a butterfly repeatedly flying into a flame and singeing itself; or a bee buzzing helplessly near a half-open window and hurling itself in vain against the impenetrable pane, becoming utterly confused when the opening is shifted ever so slightly; or a sand-hopper which the harmless walker by means of his own preceding shadow can flush again and again and chase for miles ahead, when it could simply move aside; then we can see why Descartes was able to conceive animals as machines, some kind of weird and wonderful automata. The virgin Queen Christina with her apt ironic taunt that the propagation of watches was something unheard of may well have made the philosopher see the defects of his view, warning him to be careful.

Looking more closely at the two opposite tendencies of animal life that seem so contradictory, we find both of them clearly marked in our own nature. The pupils of the eye contract mechanically with brightness and dilate with progressive darkness, without our will or knowledge, just as the functions of digestion, nutrition and growth occur without our conscious intervention. However, the arm that stretches to open the drawer of a table when we recall that in it lies a ruler we presently need seems to obey only our well-considered command without external impulse; but an accidentally burnt hand or a foot tickled on the sole will withdraw involuntarily and without reflection, even if a man is asleep or paralysed by a stroke. In the movement of the eyelids, automatically closing when an object suddenly approaches, but also movable at will, as in many other movements such as breathing and walking, the two features occur in constant alternation or combination.

6. Accurate self-observation of the processes we call weighing, deciding, willing, brings us to know a simple state of affairs. A simple experience, such as meeting a friend who invites us home, connects with many memories which become vivid in turn and displace one another: we remember hearing his witty conversation, the piano standing in his room and his excellent playing of it; suddenly we remember that today is Tuesday and that this is the day on which a quarrelsome man generally visits our friend, and so we thank him but decline his invitation. Whatever our decision, in the simplest as in the most complicated cases, memories that become effective influence our movements just as definitely and provoke the same advances or retreats as the respective sense experiences of which they are traces. We do not control what memories come to the surface and which of them will carry the day[3]. In our 'voluntary actions' we are no less automata than the simplest organisms, but the part of the mechanism that undergoes constant small changes through experience is visible only to ourselves while remaining hidden to others; indeed, we ourselves, however attentive, may miss its finer features. Thus what emerges in our voluntary actions is a much larger and less perspicuous or ordered section of the universe and a spatio-temporally much more far reaching set of connections: it is this that makes such actions appear unpredictable. The organs of lower animals react fairly simply and regularly to the overt stimuli. All relevant circumstances seem to be concentrated in a single spatio-temporal point. Here the impression of automatic behaviour arises very readily indeed. Yet more refined observation here too reveals individual differences, some innate, some acquired. Animal memory differs greatly according to genus and species, and also between individuals, though less markedly. From Ulysses' dog on the point of expiring and no longer able to stand up yet still recognizing his master returning after twenty years and greeting him with wagging tail, to the pigeon that hardly remembers a kind action for a day, to the bee that barely finds its way back to a source of food – what a tremendous gap! Are the lowest organisms completely devoid of memory?

That men are inclined to think of themselves as totally different from the most simple organisms is due only to our complex and varied mental life, compared with theirs. A fly whose movements seem to be determined and guided by light, shade, smell and so on, may have been chased off ten times, it will settle back on the same part of the face and cannot give in until it drops to the floor swatted. A poor beggar is anxious to get a penny to secure him for the day and therefore goes on pestering a comfortably somnolent bourgeois until the latter dismisses him with a curse: both act automatically, except that they are somewhat less simple automata.

7. The basic feature of animal and human behaviour is this rigorously determined regular automatism: it is only because we see it in such different degrees of development and complexity that we seem to perceive two quite different basic traits. To understand our own nature it is vital to pursue the determined aspect as far as we can, for it is of no practical or scientific value to perceive an absence of rule. Gain and insight accrue only when we discover the rule in what until now was regarded as lawless. It will always be difficult to refute the assumption of a free soul acting without law, since experience always contains an unperspicuous factual remnant; but to assume a free soul as a scientific hypothesis, or indeed to look for it, is surely a methodological aberration[4].

What we seem to find free, arbitrary and unprecidtable especially in man hovers over the automatic like a thin veil, a mere whiff or enveloping mist. We see human individuals from too close up, as it were, so that the picture is overburdened with to many confusing details that are not immediately perspicuous. If we could observe men from farther afield, in birds' eye view or from the moon, the finer details, with influences deriving from individual experience, would disappear and we should see merely men that grow, feed and propagate with great regularity. A mode of observation that deliberately slurs over the individual, and attends only to the most essential and most strongly linked circumstances, is indeed used in statistics, and then men's voluntary actions show themselves to be just as regular and determined as any vegetative or even mechanical process, where usually nobody thinks of mental or volitional influences. The yearly number of marriages and suicides fluctuates as little or less than the birth and natural death rates, although the former greatly involve volition and the latter not at all. If however, even one lawless element were involved in determining these mass phenomena, then however great the number of cases no further rule could emerge[5].

Thus Descartes could easily have reached the position that not only animals but men too appeared to be automata. The great doubter's aim was indeed to mechanize or rather geometrize the world; but here he may have lost the courage to doubt, in deference both to the strength of the Inquisition and to the hold on him of the traditional views so pointedly expressed in his dualism. Spinoza avoided such inconsistencies. Amongst later writers we must emphasize Lamettrie[6], because of his uniform conception of men and animals, as expounded in *L'homme machine* (1748) and in the essays *L'homme plante* and *Les animaux plus que machines*. One should not look for deep philosophical thought in Lamettrie: though important at the time, his writings make dull reading today. Not so with Diderot, whose penetrating article *Entretien entre D'Alembert et Diderot. Le rêve de D'Alembert* anticipates the ideas of modern biology.

8. People who have tried to grasp nature by reason have always been tempted to simulate living things by automata or machines, thus trying to understand them at least in part. One of the oldest automata recorded beyond mere legend is the flying pigeon of Archytas of Tarentum. Hero of Alexandria[7] too was much absorbed in building automata, and these attempts were later understood more clearly than the scanty enough fragments of ancient science that his writings transmit. In the 16th century we meet ingenious clocks with moving figures of men and animals, in Strasbourg, Prague, Nurenberg and elsewhere; in the 18th century, Vaucanson's swimming and eating duck, and his flute player, as well as Droz' boys that draw and his girl pianist. Much as one might feel like dismissing such items as mere trifles, one must not forget that the knowledge gained in the course of their making can be directly used in scientific enquiries, such as for example Borelli's *De motu animalium* (1680), W. Kempelen with his speaking machine (*Mechanismus der menschlichen Sprache, nebst Beschreibung einer sprechenden Maschine*, Vienna 1791) likewise made significant scientific progress[8]. Much scientific physiology may be viewed as an extension of what the makers of automata did. Kempelen's automatic chess player, in which a man had to be concealed, does of course provide the superfluous proof that intelligence cannot be replaced in this simple mechanical way. Living things are automata that bear the influence of the whole past and always continue to change, that arise from and can in their turn produce similar automata. One is naturally inclined to imitate and reproduce what one has understood. How far one succeeds in this is itself a good test of comprehension. Considering the benefit that

modern makers of machines have derived from the construction of automata, and looking at calculating machines, control mechanisms, vending machines and the like, we may expect further progress in technology. An absolutely reliable automatic post-office clerk who will accept registered letters seems not at all impossible and would be a welcome relief to intelligent beings tortured by sheer mechanical repetitiveness.

From our standpoint there is no reason for further discussing the opposition between the physical and the mental. The only thing that can be of interest to us is the recognition of the interdependence of elements. That these relations are rigid, if complex and hard to discover, it will be reasonable to presuppose in embarking on our investigations. Past experience affords us this presupposition and every new success in enquiry reinforces our confidence in it, as will emerge more clearly from the special investigations that follow.

NOTES

[1] Cf. *A* 4, p. 159.

[2] Cf. A. Fouillée, *La psychologie des idées-forces*, Paris 1893. This correct and important notion is there presented at rather great length in two volumes.

[3] Mistaking this fact during subsequent reflection produces repentance, which is sensible and important only for future repetitions of the same or similar situations. And there it is not punishment or expiation that matter but only a change of attitude. The question of freedom and responsibility can refer only to whether an individual is sufficiently developed mentally to take into account the effect on himself and others when deciding on his actions. Cf. the views maintained by A. Menger in his remarkable book *Neue Sittenlehre* (Jena 1905). The courageous truthfulness that he evinces in all his writings deserves our respect.

[4] The views maintained in the writings of H. Driesch derive from philosophical foundations that are totally different.

[5] On this point I have already made some comments: 'Vorlesungen über Psychophysik', *Zeitschr. f. praktische Heilkunde*, Vienna 1863, pp. 148, 168, 169.

[6] Lamettrie, *Oeuvres philosophiques*, précédées de son élogement par Frédéric II, Berlin 1796.

[7] Hero's Works, published by W. Schmidt, Leipzig 1869, Vol. 1.

[8] What remains of Kempelen's speaking machine is in the physical collection of the Technical College of Vienna (communication from Prof. A. Lampe).

CHAPTER III

MEMORY. REPRODUCTION AND ASSOCIATION

1. On a walk through the streets I meet a man whose face, frame, gait and speech provoke in me a lively idea of such a set of charateristics in different surroundings. I recognize *X* who stands before me as a sense experience to be the same as forms a part of my memories from the other setting. Recognition and identification would be senseless unless *X* were given twice over. Soon I recall previous conversations with him in the other surroundings, joint excursions and so on. Similar situations are observed in the most varied circumstances, and we may gather them under a rule: a sense experience, consisting of *ABCD*, revives the memory of an earlier one consisting of *AKLM*, thus reproducing it as an idea. Since *KLM* is not generally reproduced by *BCD*, we naturally take it that the common element *A* starts the process. First *A* is reproduced, and then follow *KLM* which were directly experienced with it or with other simultaneous features already reproduced. All processes in this field can be subsumed under this one law of association.

2. Association has great biological significance since on it depends any mental adaptation to one's surroundings and any ordinary or scientific experience. If the surroundings of living things did not consist of parts that remain at least approximately constant or that recur periodically, experience would be impossible and association useless. Only if the surroundings remain constant can a bird connect the visible portion of it with the idea of the location of his nest. Only if it is always the same noise that foretells the approaching enemy, or the fleeing quarry, can the associated idea serve to trigger off the corresponding movement of flight or attack. An approximate stability makes experience possible, and the fact of this possibility in turn allows us to infer that stability. Our successes justify the stability presupposed by scientific method[1].

3. A new-born child, like the lower animals must rely on reflex movements. It has the innate drive to suck, cry when in need of help and so on.

As it grows up it acquires, like higher animals, its first primitive experiences through association: it learns to avoid touching flames or colliding with hard bodies as being painful, to link the seeing of an apple with the idea of the corresponding taste and so on. Soon, the child far surpasses all animals as to the wealth and subtlety of experience. Much can be learnt from observing how associations are formed in young animals, as C. L. Morgan[2] has done with chicks and ducklings hatched by incubator. A few hours after hatching, chicks already possess appropriate reflex movements. They walk and peck with accurate aim at noticeable objects. Partridge chicks even run off still partially covered with eggshell. To begin with, chicks pecked at everything: the letters of a printed sheet, their own claws and excrements. In the last case, however, the chick immediately rejected the evil-tasting object, shaking its head whetting its beak clean on the ground. The same happened when the chick had picked up a bee or a caterpillar of disagreeable taste, but pecking at unsuitable, or inexpedient objects soon ceases. A dish of water they leave alone, but start drinking at once if they happen to walk into a puddle.[3] Ducklings, on the other hand, fairly hurl themselves into the dish, washing themselves, diving and so on. The day after, on being presented with an empty dish, they stormed into it again and performed the same actions as in water; but soon they learnt to distinguish an empty dish from a full one. I myself once put an empty beaker over a chick a few hours old and brought a fly into its company. At once a rather amusing chase began, but without success; the chick was as yet not deft enough.

4. The behaviour of chicks and ducklings is innate and occurs without any instruction, being prepared by mechanisms of motion. The same holds for their cries: in chicks we distinguish sounds of comfort, when they creep into the warmth of an outstretched hand, cries of danger at the sight of a fat black bug, a cry of loneliness and so on. Whatever is here mechanically prepared and innate and however much anatomy may favour and facilitate the forming of certain associations, these latter as such are not innate but must be acquired through individual experience.

This will indeed be so if we confine the term 'association' to conscious ideas. If we take it in the wider sense of simultaneous processes triggering each other off, the boundary between the innate (or hereditary) and the acquired may become rather difficult to draw. This must indeed be the case if what a species has acquired is to be amplified or modified by the individual. My tame sparrow is quite unafraid, alights on the shoulder of any member of the family, tugging at hair or beard and fighting off with vigour and angry chirps the hand that would chase him from the shoulder of his choice. However, his wings flutter nervously at any noise or movement in his surroundings. Whenever at table he snatches small bits he flies off, even if only a foot, just as his colleagues in the street, even though he is not being disturbed by any companions.

Chicks raised in an incubator take no notice of a hen's clucking and show no fear of cars or hawks. If kittens that are still blind really start spitting if touched by a hand that has just stroked a dog, one would have to regard this as a manifestation of an olfactory reflex.[4] Of course unusual phenomena easily frighten young animals: chicks fed on small worms will occasionally swallow twisted wool but stop in doubt in front of a large piece of wool; my daughter observed that a tame young sparrow would not go near his food tray for a while when by way of experiment she put a fat meal-worm into it.[5] For many animals the fear of what is unusual and striking seems indeed to be one of the chief means of protection.

5. In more highly developed animals the formulation of associations is even more marked and at the same time more lasting. In the village where I spent part of my youth many dogs molested by village boys had adopted the habit of slinking off on three legs as soon as anybody picked up a stone. People naturally tended to regard this as a clever trick to provoke sympathy. Of course it was only a vivid associated memory of the pain that had at times followed the picking up of the stone. I once saw one of my father's young pointers fiercely root up an antheap and then desperately clean his sensitive nose with his paws. Thenceforth he carefully respected formic habitations. When on one occasion this dog kept on disturbing me in my work with his unwanted and excessive affection, I slammed a book shut in front of his nose and he retreated in fright: from then on the mere grabbing of a book was enough to forestall any disturbance. To judge by his muscular jerks during sleep, this dog must have had a vivid dream life. Once when he lay quietly asleep, I placed a small piece of meat near his nose and soon his muscles began to move strongly especially round the nostrils: after about half a minute he awoke, snapped up the meat and then

went on sleeping in peace. Besides, his associations lasted: when after nine years' absence I returned unexpectedly to my father's house on foot and in the dark, the dog received me with angry barking, but a single call sufficed to provoke the friendliest behaviour. Homer's tale of Odysseus and his dog is surely no poetic exaggeration.[6]

6. The import for mental development of comparing a sense experience *ABCD* with an idea of a reproduced sense experience *AKLM* cannot be overrated. Let the different letters stand for whole complexes of elements. For example let *A* be a body previously encountered in a setting *BCD*, but now in a setting *KLM*; let us say a body moving in front a of background and thereby recognized as being a separate and relatively independent thing. If now we assign separate letters to separate elementary sensations, we come to recognize these elements as independent constituents of our experience, so that the reddish yellow *A* occurs not only on an orange but in various different complexes: in a piece of cloth, a flower or a mineral. However, association is the basis not only of analysis but also of combination. Let *A* be the visual image of an orange or a rose, whereas *K* in the reproduced complex stands for the taste of the orange or the smell of the rose. We immediately associate the renewed visual image with the previously experienced characteristics. The ideas provoked by the things surrounding us therefore do not correspond precisely to our actual sensations, but usually are much richer. Whole bundles of associated ideas, provoked by previous experiences, interweave with actual sensations and condition our behaviour much more extensively than these latter could do on their own. We see not only a reddish yellow sphere, but think we perceive a soft, fragrant, refreshing and tart-tasting object; we see not a brown, vertical, shining surface, but for example a wardrobe. By the same token we may at times be deceived by a yellow wooden sphere, a painting or a mirror image. As we grow older the range and wealth of sense-experience grows and with it the number and range of associable connections between them. Hence, as we saw, the progressive analysis into components as well as the constant formation of new syntheses out of them. Once the life of ideas has become strong, complexes of ideas can be reproduced and associated like sense experiences. Here too new analyses and syntheses appear, as any novel and any scientific paper shows and as any thinker can observe in himself.

7. Although we can discover only one principle of reproduction and association, namely that of simultaneity, the run of ideas varies greatly in different cases, as the following reflection will make clear. In the course of a lifetime most ideas have become associated with many others, and these widely divergent associations partly resist and weaken one another. Unless some convergent ones happen to attain dominance or chance especially favours a particular one, none of these associations will become effective. Can anyone tell where and when he has become acquainted with, or seen in use, some particular letter, word, concept, or method of calculation? The more often he uses them, the more familiar they are to him, the less likely it becomes that he should remember. The name Smith, in whatever spelling, is linked with so many different fields and occupations, that on its own it does not trigger off any association. According to what I happen to be thinking or doing at the moment, the name can remind me of an ecomonist, a clergyman, an archaeologist, a geologist and so on. The same can be observed with more uncommon names. I used frequently to travel past an advertisement for Maggi's meat extract, but only once, when I happened to be thinking about some physical problem, was I reminded of the man of that name who wrote a book on mechanics that was of interest to me. Likewise, a blue cloth will not suggest anything to an adult, while a child may remember the cornflower it picked the day before. The name 'Paris' may summon up the collections of the Louvre, or the famous Parisian mathematicians and physicists, or the splendid restaurants, depending on whether I happen to be in the mood for art, science or gastromony. Even circumstances factually unrelated to one's particular line of thought can become decisive. It is reported of Grillparzer that a poetic sketch completely forgotten through a long illness sprang back to mind when he played again the symphony that he had been playing when first engaged on the sketch. Even unconscious mediating links can produce associations, see the case reported by Jerusalem.[7] In these cases, the principle of simultaneity stands out clear and unalloyed.[8]

8. Consider now some typical ways in which ideas run.[9] If I give free rein to my thoughts, without plan or purpose and as far as possible in isolation, as for example during a sleepless night, I soon drift into all manner of subjects: the comic and the tragic, remembered or invented, constantly alternate with scientific inspirations and plans, so that it would be very diffi-

cult to identify the small accidental features that momentarily direct this flow of 'free phantasy'. Much the same occurs when two or three people freely chat with one another, except that here their thoughts exert mutual influences. The surprising twists and turns of conversation often lead to the wondering query: how on earth did we come to talk about *that*? With several observers, their thoughts fixed in spoken words, an answer becomes easier, indeed rarely fails to emerge. It is in dreams that ideas proceed most strangely, but the thread of association is here most difficult to follow, partly because of incomplete memories left behind by the dream and partly because the sleeper is disturbed more often by slight sensations. Situations experienced in a dream, such as figures seen or melodies heard, are often extremely valuable as a foundation for artistic creation,[10] but the enquirer can use them only in very exceptional cases.

9. Lucian's adventure story *The True History* no longer quite corresponds to free phantasy. This most witty of ancient romancers here makes a principle of retaining only the most adventurous and improbable of his notions. He invents enormous spiders that span the space between Moon and Venus with webs affording a path, and playfully bids Moondwellers drink liquid air 1700 years before any was actually produced. Using an itinerary as guideline for his phantasies, he visits amongst other places the Isle of Dreams, whose indefinite and contradictory character he admirably describes by stating that it recedes in equal measure as the traveller approaches. In spite of this surfeit of rank phantasy we can nevertheless discover the threads of association, unless they have been deliberately concealed. The journey begins at the Columns of Hercules and proceeds westward. After 80 days he reaches an island with a commemorative column and an inscription of Hercules and Dionysus together with their giant footprints. Naturally there is also a river of wine containing fish that make one tipsy if eaten. The springs of this river well up near the roots of lush vines, and on its banks one meets women who like Daphne have been half transformed into vines. At this point the thread of association has grown into quite a carrier rope. Elsewhere the author has cut back the shoots and blooms of his phantasy, where they failed to meet his aesthetic and satirical ends. It is this rejection of what is unsuitable that distinguishes the play of ideas, however free, in a literary or other work of art, from the aimless abandon to one's own ideas.

10. If I arrive in a place or region where I have spent part of my youth and I simply abandon myself to the impressions of the locale, the ideas take quite a different kind of turn. Everything that now impinges on my senses is so richly linked with experiences of my youth and little or not at all with later ones, that the events of that earlier period gradually emerge from oblivion quite accurately and tightly interconnected, as to space and time. In such a case one always feels involved, as Jerusalem[11] aptly remarks. We can therefore use the person as the thread for arranging the elements of memory in temporal order. Just so, if somewhat less completely, if the picture of home comes to mind, provided only one is not disturbed but is given time for the picture to complete itself. For example, everyone knows tales the old tell of their youth, or reports about holidays and all that happened there, to the last merciless detail.

11. The previous example concerns essentially the reviving of already existing connections between ideas and memories; on the other hand, solving verbal or other puzzles, geometrical or constructional tasks, scientific problems or the execution of an artistic design and so on, involves a movement of ideas with a definite goal and purpose: we seek something new, as yet imperfectly known. Such a movement, which never loses sight of the more or less circumscribed goal, is called reflection. If someone standing in front of me proposes a riddle or problem or if I sit down at my desk on which traces of my preparations for work are already visible, this posits a set of sensations that constantly redirects my thoughts to the goal, and thus prevents aimless meanderings. Such external constrains on thought are valuable in themselves. If, with some scientific task in mind, I finally fall asleep exhausted, this external monitor and guide vanishes at once, and my ideas become diffuse and leave the appropriate pathways. That is partly why scientific problems are so rarely furthered in dreams: but if the involuntary interest in the solution of a problem grows strong enough, the external monitors become superfluous: whatever one thinks or observes will of itself lead back to the problem, sometimes even in dreams.

The idea we seek by reflecting must satisfy certain conditions, it must solve a riddle or problem or make a construction possible. The conditions are known, the idea is not. To illustrate the process that leads to the solution, consider a simple geometric construction; the form of the procedure is the same in all the relevant cases, so that one example suffices to elucidate them all (see Figure 1). Two lines a and b at right angles and cut by a skew line c form a triangle into which a square is to be inscribed with corners on a, b, their intersection and c. Let us try to imagine a square that fulfils all these conditions. The first three corners do so at once if two adjacent

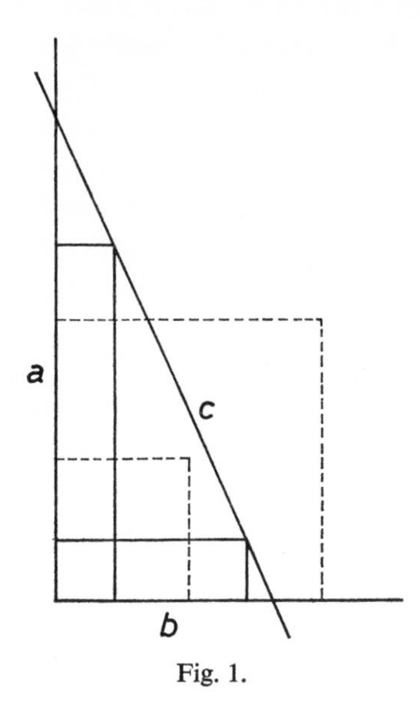

Fig. 1.

sides of the square lie along a and b. The fourth will generally fall either inside or outside the triangle. If one corner is arbitrarily taken on c, the rectangle with this and the intersection of a and b opposite corners is generally not a square. However, as the corner on c descends we pass from an upright rectangle to a level one, so that in between we must reach a square. Thus amongst inscribed rectangles we can select one arbitrarily close to the square. However, we may proceed differently, starting with a square whose fourth corner falls inside the triangle and then increasing the side of the square until that corner falls outside: in between, it must fall on c. In this sequence too we can select the required square arbitrarily closely. Such tentative soundings of the area in which the solution is to be sought naturally precede complete solution. Ordinary thought may feel satisfied with a solution that is near enough in practice. Science requires the shortest, clearest and most general solution, here obtained by recalling that all inscribed squares share as diagonal from the intersection of a and b the bisector of the angle there, which cuts c at the fourth corner required, so enabling us to complete the required square. Though the example we have just discussed in detail is simple, it nevertheless brings out clearly the essential points of problem solving, namely experimenting with ideas and memories[12] and the indentification of the well-known solution. The riddle is solved by an idea that has properties corresponding to conditions *ABC*. Association gives us series of ideas with property *A*, property *B* and so on. The term or terms belonging to all these series, the point where they all intersect, resolves the problem. We shall return to this important issue later, here we were merely concerned to illustrate the type of succession of ideas we call reflection[13].

12. The preceding establishes that reproducible and associable memory traces of sense experience are important for the whole of our mental life; at the same time it shows that psychological and physiological enquiries cannot be separated, since they are intimately linked even within the elements.

13. This ability to be reproduced and associated moreover constitutes the basis of 'consciousness'. A constant unchanging sensation can hardly be called consciousness. Already Hobbes has pointed out that to feel always the same comes to the same as not to feel at all.[14] Nor can one see any point in assuming the existence of some form of energy different from all others and peculiar to consciousness. In physics it would be otiose and superfluous, nor would it explain anything in psychology. Consciousness is not a special mental quality or class of qualities different from physical ones; nor is it a special quality that would have to be added to physical ones in order to make the unconscious conscious. Introspection as well as observation of other living things to which we have to ascribe consciousness similar to our own shows that consciousness has its roots in reproduction and association: their wealth, ease, speed, vivacity and order determine its level. Consciousness consists not in a special quality but in a special connection between qualities. As to sensation, one must not try to explain it: it is something so simple and fundamental that it is impossible, at least at present, to reduce it to something even simpler.

Besides, a single sensation is neither conscious nor unconscious: it becomes conscious by being ranged among the experiences of the present[15].

Whatever disturbs reproduction or association, disturbs consciousness, which can range from total clarity to total unconsciousness in dreamless sleep or fainting. Temporary or permanent disturbances of the connection between brain functions correspondingly disturb consciousness. Comparing anatomic, physiological and psychopathological facts we are forced to assume that the integrity of consciousness hinges on the integrity of the cerebral lobes. Different parts of the cortex retain the traces of different sense stimulations (visual, aural, tactile and so on). The different cortical areas are variously linked by 'association fibres'. Whenever an area ceases functioning or a link is severed, mental disturbances supervene.[16] Details aside, let us consider a few typical examples.

14. The idea of an orange is a very complicated business. Shape, colour, taste, smell, touch and so on are interwoven in a peculiar way. When I hear the word 'orange', the sequence of acoustic sensations calls forth the whole bundle of these ideas as though strung together. Moreover there are the memories of sensations that went with previous enunciations of the word, or with previous writing movements or previous sights of the written or printed word. If therefore there are special visual, aural and tactile areas in the brain, the failure of one of these, by suppression of its function or severance of its links with the other areas, must produce peculiar phenomena, as has indeed been observed. Suppose the visual or aural field remains active but its associative fields are severed, we find a kind of mental blindness or deafness as Munk has observed in dogs whose cerebrum had been operated on.[17] Such animals can see but cannot understand what they see: they fail to recognize the food tray, whip or menacing gesture; they can hear, but fail to follow a call, that is fail to understand it. Physiological observations are here confirmed and supplemented by psychopathological ones, especially the study of language disturbances.[18] For the meaning of words lies precisely in the mass of associations that they awaken, and correct use in turn depends on the existence of these associations: disturb these latter and marked effects must follow. Most people are right-handed and therefore adapt the left cerebral hemisphere to more delicate operations including speech. Broca recognized the importance of the posterior third of the third frontal

convolution for articulate speech, which is lost whenever this part of the brain suffers disease (apoplexy). Besides, aphasia can be determined by many other defects. For example, the patient might remember words as sounds and even be able to write them down, but unable to pronounce them though his tongue and lips can move: the motor image is missing and therefore the appropriate motion is not triggered off. Or the visual and motor images of writing may be missing (agraphia), or the ideas may be present and the acoustic image absent; or conversely, the spoken or written word may not be understood and thus fail to trigger off associations (word-deafness, word-blindness). A case of this last kind was observed in his own person by Lordat, who recorded it after recovery: he gives a vivid account of the moment when after dreary weeks he saw the words "Hippocratis opera" on the spine of a book in his library and could read and understand them again.[19] These few abbreviated summary accounts alone show how many links between sensory and motor areas must be taken into account here.[20] Lesser language disturbances such as ordinary mistakes in speaking and writing occur even in quite normal people as a result of temporary fatigue and distraction. Spoonerisms are an example of this.[21]

15. An interesting case of mental blindness is quoted by Wilbrand.[22] An educated and well-read businessman had an excellent visual memory so that the facial features of those he remembered, the shapes and colours of objects he was thinking about, stage-sets and landscapes he had seen all stood clearly before his mind in every detail. He was able to 'read off' from memory sections of letters and passages of several pages from his favourite authors: he saw the text with all its detail in front of him. His aural memory was slight and his sense for music entirely lacking. After some grave worries that later turned out to have been unfounded, he was confused for a time and then underwent a complete change in his mental life: his visual memory was entirely lost, and on revisiting a town he always thought it was new, as though it were his first visit. The features of his wife and children were strange to him and when he saw himself in a mirror he took himself for a stranger. If now he wanted to work out sums, which previously he had done through visual ideas, he had to whisper the numbers; likewise, he had to use aural ideas and ideas of speaking and writing movements in order to take note of turns of phrase or to remember written texts.

Equally remarkable is another case quoted by Wilbrand.[23] A woman suddenly collapses and is afterwards taken for blind, because she fails to recognize anyone around her. Except for a gradually improving contraction of the visual field, the fit leaves behind only the loss of visual memory, of which the patient is quite aware. She makes a striking comment on this: "to go by my condition, one sees more with the brain than with the eye, for I see everything clearly but cannot recognize it and often cannot tell what it might be".[24]

16. In view of all this we must say that there is not one memory but that memory is made up of many partial memories that can be separated from one another and be separately lost. To each partial memory corresponds a portion of the brain, some of them fairly accurately localizable even now. Other cases of loss of memory seem less readily assignable to a principle. Consider a few of those collected by Ribot (*Les Maladies de la mémoire*, Paris 1888).

A young woman, much in love with her husband, suffered a severe spell of post-partum amnesia, so that she could not remember her married life at all, while prior memories remained intact. Only by her parents' testimony was she brought to acknowledge husband and child as hers. The loss of memory remained irreparable.

Another woman fell asleep for two months and on waking fails to recognize anybody while having forgotten everything she had ever learnt. However she readily and swiftly learnt it all again without remembering that she had known it before.

In another case a woman accidentally fell into water and nearly drowned. On opening her eyes after rescue she fails to recognize her surroundings and has lost speech, hearing, smell and taste, and has to be fed. Every day she begins learning afresh and gradually she improves. Finally she remembers a love affair and her fall into the water, and the cure is effected through jealousy.[25]

17. Most peculiar of all are cases of periodic amnesia. After being asleep for a long time, a woman had forgotten everything she had learnt and had to start learning to read and reckon, and to recognize her surroundings. After some months another episode of sleep supervenes, and this time as before she remembers her youth but has forgotten what happened

between the two sleeping fits. From then on for four years her consciousness and memory is alternately in one or other of the two states. In the first she has a beautiful handwriting, in the second a defective one. People whom she is to know in both states must have been presented to her in each (cf. the often quoted case of a messenger who loses a parcel while drunk and can find it again during his next bout). If awake one finds it difficult to remember even a lively dream, just as conversely while dreaming we often lose the feel of real conditions. On the other hand, the same situations often recur in dreams. Finally, everyone, even when awake, can observe the change of mood with which memories from different periods of one's life rise simultaneously into consciousness. All these cases form a continuous transition from abrupt separation of different states of consciousness to almost complete effacement of the boundary. We may view them as examples of the formation of different centres of association, round which groups of ideas congregate, as time and mood might favour, while between these groups there is little or no connection.[26]

18. If we ascribe to organisms the property of adapting successively better to recurring processes, then we may recognize what is ordinarily called memory as part of a general organic phenomenon: namely adaptation to periodic processes, insofar as it is directly conscious. Heredity, instinct and the like may then be described as memory reaching beyond the individual. R. Semon (*Die Mneme*, Leipzig 1904) is probably the first to attempt a scientific account of the relation between heredity and memory.[27]

NOTES

[1] Experience has taught us to recognize stabilities; our mental organisation easily adapts itself to them and affords us advantages. Next, we consciously and arbitrarily introduce the presupposition of further stabilities expecting further benefits if the presupposition proves itself. The assumption of a concept given *a priori* as foundation of this methodological procedure we neither need nor would derive any advantage from. It would be a mistake in view of the obviously empirical formation of this concept.

[2] C. L. Morgan, *Comparative Psychology*, London 1894, pp. 85f.

[3] This however is also the behaviour of decorticated birds. The phenomenon thus would seem to rest on an ancestrally acquired reflex; cf. the end of this chapter.

[4] Schneider, *Der tierische Wille*, Leipzig 1880.

[5] Observation by my daughter.

[6] Next to the writings of Morgan, the following are instructive as to the psychology of lower and higher animals: K. Möbius, *Die Bewegungen der Tiere und ihr psychischer Horizont* (Monographs of the Naturwissensch. Verein f. Schleswig-Holstein 1873);

A. Oelzelt-Newin, *Kleinere philosophische Schriften. Zur Psychologie der Seesterne*, Vienna 1903. Amongst older books I should recommend H. S. Reimarus, *Triebe der Tiere* 1790, and J. F. H. Autenrieth, *Ansichten über Natur- und Seelenleben* 1836.

7 Wundt, *Philosophische Studien* X, p. 323.

8 That not all mental processes are explicable in terms of temporally acquired conscious associations will be discussed later. Here we are concerned with what is intelligible by means of association.

9 Cf. James, *The Principles of Psychology* I, pp. 550–604.

10 Well-known cases of this sort are the following: Voltaire dreams a complete variant canto of the 'Henriade'. Stranger still is Tartini's dream of the devil playing him a movement of a sonata that the violinist had been unable to achieve when awake, unless the report be a compromise between fiction and truth.

11 Jerusalem, *Lehrbuch der Psychologie*, 3rd. ed. Vienna 1902, p. 91.

12 These questions will be discussed more fully later.

13 One might be tempted to regard 'active' thinking as essentially different from 'passive' idling of thoughts. Yet just as we have no control over sense impressions and memories triggered off by a bodily action, so we have none over a notion of direct or indirect biological interest which always recurs and connects with new series of associations. Cf. *P* 3, pp. 287–308.

14 Hobbes, *Physica* IV, 25.

15 Anyone who thinks he can build up the world out of consciousness cannot have clearly grasped how complicated is the fact of consciousness. Succinct accounts eminently worth reading on the nature and conditions of consciousness are to be found in Wernicke, *Gesammelte Aufsätze*, Berlin 1893, pp. 130–145. Cf. also the lectures of Meynert cited below.

16 Meynert, *Populäre Vorträge*, Vienna 1892, pp. 2–40.

17 We can hardly doubt that the different parts of the brain perform differently. If nevertheless, as shown by Goltz, one part of the cortex can gradually function as substitute for another, we cannot think in terms of abrupt limits to functions but only of a gradual localisation in the sense of R. Semon (*Die Mneme*, Leipzig 1904, p. 160). Cf. also *A* 4, p. 155.

18 Kussmaul, *Störungen der Sprache*, Leipzig 1885.

19 *Ibid.*, p. 175.

20 *Ibid.*, p. 182.

21 On remarkable disturbances in musicians resembling aphasia and agraphia see R. Wallaschek, *Psychologie und Pathologie der Vorstellung*, Leipzig J. A. Barth 1905.

22 Wilbrand, *Seelenblindheit*, Wiesbaden 1887, pp. 43–51.

23 *Ibid.*, p. 54.

24 *Ibid.*, p. 57.

25 A. Forel, *Der Hypnotismus*, 6th ed., pp. 236f. contains the description of a most peculiar case of amnesia.

26 In the light of such periodic disturbances of memory, observations like those of Swoboda (*Die Perioden des menschlichen Organismus*, 1904) seem not at all so extraordinary as they might look at first glance.

27 C. Detto, 'Über den Begriff des Gedächtnisses in seiner Beleuchtung für die Biologie', *Naturwiss. Wochenschr.* 1905, No. 42. The author will hardly suppose that Hering or Semon would fall into the errors he criticizes. However, I think he underrates the advantage of examining the organic from two sides. *Psychological* observation can reveal the existence of *physical* processes that we should not so readily come to know by way of physics.

CHAPTER IV

REFLEX, INSTINCT, WILL, EGO

1. Before continuing with our psycho-physiological consideration, we remark that none of the special sciences required has reached the desirable degree of development to be able to serve as secure foundation for the others. Observational psychology needs considerable support from physiology and biology, but these in turn are as yet very imperfectly accountable for in terms of physics and chemistry. Therefore all our reflections are to be taken as provisional, with results that remain problematical and largely to be corrected by future research. Life consists of processes that actually preserve, repeat and spread themselves encompassing gradually increasing amounts of 'material'. Thus, vital processes resemble a fire, with which they are in any case related though in a somewhat complex way. Most physico-chemical processes on the other hand soon run to a halt, unless constantly resuscitated and kept going by special external circumstances. Even apart from this main difference in character, modern physics and chemistry remain rather inadequate at keeping track of the details of the vital process. In view of the main feature, self-preservation, we must expect that the parts of a complicated organism, or symbiosis of organs, will be attuned to the preservation of the whole that could not otherwise result. Likewise, it will not surprise us to find a similar trend towards preservation of the organism in those mental processes that are conscious (i.e. occurring in the cerebrum).

2. Consider first some facts that Goltz[1] has studied in detail. A healthy undamaged frog behaves in a way that leads us to attribute a certain 'intelligence' and 'voluntary' movement to the animal; it moves unpredictably of its own accord, fleeing from predators, seeking a new pond when the old one dries up, escaping through a gap of a trapping container, and so on. On a human scale, the intelligence is indeed very limited: a frog very cleverly snatches at flies in motion, and sometimes at small bits of red cloth, and tries to grasp the horns of a snail though failing repeatedly, but he will starve to death rather than eat fresh-killed flies. The frog's behav-

iour is narrowly adapted to his conditions of life. If his cerebrum is removed, he will move only under external stimulus, without which he simply sits still, without snatching at flies or red cloth or reacting to noise. If a fly crawls over him he brushes it off, but swallows it if put in his mouth. A weak stimulus on the skin makes him crawl away, a strong one makes him jump, avoiding obstacles, which shows that he can see them. If one leg is tied up, he will still avoid the obstacle in crawling. If he is put on a horizontal turntable, he will compensate for rotation. If put on a plank which is raised at one end, he will crawl up so as not to fall off, and overshoots the upper end if one continues turning the plank further in the same sense: an undamaged frog would jump off. Thus, what one might call soul or intelligence becomes restricted when parts of the brain are taken away. A frog possessing only a spinal cord cannot get up if put on his back. The soul, says Goltz, is not simple, but divisible, like its organ.

A frog without cerebrum never croaks spontaneously. However, if stroked with a wet finger on the back between the shoulders, he regularly croaks once, by reflex, like a mechanism. That beheaded frogs will wipe off acid with their hindlegs had been observed even earlier. Such reflex mechanisms are important for survival. Goltz has shown many important vital functions are ensured by these mechanisms, for example the mating of frogs.[2]

3. Consider next other living things to which, instinctively at least, no one attributes intelligence and volition; namely plants. Here too, we find appropriate reactive movements conducive to preservation of the whole. Amongst these we first notice sleeping movements of leaves and flowers, determined by light and temperature, and the stimuli of insectivorous plants provoked by shocks. Such movements might of course appear to be exceptional. However, it is a streak of general behaviour that the stem of plants grows upwards against gravity, where light and air allow assimilation to take place, while roots penetrate downwards into the soil, seeking water and the substances dissolved in it. If part of a stem is forced out of plumb, the parts that are still growing at once bend upwards with their convex side towards the ground, owing to stronger growth on the lower side. This exhibits the 'negative geotropy' of the stem, in contrast with the root, which is called 'positively geotropic'. The stem generally turns towards light, the growing parts being convex towards darkness; that is,

the shady side grows more strongly. Such a stem is called 'positively heliotropic', while roots generally behave in the opposite, 'negatively heliotropic' way. Research both old and new shows that the direction of gravity and of light determine geotropic and heliotropic behaviour. The opposite thrust of stem and root indicate division of labour in the interests of the whole. When we see a root splitting rocks as it tends downwards, we may still imagine that it does so in its own interest. However, this impression fades when we see the root doing the same in mercury, where it is out of place: the notion of deliberate purpose must yield to that of physico-chemical determination, while the determining factors must be thought of as arising from the combination of root and stem into a whole.[3]

4. J. Loeb[4] in a series of studies has shown that concepts like geotropy and heliotropy that come from plant physiology can be extended to animal physiology; especially, of course, where animals live under such simple conditions that no highly developed mental life is needed and thus cannot interfere. A butterfly fresh from its chrysalis crawls upwards head first and finds its bearings, preferably on a vertical wall. Newly hatched caterpillars crawl restlessly upwards: to coax them out of a reaction bottle one has to turn it with the opening pointing upwards, as with a hydrogen jar. Cockroaches prefer vertical walls. Flies with their wings cut off crawl vertically up a plank, compensating for any rotation of the plank in its own plane, while on an inclined plane they crawl along the line of steepest ascent. Even highly developed animals are geotropic, as is shown by recent findings concerning the inner ear, and its role in orientation, although here geotropy is overlaid by the intervention of other factors.

Likewise with heliotropy: as for plants, so for animals, the direction of light is important. Unsymmetrical light stimuli cause the animal to reorientate itself into the plane of symmetry relative to the light, pointing either front or back towards the source, moving towards or away from it: the animal is positively or negatively heliotropic. A moth is positive, worms and muscid larvae negative. When a positively heliotropic larva moves on a plane, it crawls along the light-vector's component in that plane. In thus moving towards the incident light it may well reach a region of smaller illumination. Without going into further detail, we observe complete agreement between J. V. Sachs' studies on plants and Loeb's on animals.[5]

5. Most recently, a strong opposition has emerged amongst views regarding insects: some see them only as reflex machines, others as having a full mental life, according to the scientists' like or dislike for mysticism, to which all mental features are attributed to be either saved or completely discarded. As for our own position, the mental is no more and no less mysterious than the physical, from which indeed it does not differ essentially. Hence we need not take sides, but remain neutral somewhat like A. Forel.[6] If, for example, touching a spider's web with a vibrating tuning fork repeatedly enables us to mislead the animal, this shows how strong is its reflex mechanism; but when in the end it tumbles to the trick and no longer comes out, we cannot deny it memory. Horseflies that buzz helplessly on the pane of half-open windows trying to move towards air and light but hemmed in by the narrow frame of the pane indeed give the impression of automata. However, if so closely related an animal as the more graceful housefly behaves rather more cleverly, we must presuppose some ability, however limited, to learn a little from experience. Thus the topochemical memory and sense of smell attributed to ants by Forel seems a happier assumption than Bethe's polarized scent[7] (i.e. towards or away from the antheap). Forel even claims to have trained a water bug to eat on land, against the animal's normal habit: this cannot be an automaton in the narrow sense. Similarly, he has demonstrated the ability to remember and distinguish colours in bees and wasps.

6. It is worth following up the large common traits of organic life in animals and plants. With plants, everything is simpler and more accessible to observation, and occurs more slowly. What we observe in animals as motion, expression of instinct or arbitrary action, manifests itself in plants as growth through a series of forms or as shape of leaves, flowers, fruit or seed, fixed for permanent observation. The difference thus rests largely in our subjective time scale. Imagine the slow movements of a chameleon slowed down even further, while the slow grasping motions of lianas are much accelerated[8]: the observer will find the difference becoming blurred. There is little temptation to treat processes in plants as psychological, so that the tendency to treat them as physical is all the stronger. With animals it is the other way round. However, the two fields are closely related, so that this change in approach is quite instructive and fruitful. Moreover, the mutual influences between animals and plants,

both as regards physics and chemistry, and morphology and biology, are most suggestive. Take, for example, the mutual adaptation of flowers and insects discovered by Sprengel in 1787 and revived by Darwin's work on orchids[9]: seemingly independent living organisms are here mutually determined and interdependent like the parts of a single animal or plant.

7. The movements that occur through stimuli without involving the cerebrum are called reflex movements: they are prepared by the connection and mutual adjustment of organs. Animals can go through quite complicated motions that seem to aim at a certain goal or purpose, although we cannot credit the animal with knowing or consciously pursuing them. Thus we speak of instinctive actions, which may be taken as a chain of reflex movements each triggered off by its predecessor.[10] Take a frog snatching and swallowing a fly: clearly, the first act is set off by optical and acoustic stimuli. That swallowing is a consequence of snatching we infer from the fact that the frog without cerebrum no longer snatches but continues to swallow flies put in his mouth. Likewise for nestlings that have not yet learned to pick up their food: at the approach of whoever tends them they open their beaks and cheep, perhaps from fear, and devour the food put in: pecking and snatching come later. The hamster's gathering food for winter may become intelligible if we consider that this is a rather voracious, quarrelsome yet shy animal that takes up more than it can eat only to put the excess down after being chased back to its burrow. The repeated instinctive action the year after need no longer be treated as independent of individual memory. But on the other hand, with more developed mental equipment an instinctive action may be modified by the intellect, which might even provoke repetition.[11] With a principle of chain reaction, it should be possible to make even complex instinctive actions more explicable. Since instinct ensures preservation of the species so long as it succeds in most instances, there is no need to regard it as completely determined and unchanging in its form as a whole or in all its single links. Rather, we must expect variations through chance factors, in the course of time for the kind as a whole, as well as for its members at any given time.[12]

8. A child of a few months grabs at everything that stimulates its senses, usually putting the object grasped into his mouth, just as a chicken pecks

at everything. He reaches out to a part of the skin disturbed by a fly, just as a frog would, but with a newborn child these reflex actions are as yet less mature and developed than those of the animals mentioned. However, the involuntary movements of our limbs are linked with optical and tactile sensations just in the same way as the processes in our surroundings: they leave optical and tactile memory pictures, and these memory traces of motions become connected by association with other simultaneous sensations, some agreeable, some not. We make a mental note that licking sugar goes with the sensation of 'sweet', reaching into a flame or knocking against a hard object or against one's own body[13] with 'pain'. Thus one learns from experience both as regards processes in one's surroundings and about those occurring in one's body, especially its movements. The latter are nearest to one and can be constantly observed, so that they quickly become the most familiar. By reflex, a child picks up a lump of sugar and puts it into his mouth, reaches for a flame and withdraws his hand. On seeing sugar or flame again, his behaviour will be modified by memory: the sugar will be grasped more readily, while the memory of pain will inhibit the reaching into the flame. For the memory of pain operates quite like pain itself, provoking the opposite movement: this 'voluntary' movement is a reflex tinged by memory. We cannot carry out a voluntary movement that has not previously occurred wholly or in part as a reflex or instinctive action, and been experienced by us as such. If we observe our own movements we notice that we vividly remember a movement previously performed and that the movement occurs while we thus remember. More precisely: we imagine the object to be grasped or removed, including its location and the optical and tactile sensations involved, and this at once brings with it the movement itself. Movements to which we are long accustomed hardly recur in conscious imagination: in thinking of the sound of a word we have already pronounced it, in thinking of its written shape we have already written it, without a conscious awareness of the mediating movements of speech or writing. A vivid imagination of the goal or result of a movement here rapidly sets off a sequence of psycho-physiological processes that end in the movement itself.

9. What we call will is only a special instance of associations, acquired in time, intervening in the pre-formed and fixed mechanism of the body. If circumstances are simple, this innate mechanism almost suffices to

assure the joint working of the parts towards the preservation of life. However, if there are large spatio-temporal variations in the circumstancies, reflex mechanisms are no longer enough: their function must cover a certain range, with certain changes from case to case. These changes, which are small enough, are obtained through association, which reflects the relative stability and limited variation in the circumstances. The modification provoked by conscious memory traces is called will. Without reflex and instinct there is thus no modification of them and therefore no will. They remain at the core of all expressions of life, and only if they cannot ensure preservation will they be modified or even temporarily suppressed, so that it may take a considerable detour to attain the goal that is not directly reachable. This is the case of an animal cunningly stalking its prey and snatching it at a bound when no other way succeeds; or of man building huts and lighting fires to protect himself against a degree of cold beyond what the unaided organism can stand. The advantage of man over animal (or of civilized over uncivilized man) as regards imagination and therefore action is only the length of the detours to the same goal, and the ability to find and follow them. The whole of scientific and technological civilization may be regarded as such a detour. If in the service of civilization intellect (and imagination) grows to the point where it can create its own needs and pursue science for its own sake, nevertheless this can be the product only of a social organization that allows the appropriate division of labour. Divorced from society, an enquirer who was totally immersed in his thoughts would be a biologically non-viable pathological phenomenon.

10. Joh. Müller[14] still thought it possible that impulses to motion from brain to muscles are directly experienced as such, just as peripheral nerve stimuli transmitted to the brain determine sensations. This view, which survived until quite recently has been shown to be untenable, both psychologically by W. James[15] and Münsterberg[16], and physiologically by Hering[17]. An attentive observer must admit that such innervational sensations cannot be experienced: one does not know how one carries out a movement, what muscles are involved and how they are stressed, and so on. All this is determined by the way the body is organized. We merely imagine the goal of the movement and experience the completed movement only through peripheral sensations of skin, muscles ligaments and so

on. Just as various imagined pictures consciously complete each other by association, so likewise memories of sensations can become complete through the corresponding motor processes which do not themselves enter into consciousness but intervene only by their consequences. Since the nervous system is the same throughout, we may assume that the principle of association (conjunction through habit) applies to the whole of it. Which links of the associative chains become conscious will depend on the particular nerve connections with the cerebrum. As examples for imagination setting off bodily processes, take vomiting, which in sensitive people can be provoked by merely imagining it. One whose hands easily perspire or who easily blushes when embarrassed must not think of these processes, or else they happen. The salivary glands of a gourmet react at once to his gastronomic phantasies. At one time, when I had a long bout of malaria I acquired the disagreeable knack of provoking shivers just by thinking about it, a skill I retained for years. Other facts further confirm our point of view: a muscular contraction set off not 'centrally' by the 'will' but through an induction current is felt as an exertion just as in the voluntary case, so that the sensation must be set off peripherally. The most interesting observations are those of Strümpell[18] on the behaviour of a boy who could see only with his right eye and hear with his left ear and for the rest lacked all sensation. If the boy was blindfolded, he would not notice his limbs being put into the strangest positions. He never felt tired. If he was asked to raise his arm and keep it there, he did so, but after 1–2 minutes it began to tremble and descend, although the patient maintained that it was still raised. Similarly he thought he was opening and closing his hand when it was being held fast.[19]

11. Movement, sensation and imagination are in any case closely related, notwithstanding the divisions and schemas necessary in psychology. When a wild cat, aroused by a soft noise, remembers the animal that might have caused it, the cat turns its eyes in the direction of the noise and gets ready to jump. The associated memory has set off movements towards a better optical sensation of the expected nutritious object, which can now be caught at a well-measured bound.[20] However, the eyes of the cat are now quite occupied with its prey and of little use for taking in impressions from elsewhere, which is why a lurking animal easily falls victim to the hunter. Sensation, imagination and movement interlink to determine

what we call a state of attention. Our own behaviour is similar to the cat's, when we reflect on something that directly affects our self-preservation or is otherwise of interest to us.[21] We do not give ourselves over to just any ideas that come into our head. First we turn our eyes away from all indifferent happenings and ignore or try to fend off noise in our surroundings. We might even sit down at our desk and design a construction or begin to develop a formula: at these we look repeatedly, and only those associations appear that are related to the task, whereas any others that might crop up are soon displaced by the former. Movements, sensations and associations collaborate in the case of reflection to produce intellectual attention, just as in the case of the lurking cat they produce perceptual attention. We think that we conduct our reflections 'voluntarily', while in fact they are determined by the constantly recurring thought of the problem which is linked in countless inextricable ways with our interests.[22] Just as perceptual attention concentrated on one object is thereby relatively switched off for all others, so the associations relating to a problem bar the way to all others.[23] The cat fails to notice the approaching hunter, Socrates immersed in speculation absentmindedly ignores Xanthippe's questions and Archimedes constructing his circles pays with his life for his defective biological adaptation to immediate circumstances.

12. The will or attention as special mental powers do not exist. The same power that fashions the body initiates also the special forms of collaboration of bodily parts that we have called 'will' and 'attention': these two are so closely related that it is difficult to define the boundary between them.[24] Both involve a form of 'choice', as in the geotropy and heliotropy of plants and in the stone falling to the ground. All of them are equally mysterious or equally intelligible.[25] The will consists in subordinating less important or only temporarily important reflex actions to the processes that have a leading role in the functioning of life, namely sensations and imagination that register the conditions of life.

13. Some movements that are essential to continued life, such as heart beat, breathing, peristaltic movement of the intestines, are independent of 'will' or at least are subject in very small measure to mental processes (as in the case of emotions). Still, the boundary between voluntary and involuntary movements is not hard and fast, but varies from person to

person: some can control muscles that others cannot. Fontana is said to have been able to narrow his pupils at will, and E. F. Weber to suppress his heart beat.[26] If the innervation of a muscle succeeds by chance and the attendant sensations can be reproduced in memory, then the contraction usually recurs and the muscle remains subject to will.[27] Thus the boundary of voluntary motion can be extended by chance trials and practice.

In morbid states, the relation between imagination and movement can suffer important changes, as a few examples will show.[28] Thomas de Quincey tells us that the use of opium so far weakened his will that he left important letters unanswered for months and in the end had to struggle hard to write even a few words in reply. A healthy and intelligent notary became melancholic: he was supposed to travel to Italy, declared repeatedly that he could not, but offered no resistance to his companion. In signing a power of attorney, he became stuck for three quarters of an hour trying to complete the final flourish of the signature. His weakness of will had shown itself in many other similar cases, when suddenly he recovered through seeing a woman thrown to the ground by horses: he quickly jumped from his carriage to help. His 'aboulia' was thus overcome by a strong feeling. On the other hand, mere imagination can become so impulsive that it threatens to turn into action. A person might for instance be obsessed with the idea of having to kill a certain individual, perhaps even himself; to protect himself from the consequences of this dreadful compulsion he allows himself to be tied up.

14. We saw earlier that the boundary between ego and world is difficult to determine and somewhat arbitrary. Let us consider that totality of connected ideas which alone is directly there for us, the ego: it consists of memories of our experiences along with the associations provoked by them. This whole complex is tied to the historical fate of the cerebrum which is part of the physical world and cannot be isolated from it. However, we cannot banish sensations from the range of mental elements. Take first sensations of organs from all parts of the body which reach into the cerebrum as hunger, thirst and the like and form the basis of instincts. These in turn, through a mechanism acquired at the embryonic stage, set off our movements, reflexes and instinctive actions, which our imagination developed later can at best modify. This wider ego is connected with the whole body, indeed even with the bodies of the parents. Finally, we

can include all sensations set off by the whole physical surrounding and speak of an ego in the widest sense, and this can no longer be severed from the world as a whole. The thinking adult who analyses his ego notices the life of the imagination marked by its strength and clarity as the most important content of the ego. Not so with the developing individual. A child of a few months is ruled entirely by organic sensations. The feeding instinct is the most powerful, gradually the life of the senses develops, and later still the imagination. Much later the sexual instinct is added and with the concurrent growth of the imagination alters the whole personality. In this way a picture of the world develops in which one's own body stands out as a clearly bounded central link: the strongest imaginations with their associations aim at satisfying the instincts, to which they are attuned as mediating agents. The central link is common to man and higher animals, though the imagination becomes weaker the simpler the organism. In social man, who is partly freed from struggle, imagination, which is connected with his calling, position, task and so on, may become so strong and valuable as to make everything else appear unimportant, though originally it was only a stage towards satisfying his own instincts and, incidentally and indirectly, those of others. Thus arises what Meynert[29] has called the secondary ego, in contrast with the primary in which the animal life of the body predominates.

15. Since organic sensations contribute so much to the formation of the ego, it is clear that disturbances in the former change the latter. Ribot[30] describes interesting cases of this. A soldier badly wounded at Austerlitz believed from then on that he was dead. Asked how he was, he would reply: "You want to know how old man Lambert is? Well, he no longer exists, a cannon ball took him. What you see here is a bad machine that resembles him, they ought to make another one." In speaking of himself he never said "I", but "this". His skin was insensitive and he often had fits of unconsciousness and immobility for days at a time. Twin monsters who partly share one body, such as the Siamese twins or the two sisters born at Szongy in Hungary, likewise partly share an ego, besides being remarkably similar or almost identical in character. Often one will finish a sentence started by the other.[31] This is merely a heightened form of the physical and mental similarity of separate identical twins which has supplied much material for comedy both ancient and modern.[32]

If organization determines the primary ego, experience has a decisive influence on the secondary ego, which can be greatly changed through sudden or permanent changes in the surroundings. This is well illustrated in the story 'Asleep and Awake' in the Arabian tales of *A Thousand and One Nights*, and in the induction in Shakespeare's *Taming of the Shrew*.

16. There are curious cases in which two different personalities manifest themselves simultaneously in one body. A man unconscious with typhus awakes one day believing he was two bodies lying in different beds: one of the bodies seems to be cured and resting, while the other feels sick. A policeman suffering a loss of memory because of blows on the head believes himself to consist of two people of different character and will, inhabiting the right and left part of his body. Here, too, belong cases of so-called possession, in which the body of a person seems to become occupied by a second person that seeks to control it and give it orders, often exclaiming with an alien voice. Small wonder that the uncanny impression of such events has provoked a demonological view of them.[33] More often different personalities manifest themselves in one body successively or in alternation. A reformed prostitute admitted to a convent fell into religious insanity followed by a phase of stupidity and next a period in which she alternately took herself to be a nun or a prostitute with appropriate attendant behaviour. There are even recorded cases of three different personalities.

To form a scientific view of such cases in the light of all the factors involved in the formation of the ego, one might take it that the changing organs are linked with closely connected areas of association that have nothing in common with one another: as the organic sensation changes, perhaps through disease, memory and personality as a whole likewise change. During the transitional stage, if this is long enough, a double personality appears. Anyone who is able to observe himself in a dream finds such states not unfamiliar and certainly not inconceivable.

17. The parts of the body are very closely related and almost all vital processes somehow reach into the cerebrum and thus into consciousness. This is not at all so with all organisms. In some animals we must assume a less close interdependence between adjacent parts than in man; witness the fact that a caterpillar with a lesion at the back begins to devour itself

from the rear[34], a wasp continues to suck for honey even if its abdomen is being cut off, a worm cut in two and sewn together by a thread goes on crawling as though undamaged. One ring of a worm does indeed stimulate the next, thus causing it to crawl on if the thread transmits the stimulus, but there is no central organization of life in a brain and correspondingly no ego.

NOTES

[1] Goltz, *Die Nervenzentren des Frosches*, Berlin 1869.
[2] *Ibid.*, pp. 20f.
[3] J. v. Sachs, *Vorlesungen über Pflanzen-Physiologie*, Leipzig, 1887.
[4] Loeb, 'Orientierung der Tiere gegen das Licht', *SB. d. Würzburger ph.-med. Gesellschaft*, 1888; 'Orientierung der Tiere gegen die Schwerkraft', *ibid.*, 1888; *Heliotropismus der Tiere*, Würzburg 1890; 'Geotropismus der Tiere', *Pflügers Archiv* 1891.
[5] Cf. the writings of Sachs and Loeb cited above.
[6] A. Forel, 'Psychische Fähigkeiten der Ameisen', *Verh. d. 5. internat. zoologenkongresses*, Jena 1902; 'Geruchsinn bei den Insekten', *ibid*; 'Expériences et remarques sur les sensations des insectes', Pts. 1–5, *Rivista de scienze biologiche*, Como 1900–1901.
[7] By means of topochemical memory a kind of spatial olfactory picture of the region traversed is supposed to emerge, as will hardly be denied in the case of dogs. From the polarized olfactory track an ant is to recognize whether the path leads towards or away from the nest. Thus, left and right on the track should be distinguishable by smell.
[8] Cf. Haberlandt, *Über den tropischen Urwald*, Schr. d. Vereins z. Verbr. naturw. Kenntnisse, Vienna 1898.
[9] H. Müller, *Befruchtung der Blumen durch Insekten*, Leipzig 1873.
[10] Loeb, *Vergleichende Gehirnphysiologie*, Leipzig 1899.
[11] The first few times, the sensation of hunger or thirst are accompanied by reflex movements that under suitable circumstances lead to the satisfaction of the need, witness the behaviour of infants. The more mature a person the clearer and more distinct the memories that help in the satisfaction of needs, starting from sensations before and after satisfaction and pointing the way. The mixture of conscious and instinctive behaviour can moreover occur in the most varied proportions. Some years ago I suffered from a severe neuralgia of the leg that set in at 3 a.m. precisely and tortured me till morning. On one occasion I observed that I found it very difficult to wait for my breakfast coffee. It occurred to me to take coffee at 3 a.m. and in this way I actually suppressed the neuralgic pain. This success, which comes near to the seemingly miraculous self-subordination of hyper-sensitive sleepwalkers, amazed even me at first. However, mysticism cannot stand up to attentive reflection. Indeed, the pains had regularly much abated shortly after breakfast and the supervening euphoria had become associated with the idea of 'coffee', without my having become clearly conscious of it.
[12] Variation in sexual instincts doubtless is based on chance circumstances attending their being first excited. It will hardly be justifiable to turn every 'perversion' into a special kind of 'sexual psychopathy' and even to regard it as physically based. We need recall merely the gymnasia of antiquity, the relative seclusion of women, and pederasty.
[13] Preyer, *Die Seele des Kindes*, Leipzig 1882.
[14] J. Müller, *Handbuch der Physiologie*, Koblenz 1840, **II**, p. 500.

[15] W. James, *The feeling of effort*, Boston 1880; *Principles of Psychology*, New York 1890, **II**, pp. 486f.
[16] Münsterberg, *Die Willenshandlung*, Freiburg i.B. 1888.
[17] Hering, in Hermann's *Handb. d. Physiol.* **III**, 1 pp. 547, 548.
[18] Strümpell, *Deutsch. Archiv. f. klin. Medic.* **XXII**, p. 321.
[19] I was myself unable for some time to break loose from Müller's views. The observations (*A* 4, p. 135) on my own apoplectically paralysed but sensitive hand, which reveals no movement whereas I seem to feel a slight opening and closing, I cannot really fit into the new theory either.
[20] Groos, *Die Spiele der Tiere*, Jena 1896, p. 210f.
[21] Cf. Ch. III, sn. 11.
[22] Cf. *P* 3, pp. 287f.
[23] Cf. 'Zur Theorie des Gehörorgans', *Sitzb. d. Wiener Akademie* **48**, July 1863, where a rather more biological view of attention is given.
[24] Cf. J. C. Kreibig, *Die Aufmerksamkeit als Willenserscheinung*, Vienna 1897.
[25] Cf. Schopenhauer, *Über den Willen in der Natur.*
[26] Ribot, *Maladies de la volonté*, Paris 1888, p. 27.
[27] Hering, *Die Lehre vom binocularem Sehen*, Leipzig 1868, p. 27.
[28] Ribot, *l.c.*, pp. 40–48.
[29] Meynert, *Populäre Vorträge*, Vienna 1892, pp. 36f.
[30] Ribot, *Les maladies de la personnalité*, Paris 1888.
[31] Vaschide et Vurpas, *Essai sur la Psycho-Physiologie des Monstres humains*, Paris (undated).
[32] Cf. Plautus' *Menaechmi* or Shakespeare's *Comedy of Errors*. Galton's 'History of Twins' is instructive as to the facts.
[33] As regards the demonological conception, see Ennemoser, *Geschichte der Magie*, Leipzig 1844; Roskoff, *Geschichte des Teufels*, Leipzig 1869; Hecker, *Die grossen Volkskrankheiten des Mittelalters*, Berlin 1865. Pathological phenomena, mental disturbances and especially hallucinations, whether chronic as in paranoia or temporarily provoked by poisons (witches' ointment), all these, with inadequate scientific criticism, helped to support belief in demons and witches, both in the victim and the observer. Cf. P. Max Simon, *Le Monde des Rêves*, Paris 1888. Further interesting data in Walter Scott, *Letters on Demonology and Witchcraft*, 4th ed., London 1898.
[34] This process is mentioned in biological writings. My sister, who for many years raised yama-mai silkworms in open oak forests, where caterpillars often sustained injury but also often healed again, denies the accuracy of the observation. The caterpillars seem to examine the wound and perhaps try to close it.

CHAPTER V

DEVELOPMENT OF INDIVIDUALITY IN A NATURAL AND CULTURAL HABITAT

1. Severed from the parent body, the animal organism begins a life of its own. Its only inheritance is a set of reflex actions to help it over its immediate needs. By adapting this set to the special surroundings, modifying and extending it and gaining experience, the animal grows into a physical and mental individual. The human young here behave exactly like the chick running off with its eggshell and pecking at everything, or like the young alligator[1] still dragging the shell attached to the umbilical cord while already snarling with open jaws and pouncing on any object brought near. The human young is less mature and less well furnished when it becomes a separate organism from its mother, whose physical and mental powers must long continue to make up for the child's lack of independence.

2. Animals gather individual experience in the same way as humans. Biology and the history of civilization are equally valid and complementary sources for psychology and the theory of knowledge. Though it is difficult to think oneself into the mental life of an insect, whose conditions of life and sensory equipment are so obscure to us, so that we are tempted to study them as mere machines and to avoid any inference to a mental life, one ought not to disregard the valuable lead from analogy with one's own psyche, the less so since all other means of research are here inadequate. We often tend to overrate this gap between man and his fellow animals, forgetting too readily how much in our own mental life proceeds mechanically. If we regard the behaviour of insects, fish and birds towards fire and glass as remarkably stupid, we fail to reflect how we should behave towards these objects, if they were totally alien to our experience and were suddenly to appear. They would seem like magic, and no doubt we should at first run into them more than once. Starting from the study of our nearest animal relatives and gradually proceeding to remoter ones, we must surely obtain a solid comparative psychology, and only then will the phenomena of the highest and lowest mental life and their real agreements and differences become clear.

3. Let us consider a few examples. Lloyd Morgan[2] let a young dog retrieve a stick. In doing so, the animal was stung by nettles and refused to pick up this particular stick, even in an open field. Other sticks he would grab without resisting, and after some hours even the first stick, when the pain and therefore the image of pain had subsided. Another dog grabbed by its middle a stick with a heavy knob, which caused him much discomfort; but through trial and error he learnt to grab the stick near the knob, at the centre of gravity. Two young dogs were carrying sticks transversely in their jaws, the ends hitting the posts of a narrow passage for pedestrians: the dogs dropped the sticks and went through: when sent back, one dog grabbed his stick at one end and dragged it through, while the other continued to pick it up at the middle, hit the post and let it fall. On returning an hour later, the apparently more intelligent dog too had forgotten how to take advantage of his chance discovery. A dog easily learns how to open a gate, by slipping his head underneath the bar and lifting it. Careful observation however shows that this procedure was found accidentally through playful and impatient attempts at getting out, and not through clear insight into the conditions for opening the gate. A dog several times ran after a flushed rabbit on a winding path through a thicket, but the rabbit escaped into his burrow. In the end, the dog went straight to the burrow and grabbed the rabbit when it arrived. Horses and dogs carrying a load up a steep hill prefer a less steep winding path to the direct route.

From these examples it seems that we can derive the following rules: 1. Animals know how to exploit associations obtained by chance. 2. Because the facts are complex, unrelated features may become associated; for example, the stinging may be attributed to the stick which happens to be the object of attention, while the nettles remain unnoticed. 3. Only those associations are maintained that are biologically important and often repeated. Surely the behaviour of most humans is intelligible in terms of the same rules.

Morgan[3] relates the case of an incredibly stupid cow whose calf died soon after birth. Since the cow would not let herself be milked except in the presence of the calf, the farmer stuffed the headless and limbless calf-skin with hay, and the cow sniffed this phantom and licked it gently, while the farmer proceeded to milk her. When after a while, these tender gestures caused the hay to appear, the cow quietly consumed it. Comparable examples of occasional human stupidity are related in Maupassant's short stories, surely not purely from invention.

4. Once biological needs have caused mental life to develop to a certain level, this mentality goes on to manifest itself independently beyond those needs. Such an excess of mentality occurs in curiosity. We know the short abrupt barking of a dog whose attention is absorbed by an unusual appearance: he calms down only when it is resolved in a way he can grasp. A sleeping cat[4] much agitated by the sound of a child's trumpet lay down again quietly once it had seen the boy who had caused the noise. A monkey[5] in a zoo caught an opossum, examined it, found the pouch and took out the young, looked at them and put them back: here the curiosity of the little zoologist goes far beyond biological needs. Romanes[6] observed that a dog became worried and frightened when a bone it was gnawing was set in motion by an invisible thread: his somewhat bold interpretation was that the dog had a tendency to fetishism. There is indeed a remote resemblance to the way certain Pacific Islanders revere a chip of wood with writing on it that conveyed a message by some means they could not understand.[7]

5. The mental life of animals is further considerably enriched by observation of the behaviour of fellow members of his species, because of their example and their albeit imperfect linguistic communications that are contained in the reflexes of warning signals or calls. In this way behaviour of older members can be transmitted to younger ones by tradition,[8] but furthermore, new modes of behaviour discovered by individuals can be spread to several or all members of the species. The life of a species thus changes with time. Although this rarely happens as fast[9] as for civilized humans, for example by invention, nevertheless the processes are similar in kind for both, and for both we can speak of a history.[10]

6. The mental differences between man and animal are not qualitative but only quantitative. Because of his complex conditions of life he has developed a more intense and richer mental life, his interests are wider, he is able to take longer detours to reach his biological goals, the life of contemporaries and forebears has a stronger and more direct influence on his life (through more perfect communication in speech and writing), and

there are more rapid changes of mental life within an individuals's time span.

7. Man acquires his cultural advances in small steps by way of primitive experience, like animals. When the fruit from trees no longer suffices, he stalks his prey like a predatory animal, using similar tricks, but here at once we notice, in his choice of means, the greater power of his imagination, strengthened through wider experience. The redskin dons a reindeer mask to stalk the herd,[11] the Australian aborigine breathes through a tube when he approaches aquatic birds, swimming below the surface to kill them easily by dragging them under, while an Egyptian does the same by holding his head in a gourd. It may be that chance discoveries led to these devices. Similarly for the catching of fish at low tide by setting a fence of stakes.[12] The highly ingenious construction of various traps stands witness to the cunning of man as of those animals that soon learn to know and shun him, thus setting man ever new tasks. Further, man had to gather new experience as the increasing numbers of his kind compelled him to turn from hunting to grazing as a nomad and finally to fixed agriculture.

The heaps of shells near coastlines show that stoneage man often had a diet that hardly differed from that of animals. Primitive man sets up camp in thickets, like birds or monkeys, or in a cave like predators. The round hut of the Red Indian[13] which arises from tying together the tops of small trees is gradually displaced by the more roomy long rectangular type. Climate and availability of material determine the transition from timber to stone, whether dressed or not.

8. Man is markedly different from animals in that he clothes himself. Sensitive crabs do indeed protect themselves by crawling into shells, and certain larvae prepare a cover of small pebbles and leaves, but such cases are very rare. Mostly, the natural cover of the body is adequate for protection. What are the circumstances that made man lose all but a rudiment of the inherited cover of hair? What were the antecedent causes that led him to meet unfavourable climatic conditions with protective clothing? Is it that man was displaced to the north from warmer climates and lost his hairy cover through using clothes? Or is our present state the result of complex prehistoric events? The first covers adopted by men in need of protection were animal skins and treebark,[14] or sometimes plaited grass

cloth as a substitute. This progressively led to the making of twisted thread from plant fibres, hair and wool, to spinning and plaiting, that is weaving. The need to combine skins or woven cloth into clothes teaches one sewing.

9. Man and animal choose somewhat different paths towards satisfying their needs. For both it is only by means of their bodily muscles that they can relate themselves to their surroundings; but whereas animals, completely absorbed by their needs, mostly aim at directly grasping the object that will bring satisfaction or removing the object that disturbs, man with his greater mental power and freedom sees indirect paths as well and chooses the most comfortable. He has found the leisure to observe how body behaves towards body, even though this hardly touches him, and occasionally he knows how to make use of it. He knows that animals are not afraid of their fellows, nor birds of gourds, and he chooses his mask accordingly. Whereas a monkey grabs at birds always in vain, man hits them by means of a missile whose behaviour and effect on colliding with other bodies he has tried out by systematic play. A monkey, too, likes to wrap himself in a blanket, if he has one, but he does not know how to obtain skin or bark. Occasionally, a monkey will pelt an enemy and use a stone to open fruit. Man, however, stabilizes every such advantageous procedure, he is more economically minded: he busies himself with the stone, shapes it into hammer or axe, grinds away at his spear for weeks; and, turning his attention to means, he invents weapons and tools that afford him invaluable advantages.

10. If lightning produces fire somewhere, monkeys use the opportunity to warm themselves, as gladly as a human. However, the latter alone observes that additional wood maintains the fire and he alone makes use of this discovery, by tending, keeping and transmitting fire for his own purposes.[15] Indeed, the new discoveries he makes in gathering readily inflammable and smouldering material, namely tinder, enable him to create fire afresh by inventing the fire drill and thus enter into permanent possession of fire. This done, his sight set beyond the most urgent needs, he may observe the formation of glass, the smelting of metals and so on. Fire is the key to the treasures of chemical technology, just as tools and weapons are to mechanical technology. Tempting and psychologically instructive though it might be to pursue the development of technology from primitive experi-

ence, it would lead us too far afield. I have elsewhere given a brief attempt at outlining the psychological inferences from such a study.[16] Much is recorded also in books on cultural history.[17]

11. Anyone who has carried out experiments knows that it is much easier to perform a purposive movement of the hand, which automatically corresponds to our intentions, than it is accurately to observe the interaction of bodies and to reproduce it in memory. The former belongs to our constantly exercised biological function, the latter lies outside our most immediate interests and cannot become one of them until there is an excess of sensation and imagination playfully at work: observation and inventive speculation presuppose some measure of well-being and leisure. For primitive man, this means relatively favourable conditions of life. In any case, very few people are inventors, most others use and learn what those few have invented: that is the essence of education, which can make good certain deficiencies of talent and thus at least preserve what has already been achieved. It lies in the nature of things that the man of vision beyond the immediately useful brings greater blessings to the community than to himself.

12. All this shows how slow and difficult it was for primitive man to rise above his animal companions. With this rise but not until then the growth of culture accelerates. A powerful impulse comes through the formation of society, division into classes, professions and trades, partly relieving each man of looking after subsistence and thus giving him a narrower field of activity in which he can attain greater mastery. Social combination further generates special inventions peculiar to it: namely collaboration of whole groups, organized in space and time,[18] towards a common goal, as found in the case of troops fighting with weapons, the transporting of heavy loads as in Ancient Egypt, and to some extent in factory work today. In such communities, those classes that because of historical circumstances have a privileged position are not slow to exploit the work of the rest for a pittance. However, since the exploiters invent new needs, they provide incentives for seeking new ways to satisfy such need smore readily; although these inventions may not be for the sake of others, they too will indirectly benefit through the generally higher level of culture, both materially and spiritually.

13. Man learns to exploit the work of animals for his own purposes, thus greatly enhancing his power. As a member of society he gains experience regarding the great value of human labour. This leads to the practice of forcing prisoners of war to work instead of killing them. This is the origin of slavery, one of the basic pillars of ancient civilizations and perpetuated in various forms to this day. In Europe and America, slavery is now abolished in name and form, but the principle of the thing, the exploitation of many by few, has remained. The subjugation of members of the same or other kinds is not peculiar to man, we find it elsewhere as well, for example in the antheap.

14. Alongside animal and human work, man gradually learnt to exploit the forces of 'inanimate' nature. Thus arose mills driven by water or wind. Work that used to be done by animals or men was increasingly assigned to moving water and air, for which one had only to install machines that needed no feeding and were generally less obstinate than people or beasts. The invention of the steam engine opened up the rich reservoir of energy concealed in the vegetation of primaeval forests stored for millions of years as coal which is now made to work for man. The newly developed branch of electrical engineering, by means of electrical transmission of power, extends the scope of the steam engine as well as that of remote sources of wind or water power. As early as 1878, that is before the great upwards surge of electrical engineering, England had steam engines with a total capacity of 4½ million horse-power, corresponding to the work of 100 million people: more than her population several times over could have performed. In 1890, the industrial machinery of England produced as much as 1200 million busy workers could have done, that is almost the whole world population.[19]

15. One might think that with such an increase in working power, the working part of humanity who merely have to operate the machines, would be considerably relieved from toil. However, a closer look shows that this is not at all the case: work remains as exhausting as before, Aristotle's dream of a future machine age without slavery has not come true. Why this is so has been well explained by J. Popper[20]: the enormous output of machines is not just used for easing human subsistence, but largely for satisfying the luxury needs of the ruling section. For example, it is very

pleasant to imagine the speed of railways and ease of communication by post, telegraph and telephone, for those who enjoy this ease. Things look different when we consider the other side of the coin and observe the pain of those who have to maintain this rush of traffic. In view of the intense cultural life, other thoughts arise: the humming of tramways, the whirring of factory wheels, the glow of electric light, all these we no longer behold with unalloyed pleasure if we consider the requisite amount of coal burned every hour. We are fast approaching the time when these hoards, built up as it were when the Earth was young, will in its old age have become nearly exhausted. What then? Shall we sink back into barbarism? Or will mankind by then have acquired the wisdom of age and learnt to keep house? Cultural progress is conceivable only when there is a certain audacity and thus can generally be set going only by those who are partly relieved from toil. This holds both for material culture and for things of the spirit. These last have the splendid property that one cannot stop them from spreading to the burdened sections of mankind: sooner or later these people will recognize the true state of affairs and confront the ruling sections with the demand for a cheaper and more appropriate use of the common stock of possessions.[21]

16. Amongst the inventions arising from social existence are speech and writing. Reflex utterances of sounds occurring when certain causes provoke emotions, automatically become memories and signs of these causes and emotions: individuals living under the same circumstances will understand them in the same way. However unspecialized the sounds uttered by animals, human speech is only a further development of animal speech: because of a greater range of experiences, the sounds become more modified and specialized, spread by imitation and preserved by tradition. The emotional element that produced the sound recedes more and more, while the sound becomes specialized and increasingly associated with the corresponding ideas. Jerusalem has clearly traced the formation of nouns from such emotional sounds in the case of Laura Bridgman[22]. In a limited way we can observe the process of changes in speech in our children. More extensive evidence comes from comparing the languages of peoples of common origin: there we can see how the division of a people into several branches living under different conditions is accompanied by a division of language. Words change, or disappear if their cor-

responding objects no longer exist, or come to be used for other related or similar objects that so far lack expressions. Since the point of comparison differs from case to case, one and the same word often ends up meaning quite different things in related languages. Thus, a German can derive some harmless fun from reading Dutch newspapers or shop signs, and no doubt the reverse holds true as well.[23] Words are important as centres of associations. Mental development is most considerably advanced through linguistic communication and transmission of experience. The importance of language for abstraction will be discussed later.[24] Only in rare exceptions does sounded language use imitation to denote something audible. The language of gestures employed by alien peoples for mutual understanding, or the natural use of gestures by the deaf and dumb (as against their artificial finger language), makes most use of imitating the visible where one cannot directly point at it.[25]

17. By using permanent visible signs instead of momentary audible ones one obtains writing, whose advantage lies precisely in this permanence[26], as against the transient, quickly dissolved and forgotten spoken word. The most obvious thing is to convey information or news about events by means of pictures, a method in fact used by North American Indians; witness a rock drawing near Lake Superior reporting about a naval campaign.[27] Other beginnings of writing are found in tattooing, where drawings on the skin gradually and of their own accord become 'totems', the signs of tribes. Conventional signs for recalling things, such as knots, notches on sticks, split lengthwise with one portion going to each of the two parties to a contract, knotted strings of the kind kept by the Peruvian administration (quipus) and 'wampum' belts, all these are other such beginnings. The further development of writing can proceed along two different paths: either the representations of things shrink through a quick simplifying script into conventional signs for concepts, as in Chinese; or the pictures, reminding one of the sounds of words in the manner of picture puzzles, turn into phonetic signs, as in Egyptian hieroglyphics. The tendency to think abstractly and the desire to enlist writing to this end leads to the former method, while the need to write people's names and proper nouns in general leads to the second, which has given rise to literal script. Each method has its special advantages. The second makes do with very slender means and easily follows every phonetic and conceptual change in lan-

guage. The first method is quite independent of phonetics, so that Chinese can be read by Japanese readers, who speak a phonetically quite different language. Chinese script is almost pasigraphic, although it needs to be changed with every conceptual change.[28]

18. Language and writing, a product of society and culture, react upon these and intensify them. It is easy to imagine that human life could not be very different from that of animals if there were no fairly perfect conveying of experience from one person to another and everybody had to start again from scratch, being confined to his own personal experience. Nor could men pass beyond savages if communication did not reach far beyond the span of one man's life. The partial material relief of the individual by the community and the intellectual support he gains from what contemporaries and forebears convey, these are the conditions that make it possible for the social product that we call science to emerge. A savage has gathered all kinds of experience: he knows poisonous and edible plants, follows the tracks of hunting animals and can protect himself against beasts of prey and poisonous snakes. he can use fire and water for his own purposes, select stones and wood for his weapons, smelt metals and work them. He learns to count and reckon with his fingers, and to measure with his hands and feet, he sees the vault of heaven as a child might, observing its rotation and the change of position of sun and planets on it. Yet all or most of these observations are made on the occasion or for the purpose of some application useful to himself. These same primitive observations bear the seeds of the various sciences,[29] but science itself can arise only when relief from material cares has yielded enough freedom and leisure and the intellect has grown strong enough through use, so that observation as such, regardless of application, has acquired sufficient interest. At that stage one starts collecting the observations of contemporaries and forebears, ordering and testing them, eliminating errors caused by chance interference, and thus eliciting the way in which the ascertained material hangs together. The role of writing in this process may be illustrated by a single noteworthy case: when after centuries of barbarism, Europeans in the 16th and 17th centuries once more took up the threads of ancient science, they had no need to repeat the same preliminaries but quickly reached the level of the ancients and overtook it.

Tracing the history of scientific development from the collecting and ordering of primitive experiences affords a tempting and pleasant study.[30] Some fields, like mechanics, heat and others, are especially instructive, for in them we see most clearly how science develops from manual skills and trades[31]: gradually, the original driving force of material and technical needs gives way to purely intellectual interests. The intellectual command of a range of facts now reacts on the instinctive technology from which it sprang, transforming it into a scientific technology that is no longer tied to accidental findings but can systematically pursue the solution of its problems. In this way theoretical and practical thinking, scientific and technical expereince, remain in constant and mutually enhancing contact.

19. Like science, art[32] is a by-product of the satisfaction of needs. The necessary, useful and appropriate is pursued first, but if on the way one finds things that please regardless of utility, this too may arouse interest and so be retained and cultivated. Thus from useful plaiting with its regular repetitive pattern derives the pleasure in ornament, and from useful working rhythms the delight in measure. From the bow as weapon there develops the bow as musical instrument[33], the harp, the piano and so on. Art and science, any ideas of justice[34] and ethics, indeed any higher intellectual culture can flourish only in the social community, only when one part relieves the other of some of its material cares. Let the 'upper ten thousand' recognize clearly what they owe the working people! Let artists and scientists reflect that it is a great common and jointly acquired human possession that they administer and extend!

20. The complex and manifold influences that arise from the natural and cultural setting of man, give him a much wider field of experiences, associations and interests than any animal could attain. Correspondingly we have higher intelligence. However, if we compare the members of a given class or even profession, they will naturally share some appropriate common features; nevertheless, each single one, according to his different hereditary talents and his peculiar experiences, will constitute a single unrepeatable individuality. The difference of intellectual individuality becomes of course much greater if we go beyond the boundaries of class or profession. Imagine that these different intelligences enter into free commerce and through close contact inspire each other to enterprises that like science, technology, art and so on, are group activities, and one can estimate how

vast is the currently unexploited intellectual potential of mankind. Through the collaboration of many different individualities there arises a powerful enrichment and extension of the scope of experience without blunting its clarity and liveliness. Appropriately organized teaching can in some measure replace this commerce. However, if education becomes too rigidly organized, and uniformly divided according to class and profession with barriers raised between them, it too can spoil a great deal. Beware of formulae that are too rigid![35]

NOTES

[1] Morgan, *Comparative Psychology*, London 1894, p. 209.
[2] *Ibid.*, pp. 91, 254, 288, 301. My sister reports that a large St. Bernard managed to get rid of an abhorred muzzle by burying it. Soon afterwards a colleague told me of a dog hiding a whip.
[3] Morgan, *Animal Life*, London 1891, p. 334. Good psychological and biological comments in Th. Zell ('Ist das Tier unvernünftig' in *Tierfabeln*, Stuttgart. Also, *Das rechnende Pferd*, Berlin). Very well observed is the distinction between visual and olfactory animals, and the law of economy. Zell assumes too much naiveté in his readers, which does not enhance his books.
[4] Morgan, *l.c.*, p. 339.
[5] *Ibid.*, p. 340.
[6] Morgan, *Comparative Psychology*, p. 259. Schopenhauer's dog knew *a priori* that every event has a cause and tried to find the latter in analogous cases without resorting to fetichism (Schopenhauer, *Über die vierfache Wurzel des Satzes vom zureichenden Grunde*, Leipzig 1864 3rd ed., p. 76). In this remarkable way does a dog's philosophy follow that of its observer.
[7] Tylor, *Einleitung in das Studium der Anthropologie*, Brunswick 1883, p. 197.
[8] It was hoped to reduce the migrations of birds to imitation, perhaps from a period when their goals were not yet separated by seas. New views and considerable puzzles in K. Graeser, *Der Zug der Vögel*, Berlin 1905.
[9] However an Australian parrot is supposed to have conceived the notion of attacking and pecking at sheep, an example then copied by the other members of his species.
[10] Cf. H. v. Buttel-Reepen, *Die stammesgeschichtliche Entstehung der Bienenstaates*, Leipzig 1903.
[11] Tylor, *Anthropologie*, p. 246.
[12] Diodorus III, 15, 22.
[13] Tylor, *l.c.* p. 275.
[14] *Ibid.*, p. 290.
[15] *P* 3, p. 293.
[16] *Ibid.*, p. 287.
[17] Cf. Tylor, *Urgeschichte der Menschheit*, Leipzig Ambrosius Abel (undated). Also, *Anthropologie*. Otis T. Mason, *The origins of invention*, London 1895.
[18] Wallaschek, *Primitive Music*, London 1893; enlarged German edition Leipzig 1903. In this book the practical significance of rhythm is discussed. Bücher (*Arbeit und Rhythmus*, Leipzig 1902, 3rd ed.) discusses this topic in a somewhat different fashion.

[19] Bourdeau, *Les Forces de l'Industrie*, Paris 1884, pp. 209–240. However Kublai Khan, a farsighted and inventive figure, was already a spectacular pioneer in this field.
[20] J. Popper, *Die technischen Fortschritte nach ihrer ästhetischen und kulturellen Bedeutung*, Leipzig 1888, pp. 59f.
[21] A programme for this is given by J. Popper in his book *Das Recht zu leben und die Pflicht zu sterben*, Leipzig 1878. His goals are close to the original social democratic ones, but differ from them for the better in that according to him organisation should be confined to what is most important and essential, and for the rest the freedom of the individual should be preserved. In the contrary case slavery might well become more general and oppressive in a social democratic state than in a monarchy or oligarchy. In a complementary work (*Fundament eines neuen Staatsrechts*, 1905) Popper works out this leitmotiv: for secondary needs, the principle of majority; for fundamental ones, the principle of guaranteed individuality. In important points A. Menger, *Neue Staatslehre*, Jena, G. Fischer, 1902, agrees with him.
[22] Jerusalem, *Psychologie*, p. 105. In more detail in *Laura Bridgman*, Vienna 1890, pp. 41f.
[23] For analogous examples from children's language, see *A*, p. 250.
[24] Of older writings on philosophy and language the following are particularly worth reading because of their originality: L. Geiger, *Ursprung und Entwicklung der menschlichen Sprache und Vernunft*, Stuttgart 1868; L. Noiré, *Logos. Ursprung und Wesen der Begriffe*, Leipzig 1885; Whitney, *Leben und Wachstum der Sprache*, Leipzig 1876. In many ways very stimulating is Fritz Mauthner, *Beiträge zur Kritik der Sprache*, Stuttgart, Cotta, 1901.
[25] Tylor, *Urgeschichte*, pp. 17–104.
[26] Since the invention of the phonograph a spoken passage may be reproduced at will, just like a written one. The phonographic archives of the Vienna Academy are an example of this. The idea of the phonograph is due to the phantasy of Cyrano de Bergerac (*Histoire comique des états et empires de la lune*, 1648).
[27] Wuttke, *Geschichte der Schrift*, Leipzig 1872, I, p. 156, illustrations, p. 10, Table XIII. Other passages too are important for the matter here discussed.
[28] Today the old philosophic problems of pasigraphy and an international language are once again under theoretical discussion and attempts are being made at a practical solution, particularly by the 'Délégation pour l'adoption d'une langue auxiliaire internationale'. If this technical linguistic problem were solved it would amount to a most vital piece of cultural progress.
[29] Tylor, *Anthropologie*, pp. 371f.
[30] We cannot here discuss the detailed history of development of the sciences. Cf. writings of more general content, such as Whewell, *Geschichte der induktiven Wissenschaften*, German ed. by v. Littrow, Stuttgart 1840. Especially instructive are works on special subjects, such as Cantor, *Mathematische Beiträge zum Kulturleben der Völker*, Halle 1863; Cantor, *Geschichte der Mathematik*, 1880.
[31] Cf. *M*, *W*.
[32] Cf. Haddon, *Evolution in Art*, London 1895; Wallaschek, *Primitive Music*; Tylor, *Anthropologie*, pp. 343f.
[33] *Ibid.*, p. 353.
[34] Lubbock, *Die Entstehung der Zivilisation*, Jena 1875; *Die vorgeschichtliche Zeit*, Jena 1874.
[35] Natural science will have emerged as a by-product of the manual trades. Since the latter, like physical work in general, were despised in antiquity, and the slaves who

worked and who observed nature, were strictly severed from their masters, who had leisure for amateurish speculation and often knew nature only from hearsay, it is in large measure clear why ancient science has a streak of naive and dreamlike vagueness. Only rarely does the urge for independent trial and experiment break through, in the case of geometers, astronomers, doctors and engineers, and every time it occasions important progress, as with Archytas of Tarentum or Archimedes of Syracuse.

CHAPTER VI

THE EXUBERANCE OF THE IMAGINATION

1. The development of the imaginative process brings with it advantages for organic life, especially the vegetative kind. However, if ideas outbalance sensation too far, drawbacks may result: the soul becomes a parasite on the body, consuming the oil of life (Herbart's phrase). This is evident in view of the chance events that may affect the associations on which adaptation of thought to facts depends, as was illustrated earlier. If favourable circumstances guide the imagination in such a way that it follows or anticipates facts, we gain knowledge. However, unfavourable circumstances can direct the attention on to inessentials, promoting thought connections that do not correspond to the facts but mislead. Thoughts repeatedly tested and found to correspond to the facts will always be constructive rules for action; but if in special circumstances one adopts untested chance connections as generally corresponding to fact, serious error will result, and disastrous practical consequences if one acts on them. Some examples from the history of culture illustrate the point.

2. Children hit at the picture of a person they detest, and even voice their ill-will in words. They maltreat the pictures of beasts of prey and try to protect the picture of the attacked animal from that of the predator. As the imagination grows in strength it occasionally outweighs the senses. It seems likely that less civilized men and savages will behave similarly. If one such maltreats and curses the picture of an enemy who then happens to fall ill or even die, he may easily get the idea that his action and desire have caused that death. This belief will maintain itself the more readily as it is very difficult to disprove it in this uncontrollable field. Indeed, there is a widespread practice of maltreating a doll representing the enemy, or hurting it in the hair, nails or elsewhere, in the belief that these moves are effective.

Dr Martius[1] reports the case of an old Red Indian woman slave, a captive from another tribe, furtively conducting magic rites aimed at wiping out her oppressor's children. This shows us the simple psychological basis for the widespread practice of magic amongst savage tribes and makes it intelligible that at that level people seek to protect themselves against witches by burning them to a cinder, as is still customary in Africa. It is well-known how this age-old belief of savages began to revive even in Europe from the 13th century and under the authority of the Church: in 1448 Pope Innocent VIII published a bull formally sanctioning this belief; during the 15th, 16th and 17th centuries the diabolic witch trials regulated by the "Malleus Maleficarum" claimed thousands of victims of all ages, classes, and sexes, particularly poor old women; towards the end of the 17th century reason at last prevailed and the last witch was executed in 1782 at Glarus. This dreadful delusion fraught with its frightful consequences and lasting for centuries, ought to warn mankind against letting its ways of life be dictated by any kind of faith.[2] That such ideas were not alien even to fairly educated sections of ancient civilized peoples can be seen for example from Petronius' satire (the werewolf story of Nicerus, Trimalchio's adventure with witches). Books I to III of Apuleius' "Metamorphoses", which admittedly are designed for entertainment, are full of such material. Lucian's biting mockery of educated people who takes such things seriously breaks through freely in his account of a conversation with the sick Eucrates.[3]

3. In general it is true that what is contiguous in sensation will link up in thought, but since ideas by association easily enter into all kinds of accidental conjunctions, one is constantly risking error if one conversely assumes that what is linked in thought must be so in sensation. Words are centres of association in which various threads run together: that is what makes words the object of a peculiar and widespread superstition.[4] In pronouncing a word one vividly remembers the thing referred to and all its connections: one sees the approach of a feared enemy in naming him, and therefore one avoids doing so. "Talk of the devil, he will soon appear": therefore one avoids talking of him. "Dii avertite omen" the Romans called out when a word of malicious significance had been spoken. Conversely, a spoken wish becomes more vividly conscious and seems nearer being realized, for a person often fulfils the wishes of others, and others have followed his word. Why should there not be a demon, such as primitive man suspects always and everywhere present, to fulfil the spoken wish? To the savage, his name is a part of him, to be concealed from the

enemy who would otherwise gain power over him and a purchase for magic. In sickness, he changes his name to deceive the demon of the disease. A dead man's name or words that rhyme with it are 'taboo' and must not be spoken. Mohammedans believe that whoever knew the great secret name of God could wreak the greatest miracles merely by pronouncing it: it is being kept secret precisely to forestall such abuse. "Thou shalt not take the name of the Lord thy God in vain!" This notion goes back to ancient Egypt, where the goddess Isis overcomes the god Ra by cunningly eliciting from him the secret of his true name.

The savage knows that his limbs will obey his will and change his surroundings according to his wish, but he deceives himself in failing to recognize where precisely lie the limits to his will. Thus we may see a village skittle player instinctively leaning to the side he wants his ball to run along after launching it; and if we look carefully, we observe a similar tendency in ardent billiards players. Failing to observe the boundary of one's body is indeed a main source of the kind of aberration discussed earlier and below.

4. A person wakes from a motionless sleep, during which he had been dreaming of a walk in a distant region that was not in fact visited by his body; perhaps he may have had a dream conversation with his father, who is long since dead. Moreover, take the case of fainting, apparent death and death. In naïve people who, like children, draw no sharp boundary between dreaming and waking, there inevitably arises the idea of a second shadow-ego that can leave the body and return to it, rendering the body lifeless and revived respectively. Hence develops the idea of a soul,[5] that leads an independent life. If the idea of a second shadowy life after death persists for some time, it will be amplified in detail. Men dream of this life in the realm of shadows of which they have heard so often, and the attendant ideas become ever richer and more numerous. Tylor[6] records the story of a Maori who told an elaborate tale of his aunt who had visited the land of the dead.

His aunt died in a solitary hut near the banks of Lake Rotorua. Being a lady of rank she was left in her hut, the door and windows were made fast, and the dwelling was abandoned, as her death had made it tapu. But a day or two after, Te Wharewera with some others paddling in a canoe near the place at early morning saw a figure on the shore beckoning to them. It was the aunt come to life again, but weak and cold and famished. When sufficiently restored by their timely help, she told her story. Leaving her body, her spirit had taken flight toward the North Cape, and arrived at the entrance of Reigna. There, holding on by the stem of the creeping akeake-plant, she descended the precipice, and found herself on the sandy beach of a river. Looking round, she espied in the distance an enormous bird, taller than a man, coming towards her with rapid strides. This terrible object so frightened her, that her first thought was to try to return up the steep cliff; but seeing an old man paddling a small canoe towards her she ran to meet him, and so escaped the bird. When she had been safely ferried across, she asked the old Charon, mentioning the name of her family, where the spirits of her kindred dwelt. Following the path the old man pointed out, she was surprised to find it just such a path as she had been used to on earth; the aspect of the country, the trees, shrubs, and plants were all familiar to her. She reached the village, and among the crowd assembled there she found her father and many near relations; they saluted her, and welcomed her with the wailing chant which Maoris always address to people met after long absence. But when her father had asked about his living relatives, and especially about her own child, he told her she must go back to earth, for no one was left to take care of his grandchild. By his orders she refused to touch the food that the dead people offered her, and in spite of their efforts to detain her, her father got her safely into the canoe, crossed with her, and parting gave her from under his cloak two enormous sweet potatoes to plant at home for his grandchild's especial eating. But as she began to climb the precipice again, two pursuing infant spirits pulled her back, and she only escaped by flinging the roots at them, which they stopped to eat, while she scaled the rock by help of the akeake stem, till she reached the earth and flew back to where she had left her body. On returning to life she found herself in darkness, and what had passed seemed as a dream, till she perceived that she was deserted and the door fast, and concluded that she had really died and come to life again. When morning dawned, a faint light entered by the crevices of the shut-up house, and she saw on the floor near her a calabash partly full of red ochre mixed with water; this she eagerly drained to the dregs, and then feeling a little stronger, succeeded in opening the door and crawling down to the beach, where her friends soon after found her. Those who listened to her tale firmly believed the reality of her adventures, but it was much regretted that she had not brought back at least one of the huge sweet potatoes, as evidence of her visit to the land of spirits.

This poetic and homely tale sounds like a fairytale by Baumbach. One is almost inclined to envy the Maori's for the comfortable nature of their imaginative life. Yet many similar tales drawn from other races might be set beside this one. We shall mention only one more, and that one because it shows how dream appearances are also at the root of the imaginative picture of animals and objects as having souls.[7] An Indian chief on Lake Superior desired that a fine gun of his should be buried with him. After a few days' illness he seemed to die.

But some of his friends not thinking him really dead, his body was not buried; his widow watched him for four days, he came back to life, and told his story. After death, he said, his ghost travelled on the broad road of the dead toward the happy land, passing over great plains of luxuriant herbage, seeing beautiful groves, and hearing the songs of innumerable birds, till at last, from the summit of a hill, he caught sight of the distant city of the dead, far across an intermediate space, partly veiled in mist, and spangled with glittering lakes and streams. He came in view of herds of stately deer, and moose, and other game, which with little fear walked near his path. But he had no gun, and remembering how he had requested his friends to put his gun in his grave, he turned back to go and fetch it. Then he met face to face the train of men, women, and children who were travelling toward the city of the dead. They were heavily laden with guns, pipes, kettles, meats, and other articles; women were carrying basket-work and painted paddles, and little boys had their ornamented clubs and their bows and arrows, the presents of their friends. Refusing a gun which an overburdened traveller offered him, the ghost of Gitchi Gauzini travelled back in quest of his own, and at last reached the place where he had died. There he could see only a great fire before and around him, and finding the flames barring his passage on every side, he made a desperate leap through, and awoke from his trance. Having concluded his story, he gave his auditors this counsel, that they should no longer deposit so many burdensome things with the dead, delaying them on their journey to the place of repose, so that almost everyone he met complained bitterly. It would be wiser, he said, only to put such things in the grave as the deceased was particularly attached to, or made a formal request to have deposited with him.

5. According to these ideas, not only every human or animal body, but every object has a soul or a kind of spirit, conceived by analogy with one's own. A savage who produces changes in his setting understands these best as effects of his will. Similarly he regards all agreeable or disagreeable events as expressions of a spiritual entity that is either friendly or hostile to him. The constantly fertile and active phantasy of a negro, avid for some enterprise or frightened by some enemy, sees traces of such spirits in the most trifling things he notices. These objects or 'fetishes' are collected, revered and preserved, doused with rum if they bring good fortune and perhaps maltreated if they seem to be obstinate.[7] There is nothing a fetish might not do or achieve, provided it is the right one. Though we tend to feel rather superior to such views, there are among us people who carry amulets, lucky charms and medallions, and that not for fun. Our scientific view of the interdependence of natural events is after all a different one from the view still alive in the popular mind around us.

6. The dualist idea of ghosts, an after-life and so on are rather harmless, so long as they remain theoretical and keep to an area quite beyond the reach of tests. However, if the views that arise from dreams lead to practical consequences or to actions that destroy the life and well-being of our fellows without being useful to anyone, and the untestable gains sufficient power to clash with something else that is testable, the most dire facts of

cultural history result. The most obvious are human sacrifice at funerals to ensure that after death the deceased has wives, servants and every conceivable comfort.[8] 'The King of Dahome[9] must enter Deadland with a ghostly court of hundreds of wives, eunuchs, singers, drummers, and soldiers.' – 'They periodically supply the departed monarch with fresh attendants in the shadowy world.' – 'Even this annual slaughter must be supplemented by almost daily murder. Whatever action, however trivial, is performed by the King, it must dutifully be reported to his sire in the shadowy realm. A victim, almost always a war-captive, is chosen'. Such customs are very widely spread and used to be even more common. We are familiar with the funeral of Patroclus and the burning of widows in India. Such rites in various forms existed right into 'highly civilized' periods.

7. Since dead men were thus lusting for murder, ghosts, demons and deities could not trail behind.[10]

The Carthaginians had been overcome and hard pressed in the war with Agathokles, and they set down the defeat to divine wrath. Kronos (Moloch) had in former times received his sacrifice of the chosen of their sons, but of late they had put him off with children bought and nourished for the purpose. In fact they had obeyed the sacrificer's natural tendency to substitution, but now in time of misfortune the reaction set in. To balance the account and condone the parsimonious fraud, a monstrous sacrifice was celebrated. Two hundred children of the noblest of the land, were brought to the idol of Moloch. "For there was among them a brazen statue of Kronos, holding out his hands sloping downward, so that the child placed on them rolled off and fell into a certain chasm full of fire".

The widespread custom of human sacrifice to deities is well-known; it was practised by the savage or half-civilized forebears of all civilized peoples. Partly we have direct historical evidence, partly we find hints in legends (Isaac, Iphigenia). On this count, no people has cause for blaming any other. Consider merely that other instance, far removed in time and place from the above, that the conquering Spaniards found in Mexico.

These demons and deities, from whom imaginary advantages were so dearly purchased at the cost of real harm, are, alas, very various in kind and extremely numerous. Herodotus (VII 113–114) tells of Xerxes' expedition against the Greeks:

The tract of land lying about Mount Pangaeum, is called Phyllis; on the west it reaches to the river Angites, which flows into the Strymon, and on the south to the Strymon itself, where at this time the Magi were sacrificing white horses to make the stream favourable. After propitiating the stream by these and many other magical ceremonies, the Persians crossed the Strymon, by bridges made before their arrival, at a place called "The Nine Ways," which was in the territory of the Edonians. And when they learnt that the name of the place was "The Nine Ways," they took nine of the youths of the land and as many of their maidens, and buried them alive on the spot. Burying alive is a Persian custom. I have heard that Amestris, the wife of Xerxes, in her old age buried alive seven pairs of Persian youths, sons of illustrious men, as a thank-offering to the god who is supposed to dwell underneath the earth.

Other races and other periods are no wiser than the Persians.[11]

In Africa, in Galam, a boy and girl used to be buried alive before the great gate of the city to make it impregnable, a practice once executed on a large scale by a Bambarra tyrant; while in Great Bassam and Yarriba such sacrifices were usual at the foundation of a house or village. In Polynesia, Ellis heard of the custom, instanced by the fact that the central pillar of one of the temples at Maeva was planted upon the body of a human victim. In Borneo, among the Milanau Dayaks, at the erection of the largest house a deep hole was dug to receive the first post, which was then suspended over it; a slave girl was placed in the excavation; at a signal the lashings were cut, and the enormous timber descended, crushing the girl to death, a sacrifice to the spirits.

Old and gruesome tales associated with European buildings as well as the more attenuated custom of killing small animals or walling in empty coffins on such occasions indicate that to our ancestors too this practice was not unknown.

Water-spirits are no less cruel. 'The Hindu does not save a man from drowning in the sacred Ganges.' The islanders of the Malay archipelago share with many European peoples the belief that a drowning man cannot be rescued with impunity. 'The lake or the river will have its victim.' Volcanoes too claim their human sacrifices, who are thrown into the crater. Thus it is that man's futile but exuberant phantasy has set itself busily to work to make generous additions to the natural evils that he in any case has to bear. Such atrocities are not confined to low level culture. Europe too, and fairly recently, still suffered from such practices. We need merely reflect that the Inquisition, after centuries of savagery, causing the death of thousands and destroying prosperous states and cultures, saw itself forced to cease its nefarious activities as late as the end of the 18th century.[12] To those affected it is all one whether they are buried alive for earth spirits or burned alive for the spirits of dogmas, whether they fall victims to the superstition and despotism of Xerxes, to the intrigues of the Magi, or to the ambition and intolerance of modern priests. Our culture remains ominously close to barbarism.

8. Let us now turn to pleasanter things. The spontaneous play of ideas and the changing conjunctions of thoughts, divorced from sensation or immediate needs, indeed far beyond them, these are what raises man above other animals. Phantasies about what has been seen and experienced, poetry, is the first step up from the everyday burdens of life. Even if such poetry at times bears evil fruit if applied uncritically to practice, and we have seen examples, nevertheless it is the beginning of spiritual development. If such phantasies make contact with sense experience and seriously aim at illuminating it while learning from it, then we obtain one after the other religious, philosophic and scientific ideas (A. Comte). Let us therefore look at this poetic phantasy that busily completes and modifies all experience.

9. The bones of large animals like the rhinoceros, mammoth and so on, which are found in the soil, regularly elicit in naïve inhabitants of the region the idea and legend of a battle of giants on that site.[13] A dust devil running across the desert, or a waterspout across the sea, becomes to the naïve observer a giant demon, the Jinn of the Arabian Nights. The Chinese even manage to discern the head or tail of the dragon that plunges from the clouds into the sea. The Flood story of the Hebrew Bible follows an older Babylonian tradition, as is evident from many details they share. However, the wide spread of analogous legends is to be described to the fact that they arise everywhere almost by necessity. If at great altitudes one finds petrified shells and other aquatic animals and sometimes even boats of a kind no longer used, discoveries that are indeed widespread, the naïve observer, innocent of geological shifts, is forced to the idea of a great flood reaching abnormal heights.[14] Volcanoes are often regarded as mountains fired by spirits and dwelt in by Titans. It is these inhabitants that hurl out embers and stones. The Kamchadals have a peculiar explanation for the whalebones found on volcanoes. They fear these latter, believing them to be the abodes of spirits. The spirits catch whales at night, cook them, and throw out their bones. 'When the spirits have heated up their mountains as we do our *yurts*, they flying the rest of the

brands out up the chimney, so as to be able to shut up. God in heaven sometimes does so too when it is our summer and his winter, and he warms up his *yurt*.' This is their explanation of lightning.[15]

10. Whatever he failed to understand appeared to primitive man in a peculiar light that we can recover only if we vividly recall our early childhood. This teaches us how the savage regards his reflection in water or the echo of his voice under strange and rare circumstances: namely, as something ghostlike.[16] Who cannot recall such childhood feelings? Indeed, even given a theoretical grasp, is there anything more curious than an insubstantial vision or this simplest of phonograms that the voice impresses on the air and the ears recapture a few seconds later? Alas, civilized man loses his sense of wonder so easily, and much to his detriment!

11. Another trait shared by savage and child is the behaviour towards animals. To the savage they are almost of his kind, his 'younger brothers' with whom he speaks, as children do. He wants to understand the animals' language to learn what they know.[17] He credits them with powers surpassing his own, for he cannot fly like a bird, dive like a fish or climb on threads like a spider. When my four-year-old saw a huge tame raven on a doorstep, he stopped in amazement and asked quite in earnest "Who is that?" The form of address does not mean much with a child, but even I could not escape the impression of a thoughtful personality, especially as I had just seen the bird 'scolding' a young cobbler's boy who had been teasing him.

12. From the shore, the sea looks like a flat disc, and so, with a wide enough horizon, does the land, which in a sense swims on the sea. The whole is covered by the 'dome' of heaven. These observations are the first foundation of primitive geography and astronomy. That this aspect is based on physiological factors can be observed on a mountain top or even in a balloon. The observer feels as though he is in a hollow painted sphere, the lower half representing the earth and the upper the sky, while the whole seems to be rolling or flowing away in the direction opposite to the balloon's motion. However, this is experienced too rarely to influence popular ideas. For the common man sea and land remain physically a disc and heaven a dome. If now he sees the sun dive into the western sea he imagines he must hear it sizzle and probably does, by attributing to it

some other noise that happens to be occurring. Thus arises the idea and legend that Strabo[18] reports as current on the 'holy promontory' (Cape St. Vincent) in Iberia and which Ellis has rediscovered in the Society Islands in the far off South Pacific.[19]

13. Children and primitive man have no occasion to go beyond such naïve ideas. The child sees the sun rising or setting behind a hill and goes there to catch it. Of course he finds that it was the wrong hill, for there are a second and third one further off, on which the sun is sitting, but some hill must surely be the right one[20]: he does not exclude the idea that he might catch the sun with a net. The widespread fairy tales about a sun-catcher allow the inference to a primitive level of culture at which what seems to us an invention affording pleasant phantasies may well have been meant in earnest. Similarly for other tales, for example Jack and the Beanstalk and the whole related group. Heaven to the naïve child seems not so high that he might not reach it by climbing a high tree; that is the element, phantastic to us, that is common to the group.[21] Only gradually as culture develops can such tales acquire a faint streak of humour and irony, until they end up as phantasies merely for amusement. The fairy tales of primitive tribes combined with observation on children, afford us the deepest and strongest conceivable insight into the beginnings of culture.

14. If phantasy completes and modifies single observations, neither does it leave untouched a whole complex of historical record. However, with some care one can dig out the factual core from the poetic husk, and need not discard the former with the latter. One example is the tradition of a central American Indian tribe concerning its immigration from the North.[22]

They travelled away from the rising sun, "but it is not clear how they crossed the sea, they passed as though there had been no sea, for they passed over scattered rocks, and these rocks were rolled on the sands. This is why they called the place 'ranged stones and torn-up sands,' the name which they gave it on their passage within the sea, the water being divided when they passed." Then the people collected on a mountain called Chi Pixab, and there they fasted in darkness and night. Afterwards it is related that they removed, and waited for the dawn which was approaching, and the manuscript says: – "Now, behold, our ancients and our fathers were made lords and had their dawn; behold, we will relate also the rising of the dawn and the apparition of the sun, the moon, and the stars." Great was their joy when they saw the morning star, which came out first with its resplendent face before the sun. At last the sun itself began to come forth; the animals, small and great, were in joy; they rose from the watercourses and ravines, and stood on the mountain tops with their heads towards where the sun was coming. An innumerable crowd of people were there, and the dawn cast light on all these nations at once. "At last the face of the ground was dried by the sun: like a man the sun showed himself, and his presence warmed and dried the surface of the ground. Before the sun appeared, muddly and wet was the surface of the ground, and it was before the sun appeared, and then only the sun rose like a man. But his heat had no strength, and he did but show himself when he rose, he only remained like (an image in) a mirror, and it is not indeed the same sun that appears now, they say in the stories."

This report is not very clear, but the characteristics of the arctic region – the long winter night, the frozen sea scattered with rocks of ice, the little strength of the sun on re-appearance – are unmistakeable.

15. Out of phantastically interwoven observations of nature and historical traditions primitive man develops ideas as to his origins, his relations to spirits, life after death, in short those views generally called religious and mythological. Their poetic value was discussed earlier. Merely by hoping for help from his gods and demons, man will bear many a hardship more easily, while by fearing strokes of ill-luck, his recklessness may receive a salutary check. An observer familiar with modern religion notices that all these primitive systems and especially ideas of an after life, have nothing to do with reward and punishment, or retribution, and least of all with ethics.

16. The ethic of primitive man which, because of his different conditions of life, is very different from ours, albeit no less rigid, is prescribed for him by public opinion, which well discerns what helps or harms the common weal. If he offends against this ethic he will have to come to terms with that public opinion and its consequences. His behaviour is regulated in a natural way according to the conditions of life actually present. It is certainly not rational to base ethics on foundations whose correctness cannot be tested; but where one section of a people is condemned to permanent slavery while the other aims at securing all the goods of life in this world to itself, an ethic of retribution after death is of inestimable consolation for the first and very convenient for the second. However, an ethic is healthier if it is based only on fact, like the highly developed Chinese doctrine. Ethics and law are parts of the techniques of social cultures and are the higher the more the element of vulgar thought has been displaced from them by scientific thought.

17. Some assert that there are tribes that lack all religious or mythological ideas. However, such reports must be taken with extreme care. We know how generally savage tribes hold beliefs in spirits and demons and how much these torment them. If the report is indeed free from misunderstandings and is a clear and genuine account of the facts, then such a tribe would have to be taken as a very rare exception.

The following report deserves mention as an example[23].

It is evident, that the Arafuras of Vorkay (one of the Southern Arus) possess no religion whatever.... Of the immortality of the soul they have not the least conception. To all my enquiries on this subject they answered, "No Arafura has ever returned to us after death, therefore we know nothing of a future state, and this is the first time we have heard of it." Their idea was, Mati, Mati sudah (When you are dead there is an end of you). Neither have they any notion of the creation of the world. To convince myself more fully respecting their want of knowledge of a Supreme Being, I demanded of them on whom they called for help in their need, when their vessels were overtaken by violent tempests. The eldest among them, after having consulted the others, answered that they knew not on whom they could call for assistance but begged me, if I knew, to be so good as to inform them.

At first sight we seem to detect in these words the ironical superiority of a free-thinker who gives an appropriate rebuff to the supposedly higher insight of the importunate European proselytizer.

18. At the primitive level religion, philosophy and a view of nature cannot be separated. Where, as in ancient Greece, there is no priest caste strongly pursuing its own interests, it is easier for a philosophy of a freer kind to develop, breaking through the barriers of traditional religious and mythological ideas. To be sure, this early philosophy is itself still phantastic, witness the experiments of the Ionians and Pythagoreans. How could it be otherwise? The vital point is to gain a world view at all, criticism cannot begin until there are several attempts, several seemingly unequivalent views that invite comparison, contradiction or approval. Here philosophy and natural science are as yet one: the first philosophers are astronomers, geometers, physicists, in short, scientists. However, if they succeed, alongside their dubious world views, in transforming the aspect of smaller sections of nature into a shape better able to stand up to criticism, these aspects collect and receive more general assent, forming the beginnings of science. Consider for example the geometrical findings of Thales and Pythagoras, and the latter's acoustic observations. This

incipient science still contains plenty of phantastic elements, and we may call large parts of it without hesitation nature mythology. By virtue of these very reasonable attempts to understand the whole of nature in terms of one part of it that is more easily intelligible to the enquirer the animistic and demonological nature mythology is gradually replaced by a mythology of matter and force, a mechanistic-atomistic or dynamic nature mythology. Often these different views co-exist; their traces reach into modern times, witness Newton's particles of light, the atoms of Democritus and Dalton, modern chemical theories, cage molecules and gyrostatic systems, ions and electrons. We may mention the many physical hypotheses as to matter, the vortices of Descartes and Euler, which revive in the new electromagnetic theories of currents and vortices, sources and sinks that lead into the fourth dimension of space, the ultra-cosmic particles that generate gravity, and so on. What an impressive witches' sabbath of adventurous modern ideas! These excrescences of the imagination fight for existence by trying to overgrow each other. Countless such offspring and flowers of phantasy must, in view of the facts, be destroyed by merciless criticism, before a single one can develop further and attain some permanence. To appreciate this, consider that the aim is to reduce natural processes to the simplest conceptual elements. But before we can understand nature we must seize it through phantasy, in order to give these concepts a living and intuitive content. Phantasy must be the stronger, the more remote from immediate biological needs the problem to be solved lies.

NOTES

[1] Tylor, *Early History of Mankind*, 1870, p. 138 (=*Urgeschichte*, p. 173).

[2] Ennemoser, *Geschichte der Magie*, Leipzig 1844; Roskoff, *Geschichte des Teufels*, Leipzig 1869; Soldan, *Geschichte der Hexenprozesse*, Stuttgart 1843. If this threatens to sour the reader's sense of humour, let him turn for relief to the articles on 'Bekker', 'Incubes', 'Magie', 'Superstition', in Voltaire's *Dictionnaire philosophique*. To restore complete cheerfulness, see Mises (G. T. Fechner), *Vier Paradoxa*, Leipzig 1846, the section 'Es gibt Hexerei'.

[3] On the pathological facts that promote the development of such conceptions (lycanthropy, vampirism and so on), that is the belief in magic, cf. Ch. IV note 33). Teratological cases no doubt will also have contributed much.

[4] Tylor, *l.c.* p. 126 (=German p. 159).

[5] Alongside the conception of the shadow soul, readily intelligible reasons from waking life give rise to the development of the idea of a soul of breath and a soul of blood. Cf. Odyssey XI, 33–154. Shadow souls recover memory by drinking blood.

[6] Tylor, *Primitive Culture*, 1871, II, p. 46 (=*Anfänge der Kultur*, II, p. 49).

[7] *Ibid.* **I**, p. 434 (=German **I**, p. 479).
[8] *Ibid.* **II**, p. 144 (=German **II**, p. 159).
[9] *Ibid.* **I**, p. 417 (=German **I**, p. 451).
[10] *Ibid.* **II**, p. 366 (=German **II**, p. 405). The facts are related in Diodorus XX, 14, who gives further examples of human sacrifice. Also Herodotus, IV, 62.
[11] Tylor, *ibid.* **I**, 96 (=German **I**, 106).
[12] F. Hoffmann, *Geschichte der Inquisition*, Bonn 1878. Lea, *A History of the Inquisition*, New York 1888.
[13] Tylor– *Urgeschichte*, pp. 104–112; *Primitive Culture* **I**, pp. 265, 266 (=German pp. 288, 289).
[14] Tylor, *Early History*, pp. 329–30 (=German pp. 409f). During a visit to lake Garda I myself once heard a local inhabitant express the view that the lake had at one time been much higher and that Monte Brione between Riva and Torbole had been an island, because you can find shells on top.
[15] *Ibid.* p. 326 (=German, p. 411).
[16] J. W. Powell, *Truth and error*, Chicago 1898, p. 348. An echo that had to leave behind a phantomlike demonic impression is reported by Cardanus (*De subtilitate*, 1560, lib XVIII, p. 527) following the experience of a friend A. L. who one night came to a river he wanted to cross. He shouted: Oh. Echo: Oh! A. L.: Unde debo passa? Echo: Passa! A. L.: Debo passa qui? Echo: Passa qui! However, since at that spot there was a fearsome vortex, A. L. took fright and turned back. Cardanus recognizes the phenomenon as an echo and points out that it was easily recognized as such by its intonation.
[17] Powell, *l.c.* p. 383; also Galton, *Inquiries into human faculty*, London 1883.
[18] Strabo, III. Iberia, 1.
[19] Up to the age of four or five I continued to hear the sun hiss as it seemed to dive into a big pond and was mocked for it by adults. Still, I greatly value the memory.
[20] As a child I too pursued the setting sun from hill to hill.
[21] Tylor, *Urgeschichte*, pp. 436f., 443f.
[22] Tylor, *Early History*, p. 307 (=*Urgeschichte*, p. 387).
[23] Lubbock, *Origin of Civilization*, 1870, p. 140 (=*Entstehung der Zivilisation*, Jena 1875, p. 175).

CHAPTER VII

KNOWLEDGE AND ERROR

1. Living beings have reached equilibrium with their surroundings by means of adaptation that is partly innate and permanent, partly acquired and temporary. However, a mode of organization and behaviour that is biologically beneficial under certain circumstances may become detrimental under changed conditions and even end up by destroying life itself. A bird is organized for life in air, and a fish for life in water, but not the other way round. A frog snatches at flying insects that are his food but falls victim to this habit if he gets hold of a piece of moving cloth and is caught on the hook attached to it. Moths that fly towards light and colour for the sake of self-preservation, may in the course of this generally appropriate behaviour finish up on the painted flowers on wallpaper, which do not nourish; or in a flame, which kills. Any animal caught in a trap or by another beast thereby shows that its psychophysical organization is appropriate only up to a point. In the simplest animals stimulus and response may lie in attack and flight, so regularly linked that the observed facts would not cause us to imagine that sensation, ideas, moods and will intervene between the two, were it not that this is strongly suggested by analogy with processes we observe in ourselves. Here, stimuli work directly and actively, as in a reflex movement, say of tendons, of which we do not learn until after it has happened. Not until, with complicated conditions of life, simple stimuli become too ambiguous to determine the appropriate adaptation, does sensation arise as a separate element, and this, along with memories and ideas, determines the state of the organism and its feelings, and thus finally sets off an action towards a conscious goal. To the more complex conditions there corresponds a more complex organism adapted to them through the interaction of many parts mutually attuned. Consciousness consists in an especially important interaction of the parts of the brain. If some element, part of a conscious process, a sensation or an idea, does not appear to us as directly active, this is because of the many and versatile connections that the element has attained in a developed individual: each single connec-

tion tends to be pushed into the background, to be called into play only when elements (sensations, ideas) are suitably combined. There is no opposition between, say, idea and will: both are produced by organs, the former mainly by single organs, the latter by the connection of organs. All the processes of a living individual are reactions in the interest of self-preservation, and changes in ideas are merely part of changes in reactions. The fact that a certain living organism exists shows that adaptations were viable sufficiently often to ensure survival. We daily observe reactions in physical and mental life that are not viable and thus amount to failure of adaptation. Physical and mental reactions are ruled by the laws of probability: whether any one of them is useful or harmful, whether the ideas that arise are biologically beneficial or misleading, in either case the same physical and mental processes are involved.

2. Consider some examples. As soon as a reaction is set off by a stimulus, harmful consequences may supervene. Many flies, attracted by the carrion smell of certain plants, are misled into laying their eggs on them, but when the larvae hatch they die of starvation. Insects often fall victim to poison that tastes like certain foods. Cattle and sheep, especially in strange pastures, sometimes suffer similarly. Circumstances with close physical links that set off different sensations occur more often together than circumstances that are linked only by mere accident; so that the sensations and ideas corresponding to the former are more strongly associated than in the latter case. Innate and acquired perception is moreover directed to what is biologically important. Nevertheless, we cannot exclude unfavourable accidental events with attendant misleading associations. If Darwin is right, vividly coloured insects of unpleasant taste or poisonous sting are shunned, and so are others that are harmless but similar in appearance (mimicry). When our retina receives the optical image of a known body, association calls up the touch and other properties. If in the dark we touch a body, its optical image arises in us. It is biologically important that these associations occur quickly and vividly, almost like illusions; yet these same processes will deceive us on occasion, even if rarely. This depends greatly on one's mood or drift of thought. A young man busy turning grassland into agricultural fields using a team of oxen is often disturbed by rattlesnakes, which he kills: when trying to pick up the whip he had dropped, he grabs by chance a stick that he takes

for a snake and therefore thinks he hears the rattle.[1] Conversely, in some cases one might be looking for a stick and by mistake grab a snake, thinking it is a stick or some other harmless object. How far this agility of associative mental completion can go in man, especially civilized man, is best seen in the way one readily gives spatial content to a flat perspective line drawing. One easily recognizes a staircase, a machine or complicated crystal shapes in their three-dimensional form, although the sketch indicates a mere hint of it. T. W. Powell[2] records that Red Indians find it hard to interpret such sketches, but they soon overcome this. With paintings in colour, it must be a representation of a familiar object before they can readily understand it. I once knew an old lady endowed with great imagination who could tell marvellous fairy tales but found a painting quite unintelligible, as an animal or idiot might, hardly sure whether it was a landscape or a portrait.[3] The fact that associations are imprecise and may interfere with each other shows itself in a child's first rudimentary attempts at drawing. What they remember and what they have observed about a person, they express in his 'picture', regardless of whether all this is simultaneously visible. Likewise for Red Indians[4] and those who originated painting in ancient Egypt: the walls of their temples combine venerable age and traits of highly developed technical skills with primitive childlike art.

3. Strong physical dependence cannot easily be quite masked by chance, and biological interest promotes the noticing of correct and important associations: these therefore tend to become permanent[5] even if there is no great mental development, and so tend by instinct alone to conduct the processes of life in a largely beneficial manner. However, where misleading associations entail painful consequences, it is precisely these that act as corrective and contribute to further mental development. Dreamlike associations will give way to attentive, conscious and deliberate notice of important agreements and differences of the various cases, and to the clear discernment of their correct and misleading features. Here lies the beginning of deliberate adaptation of ideas, of enquiry. Briefly, this aims not only at permanence in ideas but at a degree of differentiation[6] adequate to cope with the wealth of experiences. The course of ideas should adapt itself as closely as possible to physical and mental experience, following and anticipating it, remaining as constant as possible in different cases

yet doing justice to their difference. The course of ideas is to be as faithful a picture as possible of the course of nature herself. Any substantial progress here requires collaboration of men in society and communication in speech and writing.

4. A person who has felt the discomfort of mistaking a poisonous mushroom for an edible one, will carefully notice the red and white patches of a toadstool and construe them as warning signs of poison, which thus stand out from the total picture of the plant; and similarly with poisonous berries and the like. In this way we learn to pay separate attention to the more important determinant features of an experience, which we thus analyze into parts or reconstitute from them. In regarding one aspect as further defined by another that seems noticeable or important, we make a judgement. One can of course judge internally without linguistic expression, or before it. This must have been the case of the intelligent savage who first protected his gourd from burning by giving it a coat of clay; he must have judged thus: the gourd burns, clay does not, a clay-covered gourd does not burn. One can gather simple observations and experiences without speaking, as is easily observed with intelligent dogs or with children before they learn to speak.[7] However, the linguistic expression of a judgement has great advantages, for it imposes on the conveying of any experience a prior analysis into generally known and named constituents, by which the speaker himself gains much in clarity.[8] His attention must emphasize special features and he must produce abstractions and compels others to do likewise. If I say "the stone is round", I separate matter from form. In the judgement "the stone serves as a hammer", use is separated from object. The proposition "the leaf is green" contrasts colour with form. Much as thought gains through expression in language, still this involves classification into accidental and conventional forms. Whether I say "the piece of wood floats on the water" or "the water carries the piece of wood" makes no difference in thought but is pschologically the same; but in verbal expression, the role of subject has gone from wood to water. Whether I say "the cloth is torn" or "the cloth is not intact" is psychologically the same, but verbally I have transformed an affirmative judgement into a negative one. The judgements "all A are B" and "some A are B" can psychologically be taken as a sum of many judgements. Since logic has to use language it has to make do with the

historically transmitted grammatical forms, which are not at all parallel to psychological events.[9] How far a logic that uses an artificial ad hoc language can free itself from this mismatch and follow psychological events more closely will not be discussed here.[10]

5. Not every judgement can base itself on such simple sense observations or pictures as 'intuitive' judgements, such as "the unsupported stone falls to the ground", "water is liquid", "salt dissolves in water", "wood is combustible in air". Further experience tells us for example that in the last case the conditions for catching fire are much more complex than that judgement suggests; wood does not burn in any sort of air, but only if there is enough oxygen and at a sufficiently high temperature. Oxygen and temperature are not seen just by looking, the words arouse no simple clear idea; on the contrary, we have to think of the whole range of physico-chemical behaviour of oxygen, with all the experiences and observations we have of it and the judgements we have made to give proper expression in thought to the condition 'presence of oxygen'. The concept of 'oxygen' cannot be given by means of a clear picture but only by a definition which exhaustively sums up experiences.[11] Likewise for 'temperature', 'mechanical work', 'quantity of heat', 'electric current', 'magnetism' and so on. By intense occupation with the field of experience and knowledge to which a concept belongs, we gain the facility of ensuring that whenever the word embodying and denoting a concept is used all the experiences linked with that concept are allowed to resonate softly within us, without any precise and explicit idea of them. Concepts contain potential knowledge, as S. Stricker once apply remarked. Through frequent use of the word for a concept we obtain a sure and delicate feeling for the sense in which and the limits within which we may properly apply that word. Those who are less familiar with a concept will find that when the word is used some picture will arise to represent the concept by a conspicious and important side of it. Thus in vulgar thought one may, on hearing the word 'oxygen', imagine a chip that glows and breaks into flame; with 'temperature' might go the idea of a thermometer, with 'work' that of a lifted weight and so on, ideas that Jerusalem (*op. cit.*) has aptly called typical.[12]

6. A judgement, our own or conveyed to us, that we find appropriate to the physical or mental finding[13] to which it relates, we call correct and see

knowledge in it, especially if it is new and important. Knowledge is invariably a mental experience directly or indirectly beneficial to us. If, however, the judgement does not stand up, we call it an error; or, if we have been deliberately misled, a lie.[14] The same mental organization that is thus beneficial, and, for example, causes us quickly to recognize a wasp, may at other times cause us to mistake a beetle for a wasp (mimicry). Even the immediate observation of the senses can lead to knowledge and to error, when important differences are overlooked and agreements mistaken; for example by taking a faintly coloured wasp for a fly in spite of the characteristic body shape. Much more readily still such oversights lead us into error in conceptual thinking, especially the person who has no practice in it and makes do with typical ideas without exact subsequent analysis of the concepts used. Knowledge and error flow from the same mental sources, only success can tell the one from the other. A clearly recognized error, by way of corrective, can benefit knowledge just as a positive piece of knowledge can.

7. If we ask ourselves from what source erroneous judgements spring when based on observation, we must say they spring from inadequate attention to the observational circumstances. Every single fact as such, whether physical or mental, or mixed, remains as it is; error supervenes only when we take a fact as continuing to exist under other circumstances and ignore the change in circumstances, physical, mental or both. Above all, we must not ignore our bodily boundary, in that dependence is rather different according to location: on one side, on the other, or across.[15] For example, we may consider the mistaking of a genuine hallucination for a sensation, though this is rather rare in a healthy person. However, it happens daily that one mistakes a sensation for the idea called up by association, or that one fails to distinguish them adequately. The simplest example is taking a mirror image for the object, which is frequently observed in birds and other animals too. Monkeys grab behind the mirror and in line with their higher mental development, give vent to their displeasure at being thus tricked.[16] If stronger expectations are ready to complete the sensation by association, less pleasant forms of deception result, as in our previous example of snake and stick. This happens the more easily as the sensation is less intense, for example in weak light, and the imagination therefore more active. Such cases of illusions overcoming

sensations can wreak much havoc in scientific research too.[17] We have earlier discussed the role in vulgar thought of translating dream phantasies into physical fact. Many will recall the childhood experience of waking from a dream and crying because some beautiful toy that one seemed to hold in one's hand only a moment before was now gone. Young peoples behave not very differently from such a child. Hence their preoccupation with interpreting dreams, since these are taken to have a vital bearing on waking life.

8. The boundary between dreaming and waking becomes fully distinct only gradually. Let me illustrate this by a recent experience of mine. I awoke at night through hearing someone opening the door and entering. In spite of total darkness, I see a tall figure creeping along the wall and stopping in front of the faintly illuminated window. Keeping still and observing, I no longer hear the slightest noise, but see the figure perform various slow movements. Now it becomes clear to me that in front of the window there is a clothes-stand whose contours, because of the darkness, are being constantly changed by my waking phantasy, which was left over from my dream phantasy.[18] Such appearances are familiar to me from dark and sleepless nights: however dark it is, I see my bedroom windows; but, being a little uncertain as to their place, width and so on, I cover my eyes with my hand, or close them completely, and still see the windows. This is a good method for distinguishing a phantastic appearance in the dark from a physically determined sensation.

9. From the book by Powell quoted above, which I would not recommend for strictly philosophical purposes, but which contains many valuable particulars, I should like to adduce as an interesting example of 'physical' thinking the view of a certain Red Indian chief.[19]

A party of white men and Indians were amusing themselves after the day's work by attempting to throw stones across a deep canyon near which they had encamped. No man could throw a stone across the chasm. The stones thrown by the others fell into the depths. Only Chuar, the Indian chief, succeeded in striking the opposite wall very near its brink. In the discussion of this striking phenomenon Chuar expressed the opinion that if the canyon were filled up, a stone could easily be thrown across it, but, as things were, the hollow or empty space pulled the stone forcefully down. The doubts of European Americans as to the correctness of this conception, he met by asking 'Do not you yourselves feel how the abyss pulls you down so that you are compelled to lean back in order not to fall in? Do not you feel as you climb a tall tree that it becomes harder the higher you climb and the more void there is below?'

This 'primitive physics' seems in several respects erroneous to us modern men. First of all, Chuar was interpreting his subjective dizziness as a physical force that drew all bodies down. He was of course not at all troubled by the fact that the great abyss above us was not equally active, since 'downwards' was an absolute direction for him. We have no right to demand that he should be wiser in this respect than fathers of the church like Lactantius and Augustine. Chuar's ascription of forces to empty space would have horrified Descartes and his associates but after Fresnel, Faraday, Maxwell, and Hertz we ought no longer to be as amazed as the white and educated associates of Chuar. The modern physicist would above all question the alleged fact by careful measurement, to show that the empty space does not behave as claimed (for example, by using a balance to show that gravity is not increased above a hollow space, if anything the reverse would be shown assuming the balance to be sensitive enough). We no longer hypostatize our subjective sensations and feelings into physical forces. To this extent we are more advanced. However, let us not be too pround but consider that we still regard our subjective concepts as physical realities, as has been shown by Stallo[20], as well as by myself.[21] How this misleads enquiry will be discussed elsewhere.

10. By uncovering misleading motives one protects oneself against error and may indeed derive some use from it. Such motives are most obvious and clear in cases where someone deliberately misleads us. Leaving aside the verbal sophists, whose artful fallacies mislead our conceptual thinking, we also have sophists of action whose sham tricks mislead observation. It is of interest to analyse the procedures of conjurers or jugglers, which can deceive and surprise the public by quite simple means. A rather crude method is to cause the spectator to assume a non-existent identity: for example a borrowed watch is placed in a laterally covered mortar on a table, some trick or other is used to distract the public's attention so that a hidden assistant can remove the watch and put a similar but worthless one in its place, and that is the one which is now smashed; while the wreckage is shown round, another trick allows the assistant to produce the original

watch at a spot where nobody expects it to be.[22] In exceptional cases, a conjurer may be prepared to incur some expense for this joke in order to enhance his reputation. Thus Houdin[23] in a performance before Pope Pius VII smashed an expensively bought watch that had been fashioned after the timepiece of one of the cardinals and even bore his name. Houdin gives instructions for sham movements that provoke the impression that one has put an object down when this is not so; he shows how one can hold a small object with open hand and spread fingers and explains this by means of an illustration[24]. In card tricks conjurers use small signs, invisible to the unpractised, and that alone guides them.[25] By using unusual even if simple means that no one thinks of, they will be almost always successful.

11. In Europe today, using a strong electromagnet will not produce a sensation and people would soon see through the arrangement. However, when Houdin[26] gave a performance to an Arab audience at Algiers and rendered a small flimsy suitcase (with iron base) too heavy for the strongest man (by means of an electromagnet under the carpet), the spectators were naturally gripped by panic. A case reported by Decremps[27] shows how even educated and experienced people can easily be put off balance. Mr van Estin, a Dutch merchant on the island of Bourbon gave a sheet of paper to Mr Hill, then travelling there, asking him to write down an arbitrary question, and to keep the sheet on his person without showing it to anyone, or better still to burn it. When all this had been done in van Estin's absence, he returned with a folded sheet and declared that it bore the answer. So that Hill should not think this was a mere ordinary conjuring trick, van Estin asked him to sign it and instructed him to fetch the sheet thus marked from the drawer of a desk in a pavillion at the far end of a park, at the same time giving him the keys of pavillion and desk. Hill ran quickly to the pavillion and at the spot indicated did indeed find the sheet marked by him, with the correct answer to his question. Without going into detail on the mechanical, optical and acoustic devilries that Hill encountered in the pavillion, much to his annoyance and distracting his attention in all directions, let us at once consider the resolution of this apparently surprising trick. Why did Hill have to write down his question? Why not merely think it? Obviously because it must leave a trace: the sheet on which Hill wrote was lying on a black folder containing carbon paper, van Estin's folded sheet on which the answer could be written at

leisure after Hill had gone was conveyed by pneumatic tube to the distant desk. The complicated dressing up served merely to disguise the simple situation. If we ask ourselves in what a juggling device differs from a technical invention, it is just that the former cannot create anything useful.[28]

12. Another interesting example reported by Decremps[29] deserves mention. A man, before a jury, is accused of having thrown a child into the river drowning it. No fewer than 52 witnesses testify against him: some saw him throw the child into the river, others had seen him hit out at the child in great anger, and so on. The accused defends himself by saying that nobody had reported the loss of a child and no corpse had been found. The court is of course highly embarrassed. Then the accused asks that a friend of his be admitted, which is granted, and he appears with a huge package which on unwinding reveals a cradle with a child. The accused caresses the child which starts crying at once. "No, you poor mite, you shall not stay alone and helpless in the world!", the accused shouts, suddenly drawing a sword from the package and before anyone can stop him, cuts off the child's head with the yell "Now go the way your brother went!". Instead of the expected blood, a wooden head is seen and heard rolling across the floor. Now the man turns out to be a magician and ventriloquist who had used this method to ensure a living at his trade, for which he had first to acquire the necessary reputation. Whether this story be true or false, it is certainly instructive. Something may be very probable and yet not true. There is nothing that witnesses do not see, once they believe that this man or that is a murderer or a thief, and a biassed witness will testify anything. But what need is there of anecdotes when the actual judicial murders that become known every year show clearly enough how easily men are condemned when they are thought to be guilty. As though it were not much more important that no innocent man be condemned than that every guilty person be punished! The criminal law is meant to protect humanity, but often behaves towards it like the bear of legend, who used a stone to crush a fly on his sleeping benefactor's forehead.[30]

13. From the behaviour of conjurers and their audiences we can draw fruitful lessons for our own behaviour in scientific enquiry. Nature is not

indeed a juggler seeking to deceive us, but her processes are very complex. Besides the conditions to be investigated on which we fix our attention, there is a whole host of other circumstances that are involved in determining natural processes, and this additional material conceals what interests us by complicating and apparently falsifying the process in question. Therefore, an inquirer must not disregard any unintended factor that may be involved; he must take into account all sources of error. An experimenter might be investigating some new effect of electric current, using a galvanometer, but in his eagerness he forgets that the deflection might be caused by some unnoticed loop and so have nothing to do with the process being investigated. We must above all be on guard against assuming identities without making sure that they do exist. A chemist discovering a new reaction of some substance must take into account that the latter may have been prepared by a new process and contain impurities: it may not really be the substance he thinks he is examining. Finally, we must remember that even the highest probability is still not a certainty.

14. I should like to close with the story of an incident that was very instructive for myself. One Sunday afternoon, my father showed us children the experiment that Athanasius Kircher[31] described as the "experiment on the amazing imagination of the hen", only with a very slight difference. A hen resisting fiercely is pressed down on the floor and held there for half a minute, during which it quietens down. Now a line is drawn with chalk, across the hen's back and round it on the floor. On being released, the hen simply stays put, and quite strong shocks of fright are needed to induce it to jump up and run away, "for it imagines it is tied down". Many years later I once discussed hypnosis with a fellow laboratory worker, Prof. J. Kessel, and I remembered Kircher's experiment. We ordered a hen and repeated the experiment with great success. But, on further repetition, when the hen was merely pushed down and the hocus-pocus of the chalk line was omitted, the experiment was just as successful. The hen's imagination, which had continued unchallenged in my mind since childhood, was thus destroyed forever.

15. It is not advisable to regard a single experiment or observation alone as proving the correctness of a view that seems to be confirmed by them. Rather, one must as far as possible vary the conditions, both those

considered as important, and those that seem indifferent; and that with one's own experiments as much as with other people's. Newton in his *Optics* has made extensive and exemplary use of this method, thus laying the foundations for modern experimental physics, just as his *Principia* make him the founder of mathematical physics. Both works are indeed without equal or substitute as educational guides to enquiry.

The upshot of our considerations is this: the same mental functions, operating under the same rules, in one case lead to knowledge and in another to error; from the latter, only repeated and exhaustive examination can protect us.

NOTES

[1] Powell, *Truth and error*, p. 309.
[2] *Ibid.*, p. 340.
[3] Even the more intelligent amongst dogs are said at times to recognize their master's portrait.
[4] K. von den Steinen, *Unter den Naturvölkern Zentral-Brasiliens*, Berlin 1897, pp. 230–241.
[5] Cf. *A* 4, p. 248. Also this volume, Ch. III, sn 3.
[6] *A* 4, p. 248.
[7] Preyer, *Die Seele des Kindes*, Leipzig 1882, pp. 222–233.
[8] Cf. *W*, pp. 406–414; *P* 3, 1903, pp. 265f.
[9] A. Stöhr, *Algebra der Grammatik*, Vienna 1898.
[10] Boole, *An investigation of the laws of thought*, London 1854. E. Schröder, 'Operationskreis des Logikkalküls', *Math. Annal.* 1877.
[11] Here we are thinking in the first place of empirical concepts.
[12] Jerusalem, *Lehrbuch der Psychologie*, 3rd ed. 1902, pp. 97f.
[13] This finding may relate to physical or mental facts, where amongst the latter we include also logical ones.
[14] I find myself unable to favour the view that belief is a special mental act at the basis of judgement and constituting its essence. Judgements are not a matter of belief but naive findings. It is rather that belief, doubt, unbelief rest on judgements about agreement or disagreement between often very complicated sets of judgements. Our rejecting of judgments that we cannot accept is often accompanied by strong emotion causing involuntary exclamations. From such an utterance there develops the particle of negation, according to Jerusalem (*Psychologie*, p. 121). The need for an affirmative particle is much slighter and appears much later. Once of my sons at two to three years old used fiercely to pronounce the syllable 'meich' to signify rejection, at the same time pushing away with a strong gesture the object untimely offered. It was an abbreviated 'meichni' (mag nicht=don't want).
[15] Cf. Ch. I, end of sn. 6.
[16] Darwin, *Kleinere Schriften*, translated by E. Krause, **II**, p. 141.
[17] *A* 4, pp. 158, 159.
[18] There are phantasmata resting on the retina, dark spots or rings that expand and contract. Considering the impossibility of sharply focussing in the dark, the former

phantasmata can also combine with what is objectively seen and together feign movements.

[19] Powell, *l.c.*, pp. 1, 2.

[20] Stallo, *Die Begriffe und Theorien der modernen Physik*, Leipzig 1901.

[21] Cf. *M* 4, 1901.

[22] Decremps, *La magie blanche dévoilée*, Paris 1789, **I**, p. 47.

[23] Houdin, *Confidences d'un prestidigitateur*, Paris 1881, **I**, p. 129.

[24] Houdin, *Comment on devient sorcier*, Paris 1882, p. 22.

[25] Houdin, *Confidences* **I**, pp. 288–291.

[26] Houdin, *Confidences* **II**, p. 218f.

[27] Decremps, *l.c.* **I**, p. 76f.

[28] Cf. *M* 4, p. 535. Cardanus, *De subtilitate* (1560), on p. 494, in speaking of the contempt for alchemists and other jugglers, says: I think the cause is manifold, but primarily it is because one is dealing with useless things.

[29] Decremps, *l.c.* **II**, p. 158f.

[30] In Ernst Faber's translation of Lieh Tzŭ (Elberfeld, 1877), there are some passages that splendidly illustrate how suggestion and false suspicion work. On p. 207 there is a description of a rich man's gambling party. A buzzard flies past and drops a dead mouse amongst the people in the street. "Master Yu has long enjoyed prosperous and merry days and constantly harbours belittling thoughts about others. We have done him no harm and he insults us with a dead mouse. If this is not avenged, we can hardly hold our own in the world. We therefore ask all who belong to our group to be brought out with a will; his house must be destroyed!... In the evening of that day a crowd gathered and took arms, attacked Master Yu and caused great wreckage on his property." p. 217: "A man missed his axe and suspected his neighbour's son. He therefore spied on him: every step revealed the suspect as the thief of the axe, and so did the expression of his eyes, his every word and speech, movement, manner, behaviour and action. By chance the man was digging about in his ravine and found the axe. On seeing his neighbour's son next day, he no longer found that the latter's movements, action, manner and behaviour resembled those of an axe thief." – Especially valuable and instructive for the lawyer, I think, are W. Stern's 'Beiträge zur Psychologie der Aussage' (first fascicule published in 1903).

[31] A. Kircher, *Ars magna lucis et umbrae*, Amsterdam 1671, pp. 112–113.

CHAPTER VIII

THE CONCEPT

1. We must now look more closely at the concept as a mental entity. Remembering that one cannot imagine a man who is neither young nor old, tall nor short, that is, a man in general; likewise, that every triangle must be imagined either acute, obtuse or right-angled, so that there is no triangle in general: we might easily conclude that there are no such mental constructs as concepts at all, nor abstract ideas, a denial that Berkeley had defended with special vigour. However, we might just as easily be led to the view of the 'nominalist' Roscelin that general concepts or universals do not exist as things but are mere 'flatus vocis', while his 'realist' opponents regarded them as grounded in things. That general concepts are not, as a respected mathematician asserted quite recently, mere words, clearly emerges from the fact that very abstract propositions are understood and correctly applied in concrete instances, witness the countless applications of the proposition "energy remain constant". It would however be idle to attempt to find a clear, momentary conscious idea which would exactly cover the sense of the sentence as it is spoken or heard. The difficulty disappears if we recognize that concepts are not momentary entities, like a simple concrete symbolic idea: every concept has its sometimes long and eventful formative history, and its content cannot be explicitly expounded by a transient thought.[1]

2. We may take it that a hare soon acquires the typical ideas[2] of a cabbage, a man, a dog, a cow; because of the immediate associations linked with the respective perceptions or their typical attendant ideas, the hare will be attracted to the first, run away from the next two and remain indifferent to the last. As an animal's experience becomes wider, it will become more familiar with the common reactions of the objects belonging to each single one of these types, and these reactions cannot vividly arise in the imagination all at the same time. If the animal is attracted by an object similar to a cabbage, a testing activity is set off at once; the animal will sniff and nibble at the object to ascertain whether it really offers the

expected reactions of smell, taste, consistency and so on. Initially scared by a man-like scarecrow, the animal soon sees by attentive observation that important reactions of the type man are missing here; for example movement, change of position, aggressive behaviour. Starting from a typical idea, the testing animal begins tentatively to connect with it the gradual store of remembered experiences or reactions; these can only gradually enter consciousness. This seems to me what is characteristic of concepts as against momentary ideas: the latter, by gathering associations, very gradually develop into the former, in a continuous transition. I therefore think that one cannot deny that higher animals show the beginnings of concept formation.[3]

3. Man forms his concepts in the same way as animals, but gains enormous support through language and social intercourse, both of minor help to them. Words afford him generally seizable labels for concepts even where typical ideas become inadequate or defunct. Still one word does not always cover one concept. Children and peoples at the youthful stage who dispose as yet of few words may use one of them to denote a thing or process but also on the next occasion to denote another thing or process that in its reactions exhibits some resemblance to the first.[4] Hence the meaning of words vacillates and changes. Under given circumstances, there is however only a small number of biologically important reactions that the majority take notice of, and this restores stability in the use of words. Each word then serves to denote one class of objects (things or processes) associated with definite reactions. The multiplicity of biologically important reactions is much smaller than that of reality. This first enabled man to classify the real in terms of concepts. Such conditions persist where a social class or profession considers an area of reality that no longer holds any immediate biological interest. There, too, there are fewer reactions important for the special purpose than factual features, but here we have different reactions from the previous ones, so that each class or profession undertakes a conceptual classification of its own. The mechanic, doctor, lawyer, engineer, scientist, form their own concepts and, by circumscribing limitation (definition), give words narrower meanings, different from the vulgar ones; a concept word, let us say in science, has the purpose of reminding us of the combination of all the object's reactions denoted in the definition, in order to draw these mem-

ories into consciousness as though by a thread. Take for instance the definitions of oxygen, of momentum of a mechanical system, and of potential at a point; of course every definition may contain further concepts, so that only the last and ultimate conceptual building bricks can be resolved into characteristic sense reactions. How fast and easily one succeeds in such a resolution depends on exact knowledge and on the familiarity of the concept, and the purpose aimed at determines how necessary it is. Seeing that these concepts were formed over hundreds of years it cannot surprise us that their content cannot be exhausted by an individual momentary idea.

4. What concepts should be formed and how they are to be mutually delimited is to be decided only by practical or scientific needs. One takes into the definition those reactions that are sufficient to determine the concept. One need not specially mention other reactions, of which it is already commonly known that they are inevitably tied to those in the definition. This would only burden the definition with inessentials. However, it may happen that finding such further reactions constitutes a discovery. If the new reactions on their own likewise determine the concept, they can be used as an alternative definition. We define the circle as a plane curve whose points are equidistant from a given point. Other properties of the circle are not enumerated, for example the equality of all peripheral angles on a given arc, the constant ratio of any curve point from two special points in the plane, and so on. Each of these two properties alone defines the circles too. The same fact or group of facts may, according to circumstance, lead one's attention to various reactions and concepts. A circle may be regarded as a section of a projective pencil, as a curve of constant curvature, a circular thread can be regarded as a curve of constant tension, as periphery of the area enclosed, and so on. A piece of iron may be regarded as a complex of sensations, as a weight, mass, thermal or electric conductor, magnet, rigid or elastic body, chemical element, and so on.

5. Every profession has its own concepts. A musician reads his score as a lawyer his statutes, a pharmacist his prescriptions, a cook his cookery book, a mathematician or physicist his treatise. What the layman sees as an empty word or sign has a very precise meaning for the expert; it offers

him instructions for clearly defined mental or physical activities that, if performed, can evoke in the imagination or present to the senses an object of equally circumscribed reactions. For this it is however essential that he has practised these activities and become fluent in them: he must have lived with his profession.[5] Mere reading cannot educate an expert, as little as mere listening to lectures: what is quite lacking here is the compulsion to test for correctness the concepts acquired, which at once supervenes when there is direct contact with the world of fact in the laboratory, because mistakes committed make themselves uncomfortably felt. Concepts based on incomplete and superficial facts learned from hearsay are like buildings made form decayed material that collapse at the slightest disturbance. For this reason it can only be detrimental to teaching if one impatiently forces pupils to premature abstraction[6]. Concepts formed in that manner potentially contain only ill-defined and shadow-like individual pictures that will mislead with special ease.

6. The nature of concepts reveals itself most clearly to a person who is just beginning to gain mastery in a field of science. He has not acquired the subjacent factual knowledge by instinct, but by careful and deliberate observation. He has often traversed the path from fact to concept and vice-versa, so that he has a lively memory of it and can readily retrace it and tarry at any point of it. Not so with the less definite concepts which are denoted by the words of ordinary language.[7] Here everything occurs by instinct and without our deliberate intervention, both as regards knowledge of facts and the definitions of the meanings of words. Much use has made speaking, hearing and understanding our language so familiar that everything proceeds almost automatically. We no longer stop to analyse the meaning of words, while the sense pictures that lie at the basis of speech scarcely appear in consciousness as hints, if at all. No wonder, then, that someone who is asked point blank what he finds in his consciousness when using a word, especially if the meaning is rather abstract, is apt to reply "nothing except the word"[8]. However, as soon as a phrase occasions doubt or contradiction, we bring out from the depth of memory the potential knowledge linked with this word. We learn to speak and understand our native tongue as we learn to walk. The separate phases of a familiar activity become blurred for consciousness. If now a solid scholar utters the pronouncement "a concept is merely a word", this must surely

rest on faulty self-observation. Because of long practice, he uses concept words correctly, just as we correctly use spoon, fork, key and pen, almost without becoming aware how slowly we acquired these skills. He can arouse the potential knowledge of a concept, but is not always compelled to do so.

7. Let us take a closer look at the process of abstraction by which concepts are formed. Things (bodies) are to us fairly stable complexes of connected and interdependent sensations, but not all elements of one of these complexes are of equal biological importance. A bird, for example, feeds on sweet red berries. The biologically important sensation 'sweet', to which his organism is innately attuned, causes the organism to become attuned by association to the characteristic 'red', which is conspicuous even at a distance. In other words, the organism is furnished with a much more sensitive reaction for the elements sweet and red which attract selective attention at the expense of other elements of the complex of the berry. Abstraction essentially consists in this division of attunement, interest and attention.[9] As a result of this process, the sense characteristics in the memory picture of the physical complex of the berry are not all of them equally intense, so that the picture is already assuming the peculiar features of a concept. Even the two sense characteristics sweet and red can vary greatly in the physical complex of the berry (think for example of the range of wavelength and colour that we call red) without the mental construct berry taking any notice. Presumably the whole range or mixture of sensations denoted by the word 'red' are selectively characterized by a simple basic sense process red, which, perhaps one day we may be able to isolate.[10] Even in such primitive cases therefore, to the inexhaustible range of physical sense features there corresponds a very limited range of uniform mental sense reactions and thus a marked tendency towards conceptual schematizing.

8. If we suppose that the edible and inedible kinds of berries growing in some area are more numerous and more difficult to distinguish, then the main memory pictures must become richer and more varied in their characteristics. Even for primitive man the need may arise to keep in mind special and clearly conscious and deliberate tests to distinguish usable from useless object, in cases where mere sense observation is no

longer sufficient. This occurs especially as soon as the less simple immediate biological goals, such as obtaining food and so on, are replaced by the much more numerous and varied technical and scientific intermediate goals. Here we see concepts developing from the simplest rudiments to the highest level of scientific concepts, each higher level using the lower ones as a foundation.

9. On the highest level of development concepts consist in consciousness tied to a word of the reactions to be expected from the class of objects or facts denoted. However, these reactions and the often complicated mental and physical activities they provoke can only gradually and successively appear as clear pictures. One can recognize edible fruit by colour, smell and taste, but the fact that whales and dolphins belong to the class of mammels cannot be discoverd just from appearance, but only by detailed anatomical examination. A glance will often decide the biological value of an object, but whether a mechanical system represents equilibrium or motion can be decided only through complex activities: one measures all the forces and all the mutually compatible displacements in the direction of the forces, multiplying the measure of each force with that of its corresponding displacement and adding all the products with due regard to signs. If this sum (that is, the work done) is zero or negative, we have a case of equilibrium; if not, a case of motion. Of course the development of the concept force has a long history which begins with the study of the simplest cases (levers and the like), which starts from the obvious observation that not only weights but their displacements too influence the process. However, if you are conscious that you can always carry out this test knowing that a case of equilibrium will yield zero or a negative sum while a case of motion yields a positive sum, then you have a concept of work and can by means of it distinguish static from dynamic cases. In this way every physical and chemical concept may be expounded. The object corresponds to the concept, if it yields the expected reaction when tested in the way intended: according to circumstances this may be merely a matter of looking, or a complicated mental or technical operation, and the corresponding consequent reaction a simple sensation or a complicated process.

10. For two reasons, concepts lack immediate clarity. Firstly, they

encompass a whole class of objects or facts, whose individual members cannot be imagined all at the same time. Secondly, the common features of the individuals (those features that alone are concerned in the concept) are usually such that we come to know them in temporal sequence and presenting them clearly to ourselves likewise takes considerable time. The feeling that they are familiar, reliably reproducible and potentially clear must here replace actual clarity.[11] However precisely these two factors make concepts so valuable and useful for science, in that they can represent and symbolize in thoughts large areas of fact. The purpose of concepts is to allow us to find our way in the bewildering tangle of fact.

11. Just as it is biologically important to observe how reactions hang together (say, the appearance of a piece of fruit and its food value), so every science aims at finding constancies of connection and of combination and interdependence of reactions. A class of objects (an area of fact) A for example yields reactions a, b, c. Further observation perhaps reveals reactions d, e, f. If now it turns out that a, b, c on their own can unambiguously characterize the object A, and d, e, f likewise, this establishes the connection between the two sets of reactions in object A. This is somewhat as in a triangle, which may be determined by two sides a, b and the angle γ between them, but equally well by the remaining side c and the adjacent angles α, β, which shows that in a triangle the second triad is tied to the first and derivable from it. The state of a given mass of gas is given by the volume v and the pressure p, but also by v and the absolute temperature T. Hence there is an equation relating the three determining factors p, v, T (namely $pv/T = \text{constant}$), which yields each in terms of the other two. Further examples are these: in a system that allows only conduction, the quantity of heat remains constant; in a mechanical system without friction, the change in kinetic energy in a given time is determined by the work done in that time; the same body which with chlorine forms cooking salt, forms Glauber's salt with sulphuric acid.

12. It is easy to see how conceptual formulation is important to science: by bringing it under a concept we simplify a fact, by leaving out of account those factual features that are irrelevant to our purpose, while at the same time amplifying the fact, by including all the characteristics of the class.[12] The two economic principles of permanence and sufficient differentiation,

which order and simplify, come into their own only at the stage where the material is conceptually articulated.[13]

13. Those who see concepts as an airy ideal construct without factual correlative should remember that, while concepts do indeed not exist as physical 'things', our reactions to objects of the same class of concept are psycho-physically similar, and those to objects of different classes dissimilar, as becomes quite clear in the case of biologically important objects. The elements of sensation to which conceptual features can in the last analysis be reduced are physical and mental facts. The constant conjunction of the reactions expounded in the propositions of physics represent the highest degree of substantiality that inquiry has thus far been able to reveal, much more constant than what has traditionally been called substance. Yet the factual elements contained in concepts must not mislead us into identifying these mental formations, always requiring correction, with the facts themselves.

14. Our body, and particularly our consciousness, is a fairly closed and isolated system of facts, whose responses to the physical surroundings move within narrow limits and in few directions. It is as with a thermometer that reacts only to thermal processes, or a galvanometer to electric currents, or any moderately imperfect physical apparatus. What at first blush seems a lack, namely scant variety of reaction to large and many-sided variations in surroundings, is the very thing that allows a first rough conceptual classification of processes that can then be constantly corrected and refined. In the end we learn to take into account the peculiar constancies and sources of error of our conscious equipment, as with any other apparatus. We are the same sort of thing as those in our physical surroundings and it is through ourselves that we become acquainted with them.

15. The decisive role of abstraction in enquiry is obvious. We can neither keep track of all the details of a phenomenon nor would it be sensible to do so. We take notice of those features that are of interest to us, and of those that depend on them. The enquirer's first task is thus to compare different cases in order to emphasize the mutually dependent features and to set aside as incidental or irrelevant for the purpose in hand

all the rest that have no bearing on the situation examined. This process of abstraction can yield highly important discoveries; as Apelt[14] points out, in consciousness the compound and special always precedes the simple and general: the latter is secured only by abstraction which is thus the method for seeking principles. He holds this view especially as regards the laws of inertia and relativity of motion, which may here serve us as examples of discovery by abstraction. Apelt relates how late in life and by what circuitous route Galileo came fully to grasp the law of inertia, and maintains[15] that it was by abstraction, not, as Whewell would have it, by induction. Whewell[16] does indeed speak of the induction on which the law is based, but he goes on immediately to mention Hooke's experiments with spinning tops under progressively diminished resistance, as the concrete experiments from which the general rule was drawn. In spite of the inappropriate term, Whewell thus seems to be of the same opinion as Apelt, except that Whewell is better in emphasizing the need to know various cases before abstractions can begin. For the rest, both assume a priori concepts of the understanding, which compels both of them to strange, superfluous and artificial attitudes. Apelt[17] finds the law of inertia obvious and self-evident if one brings to it the 'right' concept of matter whose basic character is 'lifelessness' and that excludes change through influences other than 'external' ones. Whewell[18], too, bases the law of inertia on the notion that nothing can happen without a cause. If man were by preference logically instead of psychologically inclined, the abstraction leading to the law of inertia would have arisen very simply. As soon as forces are known as factors determining acceleration, it follows at once that without forces we can think only of unaccelerated motions, rectilinear and uniform. History and even current discussions tell us almost redundantly that thought does not of itself run on such smooth logical paths: given the clusters of different cases and all kinds of difficulty with mutually crossing and contradictory considerations, abstraction must be imposed almost by compulsion. Whewell[20] rightly notices that a case of motion without forces never really occurs. In thus abstracting, science idealizes its objects. As for Apelt[21], he thinks that although no one had come nearer to grasping the principle of the relativity of all motion than Kepler, with his countless constructions to go from one system to another, it was Galileo who first recognized the law, not by proving it from facts but by reflecting on the nature of motion and the relation its observation bears to space, an object of pure perception but not of observation. The principle can only be intuited, not proved: its truth convinces directly, as soon as one has grasped and understood it in the abstract, no other explanatory proposition being required. That is why Apelt thinks Galileo practising abstraction was able to find the principle, while Kepler practising induction could not. My own view is that Galileo did indeed recognize the principle by abstraction, but nevertheless by comparing many observed cases. Having grasped and analysed the motion of free fall, he must have noticed that a falling motion near a stationary tower goes on in the same way as one near the mast of a fast-moving boat for an observer on it, which gave rise to the well-known conception of projectile motion as combining a uniform horizontal motion with an accelerated falling one. The remaining stages of generalisation and application offer no further difficulties. Apelt[22] tends even to think that Galileo's discovery of free fall was deductive, but Galileo's writings show clearly that the law of free fall is set up as an hypothesis, on a correct conjecture subsequently tested by experiment. It is by relying on observation that Galileo became the founder of modern physics.

16. Newton's laws of motion, but forward in his *Principia*, are indeed splendid examples of discovery by abstraction. The first law or principle of inertia has been mentioned above. Leaving aside the tautology in the second law (change of motion is proportional to moving force impressed) there is here a content that has not yet been made explicit, which is precisely the most important discovery due to abstraction. This is the presupposition that all factors that determine motion ('forces') determine acceleration. What led people to this abstraction, when Galileo's direct proof applies only to gravity? How did anyone know that this would hold equally for electric or magnetic forces? Perhaps one proceeded as follows: what all forces have in common is the pressure felt when one tries to prevent motion; pressure will always have the same effects, whatever its cause, what holds for one pressure holds for any other. This double view of force as determining acceleration and as pressure seems to me the psychological source of the tautology in the Second Law. To appreciate such abstractions properly one must consider them as acts of intellectual daring justified by success. Who will guarantee that in thus abstracting we notice the right circumstances and ignore the irrelevant ones? An intellect

of genius differs from the average precisely by quickly and surely foreseeing success for an intellectual operation. This trait is common to great inquirers, artists, inventors, organizers and so on.

To add some examples other than from mechanics, consider Newton's discovery of the dispersion of light. Besides distinguishing the finer shades of different colour and refractive index in white light, Newton was the first to recognize white light as consisting of several mutually independent radiations. The second part of his discovery seems to have been made by abstraction, the first through the opposite process, but both rest on the ability and freedom to consider or ignore factors at will and convenience. Newton's independent light radiations have the same kind of importance as the mutual independence of motions, Prevost's independent heat radiations that led to the knowledge of the dynamic equilibrium of heat, and many other conceptions that Volkmann[23] has called isolation. Such conceptions are essential to the simplification of science.

17. Although concepts are not mere words, but are rooted in fact, one must beware of regarding concepts and facts as equivalent, confusing one with the other. Such confusions provoke errors that are as grave as those arising from confusing ideas and sensations, indeed the former are much more generally damaging. An idea is a formation to which the individual's needs have essentially contributed while concepts, influenced by the intellectual needs of humanity as a whole, bear the imprint of the culture of their period. If we mix up ideas or concepts with facts, we identify what is poorer and subservient to special purposes with what is richer and indeed inexhaustible. Once again we ignore our bodily boundary, which in the case of concepts must be thought of as including all the people involved. Logical deductions from concepts remain intact so long as we retain those concepts; but the concepts themselves must always be expected to incur correction by the facts. Finally, one cannot assume that our concepts correspond to absolute constancies, since enquiry can find only constantly conjoined reactions.[24]

18. J. B. Stallo, independently and in different form, has expressed views that essentially agree with what has just been said.[25] In brief, he holds that thought does not occupy itself with things as they are in themselves, but with our concepts of them; we know things only through their relations with other things, so that all our conceptual knowledge of things must be

relative; a specific act of thought never contains the totality of an object's knowable properties, but only the relations belonging to a specific class. By not attending to these propositions, we commit several natural errors that are very common, grounded in our mental organization, as it were. Amongst these are the following: every concept is the counterpart of a distinguishable objective reality, so that there are as many things as concepts; the more general or wider concepts and their real counterparts exist before the less general, which develop from the former by the addition of distinguishing features; the sequence in the rise of concepts is identical with the sequence in the rise of things; things exist independently of their relations.

In the opposition of matter and motion, mass and force as special realities, Stallo sees the first of these errors, in the adding of motion to inert matter the second. The dynamic theory of gases is based on the theory of rigid bodies, because we became familiar with the latter before we knew gases. However, if we regard rigid atoms as the pre-existent item from which everything is to be derived, we commit the third error. The properties of gases are indeed much simpler than those of liquids and rigid bodies, as J. Fries[26] has observed long ago. Finally, as examples of the fourth error, Stallo discusses the hypostasizing of space and time, as it occurs particularly in Newton's doctrine of absolute space and time.

19. In the preface to the German edition of Stallo's book I have described where we agree and differ. Stallo's arguments, as well as mine, are never aimed against physical working hypotheses, but only against epistemological absurdities. My exposition always starts from physical details and from there rises towards more general considerations, while Stallo goes about it in exactly the opposite way. His account is aimed more at philosophers, and mine at scientists.

NOTES

[1] A psychological theory of concepts is attempted in my *A* 4, pp. 249–255; *P* 3, pp. 277–280; *W* 2, pp. 415–422. Cf. further H. Rickert, 'Zur Theorie der naturwissenschaftlichen Begriffsbildung', *Viertelj. f. wiss. Philosoph.*, Vol. 18, 1894 p. 277; H. Gomperz, *Zur Psychologie d. logisch. Grundtatsachen*, Vienna 1897; Th. Ribot, *L'évolution des Idées générales*, Paris 1897; M. Keibel, 'Die Abbildtheorie u. ihr Recht in d. Wissenschaftslehre', *Zeitschr. f. immanente Philos.*, Vol. 3 1898. Finally I wish to refer to A. Stöhr, *Leitfaden der Logik in psychologisierender Darstellung*, Vienna 1905, which appeared at the same time as the first edition of the present work. Its very

first pages contain an original elucidation of the theory of concepts from the standpoint of neuron theory.

[2] Cf. Ch. VII, end of sn. 5.

[3] Cf. *W*, p. 146.

[4] Cf. *A*, p. 250.

[5] Cf. *ibid.*, p. 253.

[6] I myself have had occasion to convince myself how futile it is to urge people towards abstraction. Children will readily grasp and distinguish small sets or groups of objects and quickly give the correct answer to the question 'how many nuts are three nuts and two nuts?' but be embarrassed by the question 'how much is two and three?'. A few days later the abstract formula will come of its own.

[7] When my boy was between four and five I gave him a small box of wooden models of geometrical bodies which I named without of course defining them. His visual imagination was greatly enriched by this and his phantasy strengthened so much that he could for example count the corners, edges and faces of a cube or tetrahedron without seeing the model. He even used the new objects and their names to describe his own small observations. Thus he called a sausage a curved cylinder. Nevertheless, he had as yet no geometrical concepts. The definition of a cylinder would have to be quite different from the usual one if the shape of a sausage is to count as a special case of it.

[8] Cf. the statistical data collected in Ribot, *l.c.*, pp. 131–145. As regards the 'type auditif' p. 139 he advances the attractive hypothesis that in the age of mediaeval oral instruction and the then current oral disputations this type was perhaps preponderant and that the origin of the expression 'flatus vocis' may be due to this circumstance.

[9] Let me again refer to the book of Stöhr (note 1 above). Note what he calls 'Begriffszentrum'.

[10] One might thus well say that simple sensations are abstractions, but one must not therefore assert that they are based on no actual process. Consider pressure and acceleration. Cf. *P* 3 & 4, p. 122.

[11] Cf. Ch. VII of this volume.

[12] *A* 4, p. 253.

[13] *Ibid.* p. 248 and Ch. VII sn. 3 of this volume.

[14] Apelt, *Die Theorie der Induktion*, Leipzig 1854, p. 59.

[15] *Ibid.* p. 60.

[16] Whewell, *Geschichte der induktiven Wissenschaften*, German translation by J. J. v. Littrow, Stuttgart 1840, II, p. 31.

[17] Apelt, *l.c.*, pp. 60, 61.

[18] Whewell, *The Philosophy of the inductive sciences*, London 1847, I, p. 216.

[19] *M* 5, 1904, pp. 140–143.

[20] Whewell, *Geschichte* II, p. 31; Wohlwill, *Galilei und sein Kampf für die Kopernikanische Lehre*, Hamburg 1909.

[21] Apelt, *l.c.*, pp. 61, 62.

[22] *Ibid.*, pp. 62, 63.

[23] Volkmann, *Einführung i. d. Studium d. theoretischen Physik*, Leipzig 1900, p. 28.

[24] In *Erhaltung der Arbeit* 1872, *M* 1883 and *W* 1896 I have explained these views in detail as regards physics.

[25] J. B. Stallo, *The Concepts and Theories of modern Physics*, 1862, German ed. entitled *Die Begriffe und Theorien der modernen Physik*, by H. Kleinpeter with a preface by E. Mach, Leipzig 1901. Cf. especially pp. 126–212.

[26] J. F. Fries, *Die mathematische Naturphilosophie*, Heidelberg 1822, p. 446.

CHAPTER IX

SENSATION, INTUITION, PHANTASY

1. From sensations and their conjunctions arise concepts, whose aim is to lead us by the shortest and easiest way to sensible ideas that agree best with the sensations. Thus all intellection starts from sense perceptions and returns to them. Our genuine mental workers are these sensible pictures or ideas, while concepts are the organizers and overseers that tell the masses of the former where to go and what to do. In simple tasks, the intellect is in direct touch with the workers, but for larger undertakings it deals with the directing engineers, who would however be useless if they had not seen to the engagement of reliable workers. The play of ideas relieves even animals from the tyranny of momentary impressions. If civilized man provides for the future more than the savage and works for goals that far transcend his own life, he is enabled to do so by his concepts and their wealth of ordered ideas. However we experience often enough how much less immediate it feels to deal with ideas than with sensible ideas. We will not easily refuse to help some unfortunate whom we actually meet, while a printed appeal for help is read much more reflectively. The platonic Socrates occasionally declares virtue to be knowledge. Yet it must be a kind of knowledge that is not always very lively. Few crimes would be committed if the consequences were vividly and accurately imagined. We should not hide want behind luxury, dance or hold a flower festival for the sick, if there were no difference between concepts and sensible ideas. A greedy rentier orders a poor beggar to be shown the door because the poor man's tale is so distressing: he is better able to cope with the concept of misery.[1] Sense perceptions are indeed the genuine original motors, whereas concepts rely on them often only through the mediation of further concepts.

2. All that men prior to the use of tools were able to learn from nature was revealed to them directly by the senses. This remains clearly visible in the traditional divisions of physics, which were conditioned by history but are no longer consistent or adequate. As soon as tools are used, every

device for observing may, in Spencer's view[2] be regarded as an artificial extension of the senses, and every machine as an artificial extension of the organs of movement. This natural thought seems to have occurred to others too: long after Spencer, and doubtless independently, E. Kapp[3] has expounded it at length, though unfortunately in rather phantastic form. An interesting and instructive account is due to O. Wiener.[4]

3. Let us consider some of Wiener's important points, without following him in detail. Sense organs are generally very sensitive, because they take up physical stimuli not only in the way inanimate objects do, but as releasing agents of ready energy stores, as happens only exceptionally in instruments like microphones, telegraphic relays and so on. The eye and ear are remarkably stimulated by energies as low as one part in a hundred million of one erg,[5] and the same energy will cause a visible deflection in the most sensitive balance. The eye is a hundred times more sensitive than the most sensitive photographic plate. If we have a weight of from 100 to 1 000 g lying on our hand, we can discern a diminution of about 30 per cent directly through the sense of pressure, while in moving the hand up and down the sensitivity can be improved to about 10 per cent. However the most sensitive balance at 1 kg load can show a difference of 1/200 mg, or $\frac{1}{2} \times 10^{-8}$ atmospheres. At 10 cm distance, the human eye can just discern two lines 1/40 mm apart, whereas with a microscope resolution can be obtained down to 1/7000 mm. Using light waves we can estimate much smaller distances still. The unaided eye can discern the gap of 1/500 sec between two electric sparks, the Wheatstone-Fedderson rotating mirror method gives us an optical means for measuring periods as small as 10^{-8} sec. Our sense of warmth feels a temperature difference of about 1/5 degree Centigrade, with the bolometric method of Langley and Paschen we can demonstrate differences of 10^{-6} degrees Centigrade. Thus the sensitivity of the sense organs in many respects can be reached and often exceeded by physical apparatus. Physicists thus learned about such delicate gradations of reactions as they could never have attained without these means.

4. Physics further knows methods of letting one sense stand for another. Optical devices can make sound waves visible and light waves audible, witness the various vibroscopic methods, the visibility of air waves by Schlieren methods and so on. Heat is felt directly only by touch[6], but a

thermometer conveys it to the eye. Even processes that do not directly affect any of the senses, such as very weak electric currents or oscillations of magnetic intensity, can be made visible by galvanometers and magnetometers; sight indeed generally intervenes where we are dealing with very delicate reactions. However, we must not forget that processes that would strictly escape all our natural senses would remain for ever undetected and undetectable. In using artificial means we are thus always seeking more versatile and more delicately graduated reactions that reach into the area of one of the natural senses.

5. Consider finally an orange, a lump of salt, platinum and air. The first of these bodies reacts without artificial operations on all senses, the second lacks smell, the third taste. Air we cannot even see, at best we feel it as hot or cold and when it is strongly moving it impresses the sense of touch as wind. That it is corporeal we are not convinced until we artificially trap some of it in a hose, which is indeed amongst the oldest physical experiments. With all these bodies, an artificial device can provoke further characteristic reactions. Thus, bodies are no more than bundles of reactions connected in a lawlike manner. Likewise for processes of all kinds, which we classify and name according to the needs of clarity. Whether it be water waves that we follow by eye and touch or sound waves in air that we hear but can make visible only by artificial means, or an electric current that can be followed almost only by artificially devised reactions, the constant factor is always the lawlike connection of reactions and that alone. That is the critically purified concept of substance that must replace the vulgar one in scientific contexts. The vulgar concept, which in ordinary life is not only quite harmless but often positively useful in manipulation (otherwise it would never have developed instinctively), plays the same treacherous role in scientific physics as the 'thing-in-itself' does in philosophy.

6. Wiener is led to the fiction of intelligent beings with senses different from ours. Nerve organs surrounded by sufficiently strongly magnetic bodies would for example represent a magnetic sense of the kind that has been artificially shown in the case of crabs by Kreidl.[7] The eye might for example be sensitive to infra-red instead of to shorter light waves. In that case one could use telescopes with lenses of hard rubber, and so on. By

such appealing considerations, Wiener thinks he can make himself independent of the special nature of our senses, in order to gain perspective on an uniform physical theory. As to that, it seems to me that all organic beings, at least on earth, are very closely related and therefore the senses of one mere variants of those of another. The perceptions of our present natural senses will doubtless remain the basic elements of our mental and physical world; but this does not prevent our physical theories from becoming independent of the special quality of our sense perceptions. We do physics by excluding variations of the observing subject, or by abstracting from them in some way. We compare physical bodies or processes, so that only sameness or difference in sense reaction counts, while the special character of the perception is no longer important to the discovered relation, which is expressed in equations. Thus, the result of physical enquiry becomes valid not only for all men, but for all beings with other senses, as soon as they regard our sensations as signs of a kind of physical apparatus[8]; except that such signs would not be directly intuitive for these beings, but would have to be translated into their sense perceptions, perhaps in the way in which we make things more intuitive by graphic representation.

7. So far we have concentrated on individual sensations and their importance. We give the name 'intuition' primarily to the whole system of spatio-temporally ordered sensations which enables us for example to recognize at a glance the whole disposition of bodies or their relative motions. The word clearly shows its origin: to one who can see, visual intuition is the most important, conveying the bulk of information and much of it at once. However, highly intelligent blind people, such as the geometer Saunderson, tell us that touch can similarly convey a rapid ordered survey, which might be called tactile intuition. Experienced musicians must be allowed to have a kind of intuitional survey of rhythmic movements in time and of the distribution and progress of vocal parts in the tonal range. Of the two calculating prodigies Inandi and Diamandi the former belonged to the aural type, the latter to the visual.[9] The former had begun to practise before learning to read, imagining numbers as sounds. The other had been to school first and had learned to write: if numbers were ordered in horizontal rows one below the other so that the digits formed vertical columns as well, Diamandi could at once quote the figures

in any columns, seeing the whole thing as a spatial arrangement. Inandi could do this only with difficulty, for he mentally heard the numbers one after another and had to cut this temporal sequence into sections and then as it were put them one under the other. Diamandi had a visual, spatial intuition, Inandi an aural, temporal one. We leave it an open question whether other fields of sense could show something analogous, for example, as Forel avers, with a highly developed sense of smell (dogs, ants).

8. It is beyond doubt that, following single sensations, intuition first set in motion ideas and actions, at a stage when conceptual thought was still rather backward. Intuition is organically older and more strongly based than conceptual thought. At a glance we see the whole lie of the land and readily move accordingly, avoiding a rolling stone, reaching out for a stumbling companion, picking up an object that interests us, without having to reflect about it. It is from intuition that the first clear ideas, concepts and reflections develop. Wherever it is possible to strengthen conceptual thought by intuition, this will happen with profit, by giving the new individual acquisitions the support of the old and well-tried findings of the species.

9. The graphic arts, and especially photography and stereoscopy, today enable us to gain a wealth of intuitive pictures that would have cost immense effort fifty years ago. Distant lands, their peoples and architecture, scenes from tropical forest and polar ice all come equally alive. Colour photography and cinematography will further enhance the natural look and the phonograph will do the same on the acoustic side. Science has found means to make accessible to intuition even objects that are not so by nature. Snapshots fix every phase of a motion too rapid for direct observation, by cancelling speed and freezing the object as it were. Marey, Anschütz and Muybridge have thus fixed the motion of animals. There are even refined methods for recording the pictures of sound waves, moving projectiles and so on. The method of picture series, long used in the special form of stroboscopic observation of rapid oscillations, allows threefold application: there are motions within the speed range of our natural intuition, which the cinematograph reproduces at the proper speed; motions that are too fast to be seen, such as the wingbeats of insects, sound oscillations and so on, can be slowed down at will; finally motions

that are too slow to be seen, such as the growth of plants or embryos, or of a town, can be suitably speeded up. Think of the changes of a growing plant with all its geotropic and heliotropic movements at increased speed, and animal movements correspondingly slowed down, and the impression given by the animal and plant worlds almost change place. No penitential sermon however gripping could surpass in impact a motion picture of a child growing, maturing and finally dissolving in old age.

10. The contrast between temporal extension and contraction is analogous to spatial magnification and reduction. The microscope is highly respected, but just as important, if less noticed, is the pictorial reduction of objects too large for our field of vision, as in a geographical map. Here too, objects clumsily grasped by concept are brought within the compass of easy and familiar intuition. The strengthening of abstract thought by means of curve-drawing devices is used in experimental work in setting out findings already to hand, in curves, geometrical constructions and so on.[10] A single example is enough to show the value of making a field accessible to intuition: Kepler had immense trouble in constructing elliptical planetary paths from individual conceptual data, whereas a mere glance would almost have sufficed to guess the answer if the motions had been given intuitively on a reduced spatio-temporal scale.

11. Memory draws on intuition. If on some chance occasion, there arises in me a mental picture of the small, cleanshaven grey-haired man, who with a friendly nod in all directions enters the dining-room, and I hear whispers from all sides in various tongues: "A German professor!", and if everything in the imaginary process occurs essentially as connected in my real experience, I shall call it a memory. If through many different experiences a great variety of associations between intuitive elements has arisen, thereby loosening the individual elements, then other influences can combine several of these connections in a way that had never happened in previous sense experience, so that this combination first exists in the imagination. Such ideas we call phantasies. If I had only ever seen one dog and now imagine one, the picture would probably have all the marks that had not escaped my attention in observing the dog. However, I have seen countless different dogs and doglike animals: therefore, the imagined dog is likely to be different from any that I ever saw. A publican may

choose the sign of "The Blue Dog". First, he takes a wooden figure of a dog, which he takes to the paint shop where he sees tins of various colours and wishing to use something conspicuous: his 'phantasy creation' thus emerges from a combination of associations that belong to different experiences. Such simple examples show that we cannot draw a sharp boundary between memory and phantasy. No experience stands so alone that other experiences might not call up the memory of it. Every memory is "Dichtung und Wahrheit".

12. A child sees a limping man. "He must have fallen off a big horse and hurt his leg hitting a stone." This phantasy story of a three-and-a-half-year-old is easily assembled from his memories. Another youngster wants to live like a fish in water or a star in the sky, with as much phantasy as a third who picked up a stone at random and peoples a cavity in it with fairies. What is more dubious to me, especially in view of observing my own children, is whether we can regard it as a matter of phantasy if a child calls a bottle top a 'door' or a small coin a 'child dollar', or when the sight of a dewy lawn provokes the statement "the grass is weeping".[11] Just as a child his phantasy, so a savage builds up his cosmogony out of elements familiar from memory. Giant frogs, toads, spiders and grasshoppers figure largely in this. With tribes that live near the sea or big rivers, giant fish or turtles rising from the depth help to fashion the present world order. If a farmer's small daughter familiar with the chicken run, asks whether stars are eggs laid by the moon, we have a beautiful example of primitive cosmogony making.[12] In Egypt where pottery early reached a high degree of perfection, we see the god Ptah using the potter's wheel to fashion the egg from which the world develops.[13] We need only recall the days of our own youth to grasp that without any solid basis of experience for understanding the world, phantasy must fill the gap and meet the need as best it may.

13. If one knows the historical development of science or has taken part in scientific enquiry, one will not doubt that scientific research requires a fairly robust phantasy, though not quite like that of the artist, to be discussed later. Consider first the experimenter. Every contemporary of Galileo knew that sound travels more slowly than light, since the impact of a carpenter's hammer can be seen at a distance before it is heard: the

very much greater speed of light is used to mark the departure of the sound. To determine the speed of light this method is useless: how are we to mark the time of its departure? Galileo imagines two observers *A* and *B* with lamps: *A* suddenly removes the shutter and when *B* sees the flash he removes his, so that *A* can observe both the departure and the arrival after a distance of 2*AB* has been traversed by the light. This ingenious arrangement arises from phantasy taking into account and combining all the relevant conditions. Perhaps the memory of echoes helped. Although Galileo recognized the operation as impracticable because of the great speed of light, Fizeau, over 200 years later, was able to continue this effort of the phantasy. He replaced *B* by a reflecting mirror and *A* by a uniformly rotating toothed wheel that marks departure and return with equal accuracy, while at *A* and *B* telescopes were added to reduce loss of light. It is a lively interest in the goal that keeps associations moving, and the presentation of the conditions to be fulfilled leads to the choice of associations usable for the purpose and from their combination arises the product of phantasy.[14] Observing lightning and the cracking of a spark, Franklin is led to suppose that lightning and thunder are electric in nature. A vivid desire arises to capture this supposed electricity, but how? A conducting rod is not enough, a tower of Babel cannot be built. Then he remembers paper kites that climb into the wind, makes one with a metal tip and attaches it to a string of hemp with a key at the lower end. He launches the kite into an approaching thunder storm, introducing a piece of silk thread between the string and his hand. Because of the rain, the string becomes a conductor, and Franklin can draw sparks from the key and charge bottles with them, thus filling the bottles with 'electric' fire. Today a captive balloon might replace the kite. Other experimental devices resulting from the exercise of phantasy include Newton's combination of a convex lens with a flat glass plate to exhibit the colours of thin plates and determine the thickness corresponding to each colour, Sauveur's rider, to determine nodal points of a vibration, Wheatstone's rotating mirror, König's acoustic flame indicator and so on.

14. In the above examples of the solution to experimental problems, we have to do not only with sensible ideas but also with concepts. Once one has acquired familiar concepts fixed by words, signs, formulae and definitions, these concepts constitute objects of memory and phantasy. One can

exercise phantasy in concepts too, searching the field by means of the thread of association until one finds a combined selection that meets the conditions of the problem. This happens especially in solving theoretical problems, if one perceives the assembly of concepts that makes everything perspicuous and gives the key to the solution. In his hydrostatic researches, Stevin noticed that when any piece of a fluid in equilibrium becomes solid, the equilibrium is not disturbed, thus reducing the solution of certain hydrostatic problems to solved problems in the statics of rigid bodies.

Kepler's laws being established, Newton proceeds to unravel their mystery. The curved path of the planets (first law) points to a centre of attraction inside the curve. The second law, concerning sectors covered by the radius vector from the sun determines the latter as this centre. The third law, stating that the cube of the distance varies as the square of the time of revolution, $r^3/t^2 = \text{constant}$, coincides with Huygens' expression for centripetal acceleration, $\phi = 4r\pi^2/t^2$, provided $\phi = \mathrm{k}/r^2$. Thus an inverse square force solves the whole Keplerian mystery.[15] The laws of reflexion and refraction of light become clear to Huygens in terms of the idea of the combined action of elementary waves moving with a speed determined by the medium. Malus' quantitative laws of polarization of light, the analogy of the colours of doubly refracting crystal plates with the colours of thin plates, Biot's formulae for the former, all these are explained and related by the Young-Fresnel conception of light as a transverse vibration together with the concept of coherence.

15. The law of association has shown itself sufficient to explain the workings of scientific phantasy here discussed. Artistic phantasy on the other hand shows some peculiar features which require some further considerations. Association is not confined to consious processes and the imagination. All processes of an organism that have repeatedly occurred together show a tendency to link up permanently. Thus movements become associated with one another through practice, and even secretions occur through association: it is a connection, acquired in time, between different organic functions, or stimulation of one organic activity by another, the temporal adaptation, through the circumstances of life, of parts in the service of the whole. However, the connection between organs that makes such interaction possible is not created by individual experience alone, indeed a great portion of it at least is given to the organism as its hereditary

property. This predetermines a set of interactions such as reflex movements; this set becomes larger in the course of organic development, and merely modified by what is acquired in the individual's life time. Thus, psychology cannot cope with all cases by means only of associations acquired in time[16]. Life would be impossible on the basis of mere associations in the ordinary sense. Further, we must consider that while organs exist for one another and serve one another, every one of them also has a life of its own. This life manifests itself in the specific energies[17] that may be modified by stimuli from without or by other organs, but still on the whole retain a definite character and sometimes make themselves independently felt. Thus, the organ of sight, or hearing or any other sense can produce as hallucinations the sensations normally provoked by physical stimuli. This happens under peculiar conditions that have as yet to be investigated. Again, the cortex can produce fixed ideas, a muscle may contract without voluntary innervation and a gland may secrete without the normal cause. It is indeed hallucinations that teach us to recognize sensations as states of our own bodies. A one-sided overestimate of this knowledge in turn becomes the basis of equally one-sided philosophic systems (solipsism).

16. Visual hallucinations, which express the independent spontaneous life of the sense of sight, have been studied in detail and well described by Johannes Müller[18]. Vividly coloured figures, for example of plants, animals men, may appear in the visual field and gradually change without our intervention. These figures are new formations, not memory pictures of previously seen objects and not provoked by thinking about them. The will has no demonstrable influence on this. Müller uses the occasion to emphasize the uselessness of the laws of assosiation, but in this he goes much too far. Of course, what appears spontaneously may alter spontaneously, but phantasms do not contradict the laws of association, even if those laws alone cannot make them intelligible: they simply belong to a different class of phenomena. In many fields the laws of association are however valuable guides. Besides there is a kind of phantasm that connects more closely with what went on immediately before, namely the phenomenon of sense-memory described especially by Fechner[19]. When we have been constantly occupied with one kind of visual objects, their images suddenly appear to us, especially in half darkness, but without change and quite objectively. These images are very similar to the previ-

ously seen objects even if perhaps no longer quite the same.[20] If in bad light we see objects modified by illusion, this points to the fact that the two extremes of spontaneous phantasms and pictures determined by physical stimuli can occur combined in all proportions. Similarly, it seems that there are all intermediate steps between sensation and idea. If then an idea is usually excited by another but may in special circumstances appear spontaneously, this agrees well with the facts known to date.[21]

17. So-called free floating ideas, sudden vivid memories of previous events or melodies once heard and so on, without any obvious point where associations might connect with immediately prior thoughts or the currently present setting, all these have no doubt been observed by everybody. Herbart was familiar with the phenomenon and tried to explain it in his own way. It seems to be related to hallucination. If, however, one takes association in a wider sense, so that a series of them might begin or end with unconscious processes, one need not regard every apparently free-floating idea as breaking the laws of association. The same bodily states, whether conscious or unconscious, can be accompanied by the same ideas. This approach seems to me to throw new light on the interesting observations of Swoboda[22] and to accord well with the views of R. Semon.[23]

18. What is usually regarded as the mark of an artistically productive phantasy is the spontaneous effortless novelty of its creations, which rules out simple imitation of experience. Besides, there is the sudden way in which at least the basic features of the creation present themselves to the artist, either as sheer hallucination or in some closely related form. In writings on phantasy we find many examples of this.[24] However, in order not to regard the exceptional as usual or to put exaggerations in the place of a sober scientific approach, consider whether it is conceivable that a Beethoven or a Raphael should appear amongst savages. One feels at once that the whole character of such artists' work greatly depends on previous art and on their experiences.[25] Granted the hallucinatory form of their inspiration, it too must be regarded as dependent on experience. Next comes the detailed work, which will hardly differ from detailed scientific work except by the more sensuous, less abstract character of art. In enjoying a Schumann symphony or a Heine poem, one recognizes the

traces of earlier art. Indeed, one might admit that much of the attraction in these works consists in the surprising variation on old themes, thus affording us agreeable reversals of our expectations. Without the older and more trivial, they could neither arise nor be understood.[26]

19. Can a scientific discovery begin with a hallucination? Perhaps Goethe's metamorphosis of plants arose in this way. Rare exceptions cannot be excluded, but in general it will be as in the case of dream phantasms. I am well acquainted with hallucinations and dream phantasms from personal experience and I have been subject to many optical and musical phantasms that might well have been used for artistic ends. However, I know of no instance of a scientific discovery based on hallucination, either amongst the great classical examples or in my own experience.[27] It is indeed not rare that a new perspective of solving some problem suddenly opens up, and this I have experienced myself. Yet on closer scrutiny, one always finds that long and painful labour and digging through the whole field have gone before; or one has collected data, idly perhaps but ruled by a specific directed interest, and a final discovery connects them together into a whole. Why, then, are science and art so different in this particular? The reason seems obvious: art remains predominantly sense-directed, appealing mainly to one sense. Each sense can hallucinate on its own. Science, however, requires concept. Are there conceptual hallucinations? How could they arise? Surely it would be pointless to expect that the ultimate acquisition of the human intellect, namely scientific consepts, which of their very nature resulted from deliberate conscious effort, should come as a gift from an unconscious organic source.

20. Let us conclude with another look at the relation of concept to intuition and sensation. The advantage of familiar concepts, acquired for oneself and not just conveyed through words or reading, is that one can easily arouse the intuitions and sensations potentially residing in them, while these contents can in turn easily be stored in concepts. Take a trivial example. Consider the period of the pharaohs some 3600 years ago, from which historical evidence is extant. These 3600 years are almost mere 'flatus vocis', so long as we cannot transform them into something more intuitive. If, however, we imagine an old Egyptian of 60 who sires

a son; the son at 60 does likewise, and so on; then, the sixtieth descendant of this line is our contemporary. Such a line is easily marked on the wall of a room. Thus, the time of the pharaohs comes rather close to us and we no longer wonder that so much barbarism still remains. If one thinks of one's worthy ancestors or imagines the attractive future of one's descendants, one does the opposite: transforming intuitive ideas into concepts. Everyone has two parents, four grandparents, eight great-grandparents, and proceeding in this way we soon reach numbers that no land can carry. Since therefore no one can have worthy ancestors peculiar to himself, everybody must be content to allow that amongst our common ancestors there are hordes of thieves, murderers and so on, who must be reckoned amongst his kin and whose mental heredity he must cope with. If someone contents himself with leaving three children, each of whom do likewise and so on, then in a few centuries his descendants would fill the earth. It follows that most of them must perish in the struggle for life, which will not always be waged with the most noble means. This simple example of the transformation of concepts into intuition and vice-versa may suggest that excessively inconsiderate and self-centred care for one's own descendants rests on an illusion, and that it would be better to care for mankind instead.

21. The owner of a well-articulated system of concepts that meets his interests, a system that he has made his own through language, upbringing and education, such a man has great advantages over one who has only perceptions to rely on. However, if a man lacked the ability quickly and surely to transform the ideas of sense into concepts and vice-versa, he would be liable on occasion to be misled by his concepts too, in which case they could become for him a mere burden of prejudice.

NOTES

[1] How much concepts lag behind sensation and imagination as regards immediacy is shown by the following occurrence. In a university town in which two nationalities *A* and *B* lived in a state of mutual tension a professor of nationality *A* had his apartment on the second floor of the institute for pathological anatomy and occasionally held a dance at home. At once a newspaper championing the interests of the *Bs* published an article entitled 'Dancing over corpses', which provoked a street riot against the professor. The impulsive mob may have felt that a professor who daily associates with corpses should not enjoy another happy hour unless he were quite depraved and heartless; at

least that was what the newspaper men pretended to believe. Yet who allows his pleasure to be disturbed by the thought that men die every minute, or that his own relatives lie buried?

[2] Spencer, *The Principles of Psychology*, London 1870, **I**, 164, p. 365.

[3] E. Kapp, *Grundlinien einer Philosophie der Technik*, Brunswick 1877. All instruments, tools and machines are regarded as unconscious projections of bodily organs. This seems rather to obscure Spencer's idea, and I think that what will be reached in this way will be only a dreamlike 'philosophy of technology'. Ask yourself which organ is projected in a screw or wheel, in a dynamo or interference refractometer and so on. What is correct is merely that the study of technology did help us to gain understanding of some bodily organs as well.

[4] O. Wiener, *Die Erweiterung der Sinne*, inaugural lecture, Leipzig 1900.

[5] I have myself occasionally tried such an estimate of the sensitivity of a sense organ. Cf. *Beweglichkeitsempfindungen*, Leipzig 1875, pp. 119f.

[6] Strictly speaking the sense for heat which is spatially allied to the sense of touch.

[7] *P* 3, Leipzig 1903, p. 398.

[8] *A* 4, 1903, p. 209.

[9] *Revue générale des sciences* 1892.

[10] *P*, pp. 124–134.

[11] Ribot, *Essai sur l'imagination créatrice*, Paris 1900, pp. 89–97. Cf. *A*, p. 250.

[12] Observation by my sister.

[13] Erman, *Ägypten* **II**, pp. 352, 605f.

[14] The theme of a periodic covering and uncovering of a lantern is indeed found also in Roemer.

[15] *M* 5, 1904, pp. 88, 195.

[16] *A*, p. 185.

[17] The theory here referred to is that put forward by Johannes Müller and further developed by Hering.

[18] J. Müller, *Über die phantastischen Gesichtserscheinungen*, Koblenz 1826. F. P. Gruithuisen, *Beiträge zur Physiognosie und Eautognosie*, Munich 1812, pp. 202–296.

[19] Fechner, *Elemente der Psychophysik*, Leipzig 1860, **II**, p. 498. Cf. also *A*, p. 157.

[20] Oelzelt-Newin, *Über die Phantasie-Vorstellungen*, Graz 1889, p. 12, reports that having killed many snakes that had pestered him he spent the following sleepless night being constantly pursued by what seemed to be their actual appearances and movements. The same happened to me after several days of experiments with spiders: I saw them creeping round me in my dreams. Once, when I raised a young sparrow on grasshoppers, I was confronted in a dream by a man-sized grasshopper crawling towards me as though it would menace me with Schiller's words: "Earth for all has ample space, wherefore persecute my race?"

[21] *A*, p. 159.

[22] Swoboda, *Die Perioden des menschlichen Organismus*, Vienna 1904. I was unable to observe a precise periodicity in myself, although the appearance of freely rising imaginations often comes to me. Perhaps a sharp periodicity shows itself only in very sensitive individuals.

[23] Semon, *Mnene*, Leipzig 1904.

[24] See note 20 above.

[25] Very sane and sober views on this in R. Wallaschek, *Anfänge der Tonkunst*, Leipzig 1903, especially pp. 291f.

[26] Cf. the charming small pamphlet by E. Kulke, *Über die Umbildung der Melodie*,

Prague 1884. Analogous considerations apply to the transformation of harmony. To mention only one example, take the sequence in the Flying Dutchman, ballad scene and overture, where the chords *F* major, *E*♭ major, *D* minor follow one another, and moreover with glaring breach of the rule forbidding successive fifths. It is only a slight modification of the trivial sequence *F* major, dominant seventh and back to *F* major, which is precisely what makes it so attractive.

[27] It is related that Kekulé saw his benzol ring as a hallucination in a London fog, but his own simple report about his speculative efforts in London and Ghent does not support this view (*Berichte d. Deutschen chem. Gesellschaft* **XXIII** 1890, pp. 1306f.).

CHAPTER X

ADAPTATION OF THOUGHTS TO FACTS AND TO EACH OTHER

1. Ideas gradually adapt to facts by picturing them with sufficient accuracy to meet biological needs. The accuracy goes no further than required by immediate interests and circumstances, but since these vary from case to case, the adaptive results do not quite match. Biological interest further leads to mutual correction of the pictures to adjust the deviations in the best and most profitable way. This requirement is satisfied by combining the principle of the permanence with that of the sufficient differentiation of ideas. The two processes of adaptation of ideas to facts and to each other cannot really be sharply separated. If the first sense impressions are already in part determined by the innate attunement of the organism in time, the later are influenced by the earlier ones. Therefore the adaptation of ideas to facts is almost always complicated by the mutual adaptation of ideas. These processes at first occur quite unintentionally and without clear consciousness. When we become fully conscious what we find within us is already a fairly complete world picture. Later, however, we gradually go over to continuing the processes with clear deliberation, and as soon as this occurs, enquiry sets in. Adaptation of thoughts to facts, as we should put it more accurately, we call observation; and mutual adaptation of thoughts, theory. Observation and theory too are not sharply separable, since almost any observation is already influenced by theory and, if important enough, in turn reacts on theory. Consider some examples.

2. We never made an effort to learn that milk and bread taste good and satisfy hunger; or that hitting a solid body hurts, flames burn, water flows downwards, lightning is followed by thunder and so on. Our body and its surroundings have achieved this adaptation of ideas, almost automatically and in the biological interest of the individual. Things become different however when the adaptation of thoughts is only of mediate interest, and its results are to the profit of people in general and conveyed through expression in language. This is much more demanding on

our mental life: the new facts must be compared with many other cases, agreements and differences be noted, and elements already known and named be sought of which the new fact may be thought of as compounded. If mediate interests strong enough to cope with this are to emerge and find satisfaction, there must be a mental actitivy strengthened in the service of life. A child learns to suck liquids through a straw without knowing how or even asking, let alone being able to say: what development is required to secure water by the indirect route of a pump, how strong must be the indirect interest to guide phantasy in making an appropriate selection of memories to produce the prototype for constructing the pump! What countless comparisons, before one can finally say that the water, in spite of its weight, because of 'fear of the void', follows the piston. For the first steps of adaptation a new combination of intuitive memories by means of phantasy often suffices. We may think of the 'attraction and repulsion' of magnets, the 'emission' of corpuscles of light, the recently revived closed magnetic flux of Euler, 'caloric' that 'flows' from the warmer to the colder body like water from a wet sponge to a dry one, or even Ampère's left-hand rule. Further adaptation requires abstract conceptual operations, looking at whole classes of facts or the reactions characteristic of them. Here belong Galileo's recognition of free fall as uniformly accelerated motion, Kepler's proof of the rectilinear propagation of light and the corresponding law of intensity, Black's construction of the concept of 'quantity of heat', Coulomb's establishment of the inverse square law of electric action.

3. Consider some simple examples of the mutual conflict and adaptation of thoughts and its outcome. A sense experience often awakens various memories that partly agree and urge towards action in one direction and partly disagree and paralyse each other. This would be the case of a fox seeing a wriggling prey but scenting that a hunter is near by or suspecting signs of a trap reminiscent of painful experiences. If the fox recognizes the putative hunter as a harmless boy without dog or gun, or the putative trap turns out to be undergrowth in which the prey became accidentally tangled, then the conflict is resolved. Before any enterprise that offers prospects partly favourable and partly not, our contradictory thoughts will place us under a more or less tormenting tension which recedes only when we recognize our hopes or fears as pointless and unjustified under

the circumstances, so that we decide to go ahead or to desist. By contrast we now feel a pleasant release from pressure. In the service of life thoughts adapt to each other and to facts, and if the thinking process has become sufficiently strong, disagreement between thoughts is in itself disturbing, so that one will try to solve the conflict if only to remove intellectual unease, even if no practical interests are involved.

4. A young savage is to deliver a basket of fruit with a letter. On the way he eats some of the fruit and is astonished that the letter was able to betray this, the next time he puts the letter under a stone to stop the 'traitor' from observing him, but again he notices that he was not sufficiently on guard against the 'magicien'. Only after learning to count and represent numbers, for example, by strokes, does he gain a roughly adequate idea how the letter could betray him. In the society of memories, the original idea of the letter is transformed until it agrees with them. The first time we see a stick standing at an angle in water it looks bent. However, we noticed no resistance when dipping the stick in, nor is it bent when we take it out, but being once bent it cannot straighten itself. Thus the bend is left out of account as an inferior sham or deception as against the remaining ideas that are in better agreement with one another and therefore of greater authority. This ignoring of an unimportant experience may meet practical purposes, but certainly does not correspond to the scientific point of view, from which any fact may become important. Thus, science is satisfied only if the straight and bent optical image are recognized as equally determined by the conditions of light propagation.

5. Thought adaptations of an individual for himself alone may be made with the help of language but are not exclusively tied to it. However, a result of adaptation that is to be useful to the community must be expressed in the concepts and judgments of language, with all the attendant advantages and drawbacks. This holds especially for all scientific adaptations, which in this instance gain expression in the mutual correction of groups of concepts and judgments.

6. It is the disturbance of ideas through contradiction that must have driven the Eleatics to their philosophical experiments. They seek a solution in what seems to us a curious manner, by allowing that the unity of

language is alone valid, while disenfranchising the senses and the differences they observe. Whatever one might think of these primitive attempts, there is no doubt that the controversies they provoked did turn attention to our thinking and speaking, which thus reached a higher degree of agility and, through the feeling of release at real or deceptive solutions, taught us to take pleasure in intellectual exercise. Besides, we must not underrate the motive power of feeling superior to the less well practised. To be sure, Zeno of Elea's primary feeling was certainly discomfort at the impossibility of discretely enumerating the infinite continuum presented by sense perception, which is indeed the main difficulty; but his 'Achilles' with the infinite geometrical progression that by his method cannot be counted through to the point and moment of overtaking, is amongst other things the work of a clever debater pleased at his own superior skill.

Eleatically inspired sophists, in the bad sense,[1] who sought to make the worse appear the better case, and eristic logicians with their fallacies, who were willing to defend any view whatever, although primarily working for their own advantage, nevertheless helped indirectly to promote critical assessments of thought and language. If fallacies of the kind that Plato in the *Euthydemus* and *Gorgias* puts into the mouths of sophists seem stale and absurd today, if clever arguments like the 'Liar' and the 'Crocodile' no longer worry us and if the case of the sophist Protagoras against his pupil Eualthus gives less trouble to modern lawyers than to ancient ones, we owe this to the fact that such difficulties have already been settled by our forebears. This shows the distance between thought in its childhood and mature thinking. Fortunately, the latter allows us to put aside sophistries and attend to more serious and fruitful tasks.[2] Still, we must remember that, besides those who incidentally advanced critical thinking by misusing it, many Greek philosophers developed the proper method of mutual adaptation of thoughts and the correction of the less founded by the better established, namely by means of geometrical proof, which concerns a simple and consistent field. This is a permanent intellectual possession. The result of these efforts, Euclid's *Elements*, remains a model of logical exposition.

7. Mediaeval logic was almost entirely sterile as regards enquiry. However, in order to reconcile its views with the dogmas of the church and the theories of Aristotle, their official philosopher, it further developed and

applied the dialectic of the Ancients. The slighter the factual material, the greater the care with which it tried to squeeze out everything that a proposition held to be true might contain. What this method revealed was mostly a rather unsatisfying paper diet that the natural scientist of today can hardly take even when diluted as in the works of Kepler, Grimaldi, Kircher and others. Nevertheless, the use of this method has schooled people in the art of exploiting ideas, as becomes obvious as soon as it bears on a real field of enquiry. This is not to say that a benevolent deity has had the foresight to place scholasticism before the beginning of scientific enquiry; but once scholasticism existed, it had to exercise its influence both for good and ill. The latter it has unfortunately continued to do through centuries, until events finally forced it into at best a sham existence for those who had been artificially blinded.[3]

8. Anyone with a strong life of ideas will readily attend to playful pursuits, when no serious tasks are to hand. Such playfulness further develops and strengthens the ideas for serious business in the future. It seems to me that both these conceptions of play are justified, whereas ordinarily one or the other side only is emphasized.[4] Consider for example the intellectual puzzles in the *Thaumaturgus mathematicus* (Cologne 1651). The book was printed at the time of scientific renaissance and bears distinct traces of ancient, scholastic and modern thought. Problem 13 demands the weighing of smoke from a burning object: the given solution lies in weighing the original object and the ash after combustion, the difference being the weight of the smoke. Both problem and answer are doubtless ancient, for Lucian reports that the cynic Demonax answered it in this way. Although we know that the answer is wrong, it nevertheless reveals a distinct feeling for the more general experience that we now express in the principle of conservation of mass; there seems to be a desire to reconcile more specific thoughts with this more general one by adaptation.[5] For some of the problems, the solution requires thought experiments. In problem 15, for example, a wolf, a goat and a cabbage have to be ferried across a river in a boat that takes only one, in such a way that none devours any other. One begins by taking the goat across and the rest follows. Problem 14 is similar, with three masters and three slaves and a boat taking only two people, with the proviso that "dominorum quisque suum amat servum". Problem 9 is a pretty puzzle from number

theory: given three vessels of capacities 3, 5 and 8 units respectively, the first two empty and the third full, to divide it into two equal parts without any further tools; the solution requires only a vivid phantasy and the only slight difficulty is the indeterminacy at the start. Problem 29 is rather curious: to place a man both upright and upside down. This seems to be impossible, but only so long as we take 'upright' as an absolute direction, like those who deny that there are antipodeans. If, however, we take the concept as relative, the problem is solved by a man placed at the earth's centre.[6] Problem 49 provides an attractive test of reflective power. A bridge is built uniformly round the earth and its props later removed all at once. What will happen? "If things happened as precisely as intellectual insight is certain", the bridge would have to continue floating as a closed vault, since no part can fall before any other. All ideas are here adapted to the general thought that every process is uniquely determined by the existing conditions. Notice that Saturn's ring might be such a bridge. However this still fails to take into account the inverse square law of gravitation and the consequent unstable equilibrium of a rigidly floating ring: the real ring of Saturn can exist only if it consists of isolated circulating masses. The problems that follow serve further to bring to notice the principle of sufficient determination and sufficient reason. In problem 53 it is stated that a uniform circular spider's thread could not be broken by uniformly exerted pulls, even if all "angels and men" were tugging. On page 230 there is the question whether two people exist with exactly the same number of hairs on their heads. This at first seems to defy answer, but it serves to emphasize the value of order and clear arrangement, that is, the value of mathematics. For if one clearly grasps that the number of people is doubtless much greater than the maximum of strands on a head, one can line up that number of people in order of hair count, assuming this to be without gaps: anyone after that has to be put together with a person already lined up.

9. These examples suffice to show that people in the 17th century in terms of their practised ability for reflection as manifested in their intellectual diversions, were well armed for great discoveries in natural science. The method of thought experiments, adaptation of isolated ideas to more general modes of though developed through experience and the search for agreement (permanence, unique determination), the ordering of ideas

in sequence, all these exercised in such deversions, are the very activities that most strongly promote inquiry in the natural sciences.

10. Let us now show some instances of how in the history of science mutual adaptations of thought of the highest importance have occurred. Stevin is looking for the magnitude of the load on an inclined plane in terms of a pull along this plane. He assumes that value as correct for which a uniform closed chain laid round the wedge would remain in equilibrium, which is familiar from ordinary experience: the less certain thought is adapted to the better founded one. When Galileo began his work, the traditional notion still survived that a projectile has a gradually decreasing impressed force, which is indeed a natural expression of everyday experience. His enquiries led him to recognize the uniformly accelerated motion of free fall and the uniformly retarded motion of vertical or inclined projection. At the same time he grew accustomed, especially through his pendulum experiments, to regard resistances as retarding, reducing the velocity. By looking at uniform horizontal motion as a special case of a uniformly accelerated or retarded motion with zero acceleration or retardation, the decreasing impressed force becomes superfluous and confusing and must give way to the generally appropriate notion of inertia.[7] Newton's *Principia* begins with eight definitions (for mass, momentum, inertia, centripetal force and so on) and three laws of motions and the consequences drawn from them. These assertions are abstracted from or adapted to experience. They bear the mark of mutual adaptation, although not a complete one, since there are some redundant statements. To appreciate the account, one must remember that it was given at a time when statics was developing into dynamics, so that it contains a twofold notion of force, as pull or pressure and as what determines acceleration. Only in this way does the formulation of the second and third law become intelligible. If, viewing statics as a special case of dynamics, one starts from the fact that bodies in pairs determine their mutual accelerations and that these pairs are independent, the ratio of masses is defined dynamically by the inverse ratio of the accelerations, adding the experience that mass ratios remain the same however determined: on this basis we can develop the whole of dynamics. Here the second law reduces to the fact of mutual acceleration of bodies and an arbitrary definition of measure, while the first law becomes a special case of the second and the third

becomes superfluous.[8] Newton's account is of course quite consistent, but the redundant element shows in that some of the propositions can be derived from others.[9]

Black, using the concept of 'caloric' had already constructed the concept of quantity of heat and had formed the idea that the sum of such quantities was constant; further, he knew that heat travels from a hot body to a contiguous colder one with fall and rise of their respective temperatures. Next he observed that the temperature of melting or boiling substances was not raised by contact with a much hotter flame, so long as the two processes continue: the fact that they seem to annihilate quantities of heat is incompatible with the constancy of the sum. Black therefore assumed that melting and boiling makes a heat quantity latent, while modern thermodynamics drops the principle of constancy. Adaptation can thus proceed in different ways: of two conflicting thoughts, the one that seems less important and reliable at the time must suffer modification by the other. S. Carnot recognized that a quantity of heat at a higher temperature must sink to a lower temperature and pass to a colder body, if the latter is to do work, for example by expansion. At first he regarded the quantity of heat as constant in the sense of Black, but Mayer and Joule discovered a diminution when work is done, while maintaining an increase through work, that is creation of heat quantity (by friction). Clausius and Thomson resolve this apparent paradox by recognizing that the heat disappearing when work is done depends on the amount transmitted and the temperatures. Both Carnot's and Mayer's views become modified and combined in a new form. Carnot's principle suggests to W. Thomson to produce ice by isothermal expansion and compression of air at 0 °C, that is without work, but J. Thomson notices that because water can do work by expansion in freezing, this work would have to come out of nothing. To remove the contradictions one had to assume that the freezing point was depressed by pressure according to a precise quantitative formula, as was confirmed by experiment. Thus, the paradoxes themselves provide the strongest incentive towards mutual adaptation of thoughts and so to new clarifications and discoveries.

11. The mutual adaptation of thoughts is not exhausted in the removal of contradictions: whatever divides attention or burdens the memory by excessive variety, is felt as uncomfortable, even when there are no con-

tradictions left. The mind feels relieved whenever the new and unknown is recognized as a combination of the known, or the seemingly different is revealed as the same, or the number of sufficient leading ideas is reduced and they are arranged according to the principles of permanence and sufficient differentiation. Economizing, harmonizing and organizing of thoughts are felt as a biological need far beyond the demand for logical consistency.

12. The Ptolemaic system contains no contradictions, all its details are mutually compatible, but we are dealing with a stationary earth, a rotating celestial sphere of fixed stars and the individual motions of sun, moon and planets. In the Copernican system and its ancient precursors, all motions reduce to circular paths and axial rotations. In Kepler's three laws there are no contradictions, but how comforting it is to reduce them to the single law of Newtonian gravitation, which in addition covers the phenomena of free fall and projectiles, the tides and much else.

Refraction and reflexion of light, interference and polarization are separate but compatible theories, and yet Fresnel's reduction of all of them to transverse vibrations was a great and welcome progressive step towards ease of exposition. A much greater simplification still is due to Maxwell, who subsumed the whole of optics as a chapter of the theory of electricity. The cataclysmic theory in geology, and Cuvier's idea of creative periods are free from contradictions, but evereybody will thank Lamarck, Lyell and Darwin for attempting a simpler conception of the history of earth, animals and plants.[10]

13. Following these examples, we conclude in general that the results of adapting thought to fact is formulated in judgements that are compared and further adapted. If there is contradiction, a less fruitful judgement can be dropped in favour of one that is more so. Which are regarded as more authoritative depends of course on how far one is familiar with the field, on one's experience and practice in intellectual thought and on the customary views of the period. An experienced physicist or chemist for example will not grant authority to a thought that offends the principles of determinism, conservation of energy and mass, while the amateur who is building a perpetual motion machine is less troubled by this. In Newton's time it required much courage to assume action at a distance, even if

presented as something still awaiting explanation. Later, success made this approach so common that nobody took offence at it. Today we feel once more so strong a need for following all mutual connections continuously through space and time, that we cannot assume direct action at a distance. Immediately after Black it was an act of daring to doubt the constancy of the quantity of heat, while fifty years later there was a strong inclination to abandon his assumption. Every period in general prefers those judgements under whose guidance it has gained the greatest practical and intellectual successes. Great and far-sighed enquirers are often put in a position where they have to oppose current views, and so tend to initiate a turn in outlook: even judgements that were hitherto authoritative now have to compromise with new ones that would otherwise simply be discarded and both are usually modified as a result; witness the thermodynamic enquiries of Clausius and W. Thomson (Lord Kelvin) and the Faraday-Maxwell theory of electricity.

14. The judgements to be compared may be from the start compatible, so that no adaptation seems required. Whether there is further demand for harmonization then depends on the personality of the thinker and what he requires in the way of aesthetic presentation and logical economy. In some heads the most diverse ideas can live together in peace because they belong to areas that never meet: a man may be intellectually sober in one field and strangely superstitious in another, especially if he reflects just when in the occasional mood for it, allowing different registers to sound from case to case without bothering about the larger organic connections of whole areas of thought. In contrast we have enquirers like Descartes, Newton, Leibniz, Darwin and others.[11]

15. The ideal of economic and organic collaboration of the compatible judgements in a field is achieved when we succeed in finding the smallest set of independent judgements from which all the rest can be deduced as logical consequences. An example is Euclidean geometry. The judgements so deduced may have been originally discovered in a totally different way, indeed this is usually so. In that case the deduction serves to make the judgement more intelligible in terms of simpler and more familiar ones that is, deduction serves to explain or to base something that is doubted on something simpler that is not, in short to furnish proof. If the deduced

judgement was previously unknown but first discovered in demonstration, we have a deductive discovery.

16. The simple generally perspicuous and familiar subject matter of geometry is well suited to illustrate the fitting together of judgements. For example, let us draw any four straight lines touching a circle and forming the quadrilateral *ABCD* (Figure 2). Not everything that we can say of it would hold of any quadrilateral, for here the sides are tangents and must

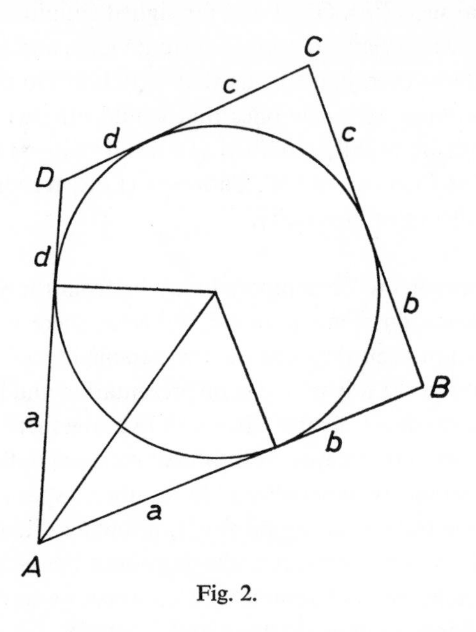

Fig. 2.

therefore be compatible with the properties of the circle: the radii to the point of contact are at right angles to the tangents. The two tangents from one vertex lie symmetrically about the join of vertex to the circle's centre and the segments from vertex to point of contact are equal.[12] Therefore the sum of two opposite sides is equal to the sum of the remaining two. This metrical property belongs exclusively to a quadrilateral circumscribed to a circle. If instead of *AD* we drew a secant to complete the quadrilateral, or a line outside the circle, the property evidently no longer holds. Equally, one cannot inscribe a circle in every quadrilateral: for that circle is determined by three tangents, or by the intersection of two bisectors of angles between adjacent tangents. The fourth side imposes requirements that are not compatible generally with the rest. Such fittings together of judgements can easily be given in the form of an explanation of a problem with its solution, or as a deductive discovery. Formulation in Euclidean or Aristotelian logical terms offers no difficulty. Examples of this are discussed at length by J. F. Fries[13] and more attractively by Drobisch.[14]

17. The forms of logic, which are not part of our account, were arrived at by abstraction from actual instances of scientific thought. However, any example, such as in geometry, shows that knowledge of these forms alone is of little use: at best it may serve to check a line of thought, but not to find a new one. Indeed, thought does not proceed in empty forms, but according to a vividly presented content, either directly or through concepts.[15] In a geometrical deduction a straight line will be considered now for its position, now for its length, or as a tangent, as a normal to a radius, as part of a symmetrical figure; in a parallelogram we must notice now the area, now the ratio of the sides or diagonals, or the angles. If we are not familiar with all the intuitive and conceptual relations and how to transform them into one another, if an interest in a putative connection does not direct our intention to the right path, then we should certainly make no geometrical discoveries. Empty logical formulae cannot replace a knowledge of the facts.[16] Nevertheless a look at algebra and mathematical symbolism in general shows that attention to thought as such and the symbolic representation of the abstract forms of intellectual operations are by no means devoid of all merit. Anyone who could not carry out these operations without such help would however gain no profit from these methods. When, however, we consider whole sequences of thought operations that contain the same or similar operations frequently repeated, symbolic representation greatly relieves the requisite mental effort, saving it for more important new cases that cannot be tackled symbolically. Indeed, mathematicians have developed in their symbolism a most valuable symbolic logic for their own purposes. Mathematical thought operations are so varied that the simple classifications of Aristotelian logic cannot encompass them. Hence mathematics given rise to its own more comprehensive symbolic logic[17], whose operations are by no means only quantitative. The beginnings go back to Leibniz[18]; in mid-19th century Germany, the only follower seems to have been F. E. Beneke.[19] It was left

to mathematicians like H. Grassmann, Boole, E. Schroeder, Bertrand Russell and others, to resume the path of Leibniz.

NOTES

[1] Th. Gomperz, *Griechische Denker*, Leipzig 1896, **I**, pp. 331f.

[2] F. E. Beneke, *System der Logik als Kunstlehre des Denkens*, Berlin 1842, **II**, p. 141. Cf. also J. F. Fries, *System der Logik*, Heidelberg 1819, pp. 492f. and the excellent and pleasing account of fallacies in W. Schuppe, *Erkenntnistheoretische Logik*, Bonn 1878, pp. 673f.

[3] According to the advice of Prof. A. Marty the best way of getting to know scholastic thought is through Francisco Suarez *Disputationes metaphysicae* (*Opera*, Tom. 22, 23, Venice 1751). Compare for example the show of ingenuity in disput. 23 'de causa finali' (T. 22, p. 442) or disput. 40 'de quantitate continua' (T. 23, p. 281), which always serves only by great detours to lead rather weakly to a doctrine of the Church or of Aristotle. Characteristic for scholasticism is what H. Reuter (*Gesch. d. religiösen Aufklärung im Mittelalter*, Berlin 1877, **II**, pp. 19f.) relates of Simon of Tournay. After a very successful lecture he said with unbridled laughter: "Oh my little Jesus, how much I have helped to make thy doctrine on this question strong and glorious! Verily, were I to appear as its malevolent opponent, I should know how to degrade, weaken and refute it with even stronger reasons and arguments!" Hardly had he spoken these words when he fell dumb: he had lost speech and memory. Of course dialectics is often the art of leading astray others and sometimes oneself. Still, it has furthered the enjoyment of thinking. The quiet happiness experienced by those who had found their way into the narrow closed circle of scholastic ideas cannot be concealed even by the '*Epistulae obscurorum virotum*', however fierce the caricature.

[4] K. Groos, *Die Spiele der Tiere*, Jena 1896.

[5] Lavoisier did not discover the law of conservation of mass, but this instinctive assumption, current already in antiquity, led him to his great chemical discoveries.

[6] This problem and its solution is ancient too, being discussed by Plutarch in the dialogue 'on the face in the orb of the moon'.

[7] Cf. *M* 5, pp. 139f. Older views of the law of inertia in Whewell, *The Philosophy of the inductive Sciences* **I**, pp. 216f. He is aware that the first source of a knowledge of inertia can only be experience. However, once recognize force as causing motion, or change in it, then without force, motion must be uniform and rectilinear. This agrees with my view, if we define force simply as that which determines acceleration. The account of D'Alembert (*Traité de Dynamique,* 1743, pp. 4–6) which is likewise discussed in Whewell p. 218, is all but unintelligible unless it were essentially changed in form. Let a body be set in motion (by an impulse?). Either the cause suffices to move the body by a foot (sic!), or constant action was needed even for this distance. In both cases the same reason for motion continues to exist for the second foot, the third and so on. Now it is clear that this consideration of the path traversed can lead to no result so long as no assumption is made about the path as a function of time. If however we assume that the motion is uniform only in a time differential after the impulse, we have already instinctively established the law of inertia and can readily philosophize it into the open. The account of D'Alembert is a pretty sophism. Playfair (quoted by Whewell, p. 219) says that we must reject the law of inertia and assume that the velocity v falls according to some function $f(t)$ of time, more simply $v=c(1-kt)$ where c is the initial velocity; but he sees no reason to give preference to any form of function or value of the constant

k. Whewell correctly comments that our lack of insight cannot decide about experience.
[8] *M* 5, especially pp. 267f.
[9] Besides what is said in *M*, let me point out that from the principle of the parallelogram of forces (Coroll. I) one can derive the proportionality enunciated in the Second Law, but the independence of the forces assumed in Coroll. I must be separately established.
[10] This approach further brings into play the Newtonian rule: as far as possible, use only an actually observed cause (vera causa) when explaining anything.
[11] Duhen (*La Théorie physique*, pp. 84–167) distinguishes two intellectual types: comprehensive minds and deep minds. The former (esprits amples) have a lively phantasy, sensitive memory, refined judgement, are able to grasp a wide variety of things but show little sense for logical accuracy and purity. Deep but narrow minds (esprits profonds et étroits) have a narrower purview and are by nature suited to conceive everything in a simplified abstract way and can estimate intellectual economy and logical connection and validity, as well as being able to apply them. The former type, he says, is found especially amongst the English, the latter amongst the French and Germans. Names of famous scientists, scientific achievement, English and French laws and so on illustrate this idea in quite an attractive way. Duhem is quite explicit that these features hold only in general and cannot simply be applied to individual cases. However I think not only that there are cases of all intermediate degrees between these two extremes, but also that each individual, according to intellectual temper and task in hand, will lean now one way now the other. William Thomson (Lord Kelvin) for example is assigned by Duhem to the first type, because of his many models, relying on the most varied principles, for illustrating physical laws; but if we look at his work in thermodynamics we should rather say that he belongs to the second type. As to Descartes, Duhem ranges him amongst type two; but if we consider Descartes' wildly unlogical attempts at finding reasons for the law of refraction, where he assumes a timeless propagation of light and yet considers times and velocities in both media, and we compare this with the splendid logical derivations in his *Dioptrics* based on that law itself, it is barely credible that it is the same author speaking. I think that we must distinguish between the intellectual work involved in derivations from given principles and the search of principles as useful starting points for derivations. If from this latter point of view we consider the writings of Maxwell, which are rather harshly judged by Duhem and Poincaré, they are supreme. We can indeed congratulate ourselves if a whole people is especially good at searching for new foundations of a field of knowledge, while another is rather better at bringing logical order, connection and unity into this field.
[12] Notice the readily evident congruence of the triangles drawn for corner *A*.
[13] Fries, *System der Logik*, Heidelberg 1819, pp. 282f.
[14] Drobisch, *Neue Darstellung der Logik*, Leipzig 1895, appendix.
[15] Cf. Schuppe, *Erkenntnistheoretische Logik*, Bonn 1878; *Grundriss der Erkenntnistheorie und Logik*, Berlin 1894.
[16] Cf. by contrast the many suggestions from an expert like F. Mann (*Die logischen Grundoperationen der Mathematik*, Erlangen & Leipzig 1895).
[17] Boole, *An investigation of the laws of thought*, London 1854. E. Schroeder, *Algebra der Logik*, Leipzig 1890–1895. Russell, *The principles of mathematics*, Cambridge 1903.
[18] Couturat, *La logique de Leibniz*, Paris 1901.
[19] F. E. Beneke, *System der Logik als Kunstlehre des Denkens*, Berlin 1842. His logic is not just formal but contains important psychological investigations, unfortunately less noticed than they deserve.

CHAPTER XI

ON THOUGHT EXPERIMENTS[1]

1. Man collects experiences by observing changes in his surroundings. However, the most interesting and instructive changes for him are those that he can influence through his own intervention and deliberate movements. As regards such changes he need not remain purely passive, he can actively adapt them to his own requirements; besides, they are for him of the highest economic, practical and intellectual importance. That is what makes experiments so valuable. If we observe how a child in the first stages of independence examines the sensitivity of his own limbs, how he is surprised by his mirror image or by his shadow in sunlight and tries out how they behave by making movements, how he practises hitting a target we are driven to conclude that man has an innate tendency towards experiment, and that without much looking about he finds within himself the basic experimental method of variation. If the adult temporarily loses these treasures so that he must as it were discover them afresh, the explanation is that his social upbringing narrows his circle of interests and confines him to it while at the same time he acquires a large number of ready opinions, not to say prejudices, that he supposes not to be in need of examination.

In experimentation, the intellect may be involved in various degrees. I was able to observe this many years ago when I was paralysed in one hand and had to do with one a great many things that one normally does with both, if I was to avoid constantly having to rely on outside help. By changing the direction of a movement towards a certain goal, though randomly and without patience, I soon made all sorts of small discoveries, not by much reflective effort but merely by retaining what was useful and adapting to it. However, it required much thought to discover a procedure to carry out geometrical drawings by means of a compass, ruler and a weight in lieu of the disabled hand; likewise for all such manipulations as exceeded the range of my hand movement alone. We can hardly doubt that there is no sharp dividing line between instinctive and thought-guided experiments. Most prehistoric inventions, such as spinning, plaiting, weaving, tying knots and so on, probably result mainly from the former: they give the impression of being thoroughly thought through and their biological antecedents may be seen in the nest-building of birds and monkeys. Most such inventions were probably made by women, half playfully, something being discovered by accident, and only later retained deliberately. Once a beginning has been made, though, comparison soon leads to more refined experiments.[2]

2. Experimentation is not exclusively a human feature. Animals too may be observed experimenting in various levels of development. The impatient movements of a hamster smelling food in a box in the end cause its lid to fall off, though there is no planning involved; this represents something like the lowest level. More interesting are C. Lloyd Morgan's dogs, who after several attempts at carrying a stick with a heavy knob at one end, no longer grabbed it at the mid-point but near the heavy end at the centre of gravity, while grabbing it at one end to drag it through a narrow gate when transverse carrying proved impossible. However, these animals show little ability to apply the experience of a previous occasion to a later similar one. I have seen intelligent horses carefully testing a dangerous path by patting the ground with their hooves, and cats trying steaming milk for warmth by dipping a paw into the pail. From mere testing by the organs of sense, turning a body over or changing position to essentially changing the conditions, from passive observation to active experiment there is a very gradual transition.[3] What distinguishes animals from men in this is above all the narrow scope of interests. A young act shows curiosity in examining its mirror image, and may even look behind the mirror, but becomes indifferent as soon as it notices that the image is not another bodily cat. The male turtle dove does not reach even this level: as I have often observed, he can start cooing in front of his mirror image for fifteen minutes at a time and perform compliments with the customary two steps, without observing the deception. What a difference of level there is when one observes a four-year-old child who spontaneously notices with amazement and interest that a wine bottle placed in water to cool will appear shortened. Another child of similar age was amazed by stereoscopic appearences that arose when he accidentally squinted in front of wallpaper.[4]

Experiments guided by thought lie at the basis of science and consciously aim at widening experience. Still, one must not underrate the function of instinct and custom in the conduct of experiments. It is impossible to gain an instant intellectual survey of all the conditions that intervene in an experiment. Those who lack the ability to hold on to what is unusual and quickly to adapt hand movements as required, will have little success in the task needed to prepare a planned experiment. In a field which has become familiar through continued concern with it, one goes about experiments quite differently. If after a gap of some time one returns to such a field, one finds that most of what one had failed to fix conceptually and the finer grasp for the bearing of subsidiary conditions must usually be acquired afresh.

3. Besides physical experiments there are others that are extensively used at a higher intellectual level, namely thought experiments. The planner, the builder of castles in the air, the novelist,[5] the author of social and technological utopias is experimenting with thoughts; so, too, is the hardheaded merchant, the serious inventor and the enquirer. All of them imagine conditions, and connect with them their expectations and surmise of certain consequences: they gain a thought experience. However, while the former combine in phantasy certain conditions that never occur together in reality, or imagine these conditions accompanied by consequences that are not connected with them, the latter, whose ideas are good representations of the facts, will keep fairly close to reality in their thinking. Indeed, it is the more or less non-arbitrary representation of facts in our ideas that makes thought experiments possible. For we can find in memory details that we failed to notice when directly observing the facts. Just as in memory we may discover a trait that suddenly reveals a man's character hitherto misread, so memory offers new and so far unnoticed features of physical facts and helps us to new discoveries.

Our ideas are more readily to hand than physical facts: thought experiments cost less, as it were. It is thus small wonder that thought experiment often precedes and prepares physical experiments. Thus Aristotle's physical investigations are mainly thought experiments, in which he uses the store of experience kept in memory and above all in language. Thought experiment is in any case a necessary precondition for physical experiment. Every experimenter and inventor must have the planned arrangement in his head before translating it into fact. Stephenson may be familiar, from experience, with carriages, rails and steam engines, but it is by first combining them in thought that he can next proceed to build a locomotive in practice. Likewise, Galileo must see the experimental arrangement for investigating free fall well represented in his phantasy before he can realise it. Every beginner learns that an inadequate prior assessment, failure to take into account sources of error, and so on, can lead to consequences no less tragicomical for him than the proverbial act first and think later in practical life.

4. When physical experience has become ampler and a given range of sense elements has entered more varied and therefore weaker mental associations, phantasy can begin the sort of play in which the associations actually arising are decided upon by the mood, conditions and drift of thought of the moment. If a physicist asks himself what is to be expected under variously combined conditions if one keeps as closely as possible to physical experience, the upshot cannot be very new and different from what is offered by particular physical experiences. Since the physicist alway turns his thoughts towards reality, his activity differs from free fiction. Yet even the simplest thought of the physicist concerning some individual physical sense experience does not coincide entirely with reality: the thought usually contains less than experience, a mere schematic representation of it with occasional unpremeditated additions. By surveying one's memory for experiences and by making up new combinations of memories one will thus be able to learn how accurately experiences are represented by thoughts and how far the latter agree with each other. We here have a process for clarifying logical economy, applied to the intellectual transformation of the contents of experience. What will determine success, what hangs together and what is independent, all this becomes much clearer through such a survey than it could become through individual experiences. This makes it plain to us how we can combine convenience with the need to do justice to experience, what are the simplest thoughts that can be most comprehensively squared with each other and with experience. We achieve this by varying the facts in thought.

The outcome of a thought experiment, and the surmise that we mentally link with the varied conditions can be so definite and decisive that the author rightly or wrongly feels able to dispense with any further tests by physical experiment.[6] However, the less certain their outcome, the more strongly thought experiments urge the enquirer to physical experiment as a natural sequel that has to complete and to determine the result. Let us first consider some cases of the former kind.

5. Conditions that have been recognized as of no account with regard to a certain result can be varied at will in thought without altering that result. By astute handling of this procedure we may reach cases that at first blush seem rather different, that is to generalisation of the point of view. Stevin and Galileo showed great mastery of this device in their treatment of the inclined plane. Poinsot[7], too, used this method in mechanics. To a force system *A* he adds two others, *B* and *C*, *C* being chosen to balance each of *A* and *B*. Since the observer's point of view is irrelevant we are led to recognize *A* and *B* as equivalent, although they might differ greatly in other ways. Huygens' discoveries about impact rest on thought experiments: starting from the knowledge that the motion of other bodies is as irrelevant to the colliding body as it is to the observer, he changes the observer's point of view and the relative motion of the surroundings: in this way he starts from the simplest special case and reaches important generalizations. Another example occurs in dioptrics, where a ray is regarded as belonging now to this now to that beam of known properties.

6. It is further profitable mentally to vary those conditions that are decisive for the result, the most fruitful approach being continuous variation, which yields a conspectus of all possible cases. Thought experiments of this kind undoubtedly have led to enormous changes in our thinking and to an opening up of most important new paths of enquiry. Even if the story of Newton and the apple must not be taken literally (though Euler still did), nevertheless it is the sort of thought process so ingeniously presented by Euler[8] and Gruithuisen[9] that gradually led from the Copernican to the Newtonian point of view, and the elements of these processes can be historically verified, though in different people and in widely separated periods.

A stone falls to the ground. Increase the stone's distance from the earth, and it would go against the grain to expect that this continuous increase would lead to some discontinuity. Even at lunar distance the stone will not suddenly lose its tendency to fall. Moreover, big stones fall like small ones: the moon tends to fall to the earth. Our ideas would lose the requisite determination if one body were attracted to the other but not the reverse, thus the attraction is mutual and remains so with unequal bodies, for the cases merge into one another continuously. Not only logical elements are at play here: logically, discontinuities are quite conceivable, but it is highly improbable that their existence would not have betrayed itself by some experience. Besides, we prefer the point of view that causes less mental exertion, so long as it is compatible with experience.

Two simultaneously falling stones move together, alongside one another. The moon consists of stones, and so does the earth, every part attracts every other, this is how masses influence each other. Nor are moon and earth essentially different from other celestial bodies: gravitation is universal. Kepler's motion is projectile, but with an acceleration of fall depending on distance. All such acceleration depends on distance, on earth as well. Kepler's laws are only ideal cases, neglecting perturbations. Here we have the logical conceptual requirement of mutual agreement between thoughts. As we see, the basic method of thought experiments, as with physical experiments, is that of variation. By varying the conditions (continuously if possible), the scope of ideas (expectations) tied to them is extended: by modifying and specializing the conditions we modify and specialize the ideas, making them more determinate, and the two processes alternate.

Galileo is a master of this kind of thought experiment. He explains the fact that particles of high specific gravity float in air and water by imagining a cube dissected by three cuts into eight smaller cubes, which leaves the weight the same but doubles the cross-section and therefore the resistance, which with repeated cuts soon becomes enormous. Similarly, Galileo imagines an animal increased in proportion in all dimensions thus preserving geometrical similarity, in order to show that the creature would have to collapse under its weight which increases according to the third power, much more rapidly than the strength of the bones.

Thought experiments on their own often suffice to reduce to absurdity what on inspection seems to be the operative rule. If a body of greater weight had the property of falling faster, a combination between a light and a heavy body, would, though heavier still, have to fall more slowly because retarded by the lighter component. The assumed rule is thus untenable because self-contradictory. Such considerations play a great historical role in science.

7. Consider a process of this kind: bodies of the same temperature do not change this through mutual interaction. A hot body A (say a glowing iron ball) heats a colder B (a thermometer) even at a distance, through radiation, for example in the well-known experiment with coaxial concave mirrors. If with Pictet we replace A by a small metal box with a cold mixture, B will be cooled. This is a physical experiment that gives rise to thought experiments. Are there cold radiations? Is the second case not simply the same as the first with A and B interchanged? In both cases the hotter body heats the colder. Let A be hotter than B, then of equal temperatures and finally A colder than B. Which body radiates heat to the other in the case of equality? Is there a sudden changeover at that point? Both bodies independently radiate and absorb from each other, which is Prevost's dynamic thermal equilibrium. Different bodies of the same temperature radiate different heat quantities, according to experiments of Leslie and Rumford: for dynamic equilibrium to continue as indeed it does, the body that radiates twice as much must absorb twice as much.

An important process consists in mentally diminishing to zero one or several conditions that quantitatively affect the result, so that the remaining factors alone must be taken as of influence. Physically, such a process is often impossible to carry out, so that we may speak of it as an idealization or abstraction. By considering the resistance to motion of a body impelled on a horizontal plane or the retardation of a body moving up a very slightly inclined plane as the angle becomes vanishingly small, we reach the idea of a body moving uniformly without resistance. In practice this case cannot be realized. Apelt therefore rightly says that the law of inertia was discovered by abstraction: thought experiment and continuous variation have led to it. All general physical concepts and laws, the concept of a ray, the laws of dioptrics, Boyle's law and so on are obtained by idealization. That is what gives them their general unspecific form, which makes it possible to reconstruct any even complex fact in terms of synthetic combination of these concepts and laws, and so to understand these facts. Such idealizations occur in Carnot's work: the absolute insulator, absolute isothermy of touching bodies, reversible processes; in Kirchhoff's notion of the perfect black body and so on.

8. Unintentionally and instinctively gained raw experience gives us fairly undetermined pictures of the world. It tells us for example that

heavy bodies do not rise of themselves, that equally hot bodies in each other's presence remain equally hot and so on. This seems meagre, but is all the more secure and broadly based. Planned quantitative experiment yields many details, but our quantitative ideas educated by experiment gain their surest support if we relate them to those raw experiences. Thus, Stevin adapts his quantitative ideas about inclined planes to that experience about the gravity of bodies by means of exemplary thought experiments, and Galileo does likewise with this ideas concerning free fall. Fourier chooses those laws of radiation, and Kirchhoff that relation between absorption and emission, that fit in with the common experience of heat mentioned above.

By means of tentatively adapting a quantitative idea to generalized experience as regards bodies under gravity (the impossibility of a perpetuum mobile) S. Carnot discovered his heat theorem that has led to so many consequences, and in so doing carried out a magnificent thought experiment. His method has grown inexhaustibly fruitful since J. Thomson and W. Thomson have taken it over.

9. It depends on the kind and extent of the experience absorbed whether a thought experiment as such can be carried through to a definite conclusion. The colder body takes on heat from the hotter touching it. A melting or boiling body is in this condition but does not become warmer. Hence Black is sure that heat becomes latent in a change of state to vapour or liquid. Thus far the thought experiment: but to determine the quantity of latent heat Black has to rely on physical experiment, even if in form this follows directly from the thought experiment. Mayer and Joule discover the existence of the mechanical equivalent of heat through thought experiments, but Joule has to determine its numerical value by means of a physical experiment while Mayer is able to derive even this from remembered numbers, as it were.

If a thought experiment is without definite issue, that is when the idea of certain conditions leads to no certain and unambiguous expectation of a result, we tend to turn to guessing, at any rate for the period between thought and physical experiment, that is we tentatively assume an approximately sufficient condition for a result. This guessing is not unscientific, but a natural process that can be illustrated by historical examples. A closer look shows even that only such guesses can give a form to the physical experiment as a natural sequel to the thought experiment. Before Galileo examined the motion of free fall of which observation and reflection told him only that the speed increases, he tried to guess the rate of increase, and by testing what follows from his assumption he was able to devise his experiment in the first place. This is because the analytic inference from the law of distance to the law of velocity that determines it was more difficult than the reverse synthetic inference. Often the analytic procedure, being indefinite, is very difficult and Galileo's position often recurred with later enquirers. Other examples of laws first guessed and later confirmed by experiments are Richmann's rule of mixtures, the sinusoidal periodicity of light, and many other important physical conceptions.

10. The method of letting people guess the outcome of an experimental arrangement has didactic value too. I have seen this method in operation both in the case of my own high school teacher, H. Phillipp,[10] and also when visiting the school of F. Pisko, another admirable pedagogue. Not only the pupil but also the teacher gains immeasurably by this method: it is the best way to get to know one's pupils. Some will guess the most obvious likely thing, while others will surmise unusual and strange results. Most people will go for what is most obvious by association: just as the slave boy in Plato's *Meno* thinks that doubling the side will double the area of the square, a primary pupil will readily say that doubling the length of a pendulum will double the period of oscillation, while the more advanced will make less obvious though similar mistakes. However, such mistakes sharpen one's feeling for the differences between what is logically, physically or associatively determined or obvious, and in the end one learns to discriminate between the guessable from that which is not. The processes here described separately and the different cases that arise in practice occur in richly varied sequence or even at once in combination. Recalling how much in the building up of knowledge is contributed by memory, we can understand Plato's view that all enquiry and learning is remembering (from an earlier life). This view of course greatly exaggerates some aspects and equally underrates others. Every present individual experience may be very important, and even if the earlier life (in modern terms, the history of the tribe), which has imposed its traces on the body, is not accounted for nothing, never-

theless the individual's memory of present life is much more important.

11. Experimenting in thought is important not only for the professional enquirer, but also for mental development as such. How does it begin? How can it develop into a method used with deliberate and conscious intelligence? Just as any movement had to occur accidentally by reflex before it could become voluntary, so here too it is essential that suitable circumstances once started an unpremeditated variation in thoughts, in order that such variation should be discovered and be turned into a lasting habit. This happens most readily through paradoxical situations: these not only give us the best feeling for the nature of a problem since paradox is what makes a problem, but contradictory elements will not allow thought to rest and so set off the process we call thought experiment. We need remember only one of the well-known trick questions when heard for the first time. A beaker with water in equilibrium on one pan of a scale has a weight lowered into it in suspension from a separate stand: does the scale pan go down or not? A fly is sitting in a closed bottle balanced on a scale: what happens if it starts hovering in the air inside the bottle? A historically important case is the seeming paradoxical opposition between Carnot's and Mayer's heat theorems; or the relation between chromatic polarization and interference which largely agree but seemed in some cases to be incompatible. The various expectations attaching to the conditions variously combined in the different cases must cause disquiet and by this very fact help to clarify and advance the subject. Clausius and W. Thomson in the one case and Young and Fresnel in the other felt the impact of paradox. By analysing one's own and other people's work one becomes convinced how all success or failure depends on whether or not the paradoxical features were tackled with the fullest energy.

12. The peculiar continuous variation occurring in some of the above thought experiments reminds one vividly of the continuous variation of visual phantasms described by J. Müller.[11] Contrary to his view, these are compatible with the laws of association, indeed one may regard them in part as a phenomenon of memory, an imitation of perspective variation in pictures. If the appearance in phantasy of sequences of notes, melodies and harmonies are not felt as strange nor as contradicting the laws of

association, it is bound to be the same for visual phantasms. One cannot deny a spontaneous, hallucinatory element in all these cases: the life and mutual stimulation of organs act jointly here. Even so we must distinguish between hallucination and the creative phantasy of the artist or enquirer. In hallucination the pictures will follow a state of excitation that depends on raw sense, whereas in creative phantasy they gather round a ruling idea that stubbornly recurs. Still, as previously stated, the artist's phantasy is nearer hallucination than the enquirer's.[12]

13. We can hardly doubt that thought experiments are important not only in physics but in every field, even in mathematics, where the uninitiated might least expect it. Euler's method of enquiry whose fruitfulness has far outpaced critical assessment, gives the impression of the procedure of an experimenter who is probing a new field for the first time. Even where the exposition of a science is purely deductive, we must not be deceived by the form. We are dealing with thought experiments, after the result had become completely known and familiar to the author. Every explanation, proof and deduction is the outcome of this process.

The history of science shows beyond doubt that mathematics, arithmetic and geometry have developed from a chance gathering of single experiences with countable and measurable bodies. By frequent mental confrontation of such physical experiences, their mutual connections first become clear. Whenever this insight happens to be absent from consciousness, our mathematical knowledge has the character of acquired experience. Anyone who has engaged in mathematical enquiry or tried to solve problems such as integrating a differential equation, will admit that thought experiments definitely precede thought construction. The historically important and fruitful method of indeterminate coefficients is really an experimental method. The series for $\sin x$, $\cos x$, e^x having been established, it was found that the attempt to expand the symbolic expressions for e^{ix} and e^{-ix} into series automatically gave the expressions $\cos x=\frac{1}{2}(e^{ix}+e^{-ix})$, $\sin x=(1/2\,i)(e^{ix}-e^{-ix})$, which had long had a purely symbolic though computationally useful meaning before it was possible to indicate their real significance.

In describing a circle one observes that for every radius left of a given initial position there is a radius at the same angular distance to the right, so that the circle is symmetrical with regard to the initial position, and therefore in all directions: every diameter is an axis of symmetry. All chords bisected by it, including the limiting one of length zero (the tangent) are at right angles to it. Two radii equally but oppositely inclined to the axis always intersect the circle in the four corners of a symmetrically inscribed rectangle. The ancient enquirer and even modern beginners may have been surprised to learn in this way that the angle in a semicircle is always a right angle. Once we have noticed the relation between peripheral and central angles, we discover by moving the vertex along the arc that at every point of it the same chord appears under the same angle, which holds even when the vertex tends to the end of the arc from either side: one side of the peripheral angle then becomes the chord and the other the tangent at its endpoint. The theorem about the proportionality of the segments of two secants from a point to a circle goes over into the corresponding theorem about the tangent, if the two points of intersection of one of the secants are allowed to move towards each other until they coincide. Depending on whether we regard the circle as drawn with a compass or generated by a rigid angle whose sides always pass through two fixed points, or whether we observe that two circles can always be viewed as similar and similarly situated, there are always new properties that emerge. The change and motion of figures, continuous deformation, vanishing and infinite increase of particular elements here too are the means that enliven enquiry, tell us about new properties and promote insight into their connections. It is to be assumed that the method of physical and thought experiment developed first in this simply accessible and fruitful field and spread from there to the natural sciences. This view would indeed be more common if elementary teaching in mathematics, especially geometry, did not move in such rigid dogmatic forms, and if the presentation did not proceed in isolated theorems out of context, which leads to criticism being monstrously twisted and heuristic methods being irresponsibly concealed. The great apparent gap between experiment and deduction does not in fact exist. It is always a matter of attunement of thoughts with fact and with each other. If an experiment does not produce the expected outcome, it may be a considerable drawback for the inventor or engineer, but the enquirer will regard it as proof that his thoughts did not accurately correspond to the facts. It is precisely this sort of clearly expressed incongruity that can lead to new clarifications and discoveries.

14. The close conjunction of thought with experience has built modern natural science. Experience produces a thought which is then spun further to be compared again with experience and modified, which produces a new conception, and so on repeatedly. Such a development may take several generations before reaching a stage of relative completeness.

It is often said that enquiry cannot be taught. In a sense this is correct: the schemata of formal and even inductive logic cannot help much, for intellectual situations never quite repeat themselves. However, the examples of great enquires are very suggestive, and practising thought experiments after their model as briefly indicated above is bound to be beneficial. By precisely this method, later generations have indeed experienced progress in enquiry, for problems that posed great difficulties to earlier enquires are now easily solved.

NOTES

[1] Parts of this chapter were previously published in *Poskes Zeitschr. f. Physik. u. chem. Unterricht*, January 1897.

[2] Mere trial will often yield quite appropriate means. I observed a servant girl who had to spread a large carpet under a dinner table too heavy for one person to carry. At once the table was standing on the carpet without being shifted, and the girl maintained she had not thought about the method. Putting the almost fully rolled carpet against one side of the table, raising the latter, holding the unrolled end of the carpet with one foot and kicking the roll with the other, it unfolded to the other side, where a similar operation completed the job. Once when able to use only one hand I wanted to open window curtains and I had to do it in several stages because of the length of the string; but suddenly I found a better procedure without having consciously or deliberately reflected on it. My hand climbed some distance up the string by alternately grasping it with thumb and index and clamping it with the remaining fingers. Having reached the maximum possible height I pulled the string down and repeated the operation.

[3] My sister's dog once jumped up in dismay at the coldness of his freshly aired cushion and since then always tested it with his paw, waiting till it reached proper room temperature.

[4] In my view it is above all the breadth of interests that makes out the superior intelligence of a three to four year old child as against that of even the most intelligent animal. I can hardly grasp how a person familiar with children and animals can really think that a horse has numerical concepts and can calculate. Cf. the book of Th. Zell mentioned in Ch. V, note 3.

[5] Cf. E. Zola, *Le Roman expérimental*, Paris 1898.

[6] Duhem (*Théorie physique*, p. 331) rightly warns against representing thought experiments as though they were physical, that is pretending that postulates are facts.

[7] Poinsot, *Éléments de Statique*, 10th ed. Paris 1861.

[8] Euler, *Lettres à une Princesse d'Allemagne*, London 1775.

[9] F. Gruithuisen, *Die Naturgeschichte im Kreise der Ursachen und Wirkungen*, Munich 1810.

[10] Unfortunately the splendid teacher in question almost completely spoiled his success by defective pedagogy and unrivalled impatience.

[11] J. Müller, *Die phantastischen Gesichtserscheinungen*, Koblenz 1826.

[12] Without underestimating the value of the laws of association for psychology one can justly doubt that they alone hold good. Alongside individually acquired temporary organic pathways the nervous system also has innate and lasting ones (at least not acquired by the individual), as is shown by reflex movements, indeed, the latter are much more important for non-individual functions. A process within an organ may be set off in both ways by a neighbouring organ, but probably it may also arise spontaneously within the organ itself. If the process is especially vigorous it will presumably spread by all available means from its place of origin. It seems to me that all this must have a mental counterpart too.

CHAPTER XII

PHYSICAL EXPERIMENT AND ITS LEADING FEATURES

1. Experiment can be described as the autonomous search for new reactions or their interconnections. Physical experiment was earlier mentioned as the natural sequel to thought experiment, and it occurs wherever the latter cannot readily decide the issue, or not completely, or not at all. Even accidental observation of something striking can instinctively provoke a special mode of motor reaction, that yields us knowledge about new reactions or the links between them. Such cases can be observed in animals, or even in ourselves, if we are sufficiently attentive: this we might call instinctive experimenting. However, if a casual observation reminds us in some unusual way of some connection already known, above all if the observation is in obvious contrast with what is known or familiar, the result is to suggest thoughts that may be regarded as the specific motive power behind the physical experiments that now follow. Amongst many cases of this kind we recall Galileo's swinging lamp, Grimaldi's coloured strips on the edge of shadows, Boyle's and Hooke's colours in soap bubbles, and in fine cracks in glass, Galvani's frog, Arago's damping of magnet needles by a copper disc and his discovery of chromatic polarisation, Faraday's discovery of induction, and so on. Every experimenter will be familiar with similar examples from his own experience, though few of them will be historically as vital and full of consequence as those mentioned. My investigations on sense organs began with the contrasting aspects of a square when the side or the diagonal is vertical. My finding an extension of the laws of contrast of brightness began with accidentally observing a phenomenon on revolving sectors with bent edges, which was unintelligible in terms of the law of Talbot and Plateau. Accidental observations can initiate not only theoretically important discoveries, but also practically valuable inventions. It is said that Samuel Brown was led to his construction of suspension bridges by observing a spider weaving his web, and James Watt to the plan of a water supply system by observing the shell of a crab.[1] I have discussed elsewhere how far such cases depend on chance and what is its function.[2]

2. Deliberate, autonomous extension of experience by physical experiment and systematic observation are thus always guided by thought and cannot be sharply limited and cut off from thought experiment.[3] That is why the requisite leading features of physical experiment, presently to be discussed, are important also for thought experiments and for enquiry in general. These leading features can be abstracted from the work of enquirers; so far they have never failed so that we can expect further success if we take notice of them. Our account is of course not exhaustive.

3. What we can learn from an experiment resides wholly and solely in the dependence or independence of the elements or conditions of a phenomenon. By arbitrarily varying a certain group of elements or a single one, other elements will vary too or perhaps remain unchanged. The basic method of experiment is the method of variation. If every element could be varied by itself alone, it would be a relatively easy matter: a systematic procedure would soon reveal the existing dependences. However, elements usually hang together by groups, some can be varied only along with others: each element is usually influenced by several others and in different ways. Thus we have to combine variations, and with an increasing number of elements the number of combinations to be tested experiment grows so rapidly (a simple calculation shows this), that a systematic treatment of the problem becomes increasingly difficult and in the end practically impossible. Deliberate experiment without some prior experience from chance observations would in most cases be powerless. The experience gained in the furtherance of biological needs makes the task much easier because it gives us a rough picture of the strongest relations of dependence and independence, which must however be considerably corrected for the new purposes of science. When therefore we start a line of experiments we know at least roughly what conditions may be ignored for the time being. A closer determination of such independences is however very important. For example, the fact that the various accelerations produced in one body by others are mutually independent, and likewise for simultaneous radiations and steady electric and thermal currents, we can proceed in terms of the principle of isolation, using the principle of superposition for their combination.

4. To determine the dependence of elements of phenomena on one other,

we must distinguish qualitative from quantitative dependence. For example, we determine a qualitative dependence if the experiment tells us that of the notes of the diatonic scale taken as discovered only by hearing, C and G are consonant, C and B dissonant. Another example is the fact that a certain red and green combine into white, while red and blue yields violet. A further qualitative case is that of the chemist examining the mutual reactions of substances that have certain specific sensible qualities, or a pharmacologist who is trying out the toxic or narcotic effect of certain vegetable substances on animal organisms. However, if we try to determine the dependence of the angle of refraction on the angle of incidence, or the dependence of the distance fallen on the time of fall, we are tackling quantitative problems. The individual angles are not so different from one another, not so incapable of being compared, as are red and green: the former can be divided into equal elements, and one angle differs from another only in the number of such elements; likewise with the distance of fall and the corresponding time and space elements. Quantitative dependence is a particularly simple case of qualitative dependence. If on top of it we can find a rule of constant form that allows us to calculate the number of space elements s from the number of time elements t, $s = gt^2/2$, or the angle of refraction r from that of incidence i, $\sin i/\sin r = n$, then we can replace or represent the somewhat clumsy table by the much more profitable rule of calculation, formula or law. An additional advantage is that by means of numbers we can carry the degree of fineness in the distinction as far as we like without having to invent a new nomenclature. Quantitative dependence presents a clear and intuitive continuum of cases, while qualitative dependence always leads to a discrete set of individual cases.[4] Wherever possible, one will try to introduce the simplicity, uniformity and clarity of quantitative treatment: we can do this as soon as we can find a set of quantitatively similar marks that completely characterize a set of qualitatively disparate elements.[5] If instead of distinguishing tonal qualities by ear, one characterizes the height by means of the frequency of oscillation, one immediately recognizes consonance, as tied to the simplest ratios of integers for the frequencies. How variously coloured light is refrected in a prism must be described in detail, but if one replaces colour by wave length (the width of interference bands under certain conditions), we readily discover a formula yielding the index of refraction in terms of the wavelength. The natural sciences show a decided tendency towards replacing qualitative by quantitative dependences as far as possible.

5. Positive examination is made much easier if we first eliminate everything that has no influence on the elements whose dependence on others are to be tested, and thus restrict the relevant area. A splendid historical example of this feature is furnished by the refraction near the edge of screens, which Newton intended to reduce to a mass effect of the screen on the corpuscles of light. S'Gravesand and Fresnel however showed that the thickness and material of the screen have no bearing on this refraction but influence merely the kind of boundary of the light. Brewster succeeded in obtaining on an impression in sealing wax the brilliance and coloured sheen of mother of pearl, which showed that the only determining factor was the shape of the surface. Le Monnier showed that hollow and solid conductors of the same shape behaved identically as regards electric charge, and thus limited the problem to examining the dependence of charge on size and shape of the surface.

6. Removing what hides or disturbs the dependence being studied is immensely important. To observe a pure case of light refraction in a prism, Newton worked in a darkened room, letting in thin pencils of sunlight so that parts of thicker pencils should not interfere and overlap. He represented the small light aperture as a lens to obtain pictures of the differently coloured rays one next to the other. In examining the errors of mirrors and lenses, Foucault and Toepler shut off the regularly reflected and refracted light so that what remains is clearly due to the errors, a method that is amongst the most refined in optics.

7. Great experimenters have always simplified their arrangements in such a way that only the factor in question remained in evidence while all other influences became negligible. Witness for example Ramsden's ingenious method for determining the thermal expansion of rods, and the equally ingenious procedure of Dulong and Petit to measure the absolute cubic thermal expansion of mercury using hydrostatic principles. The writings of great enquirers are full of such examples and are irreplaceable. Galileo demonstrates the weight of air without an air pump, measures small time spans in his experiments on fall by means of the efflux of water

and replaces free fall by bodies rolling down inclined planes. Newton tests the mutual action of magnets by enclosing them in small floating glass vessels; he also compares his calculated value for the velocity of sound with experiment, by using a bob pendulum of variable length to observe multiple echoes in an empty corridor. The apparatus of Ampère, Faraday, Bunsen are models of simplicity and purpose. However, not only should we aim at simplicity in experiment, but also learn from these giants to see more than indifferent aspects in what are quite ordinary events. If one's attention has been sharpened through a certain interest, one is able without much ado to discern traces of important connections in one's daily surroundings. Nobody who has not acquired this ability is likely to make many experimental discoveries. Huygens, observing bits of sealing wax drawn towards the axis of rotation of a vortex in water, sees in this processes that lead him to thoughts about gravitation. The perfectly sharp image of the slender feet of a fly illuminated by monochromatic light suffers no further prismatic resolution. The way a horizontally propelled hat sticks to the flat surface on which it lands is for Pascal a hydrodynamic phenomenon exhibiting the pressure of air. The traces of colours in cracks in glass observed by Hooke, led him to superimpose two spectacle lenses which show the complete phenomenon of rings that was later studied quantitatively by Newton. In the tinfoil cover of the top of a winebottle most people will not see anything special, but if one is used to observing thermal phenomena, one immediately notices the reflected radiation from a finger held in the neck of the bottle without touching it. The field of a vibrating string does not seem to show anything remarkable, but an experienced student of acoustics can see the overtones from the shading of the field. A bowed string has a uniform field, showing that every element runs through its field at constant speed: as soon as the bow is taken off, the edge of the field becomes more pronounced showing that a freely vibrating string remains relatively longer at the boundaries. A rapid look at an accidentally shiny spot on the string produces an afterimage that reveals the form of the vibration. Experiments with the most common utensils, as described for example by G. Tissandier,[6] are most beneficial, because they give us a sharper eye for noticing things that are usually overlooked.

8. If in a set of conditions one condition B is determined by another A,

we may expect that when *A* occurs or vanishes so will *B*, and similarly with relative growth and diminution. *A* might be the rise in temperature, magnetic pole strength or pressure, and *B* the gas pressure, induced current or double refraction of a transparent body. This feature of parallelism already mentioned by J. F. W. Herschel,[7] is a sure guide to the experimenter.

9. If the influence of *A* and *B* is small, making variations in *B* difficult to observe, one has to amplify these last. Galileo illustrates the process of summing effects in the case of a heavy bell which by small rhythmic impulses in phase can set the whole into vigorous motion. In this way he explains resonance. This method is at present used to obtain big galvanometer deflections for very weak currents, by means of the so-called ballistic method. By increasing the number of conducting windings the deflection is increased up to a point. Volta's electrophorus has shown how by means of two condenser electroscopes an almost unnoticeable charge can be multiplied by repeated doubling: influence machines use this process automatically to produce large charges. To make visible the influence of pressure on double refraction, Fresnel puts several prisms in a row; to obtain a noticeable path difference in dry and moist air, he uses long paths in his interference refractometer; to show clearly how the plane of polarization is rotated in his heavy glass, Faraday reflects a polarized ray to and fro many times in the direction of magnetic force; all these are examples of cumulative effects. Maxwell has observed momentary double refraction through frictional drag in viscous fluids, and I have observed the same in half fluid plastic materials through applied pressure, but both lasted only for a moment. Kundt enclosed such liquids between two long coaxial cylinders, of which one was rotating rapidly. This produces a long path and constant drag, so that the phenomenon is clearly and permanently apparent and easily measurable.

10. To determine an element whose direct evaluation is awkward, difficult or impossible, we can at times substitute another known equivalent element. For example to find electric resistance we may replace it by an amount of calibrated rheostat wire such that all phenomena remain the same. When Hirn did his tests on the heat produced by a man when working and at rest, he put the individual in a big calorimeter in which he could

mount or descend from a bicycle or remain still; but the heat produced was difficult to measure because of the heat losses from the calorimetric chamber. Hence, in other trials he replaced the man by a gas burner that produced the same calorimetric effect in the same time, and here the production of heat is easily calculated.[8] Joule compressed air by means of a pump enclosed in a pressure vessel plunged in a calorimeter. The heat of compression corresponding to the work was made more difficult to measure by the fact that the frictional heat of the pump was inevitably added; but by letting the pump run idle for the same period at the same speed, the heat of compression was easy to find indirectly.[9]

11. Another method of indirect determination is compensation. Some condition or other brings into play an element *B* that is hard to determine: by including the readily determined element $-B$, the effect of *B* is eliminated by compensation and in fact determined. If we produce a considerable path difference in two interfering rays, the system of interference bands disappears, so that the path difference can no longer be measured from the displacement of the bands. By suppressing the path difference through placing a glass plate of determinable thickness on the side that was not retarded, we have compensated and indirectly determined the path difference. In a similar way one can nullify the galvanometer deflection produced by an unknown radiation from a thermopile, by introducing a known radiation on the other side and thus determine the former.

12. The principle of compensation is important in other ways too. If a condition *A* causes the occurrence of *B*, but also of *N* which in turn influences *B* then the simple connection between *A* and *B* becomes dimmed. We must therefore compensate *N*. Jamin led two interfering beams through water-filled pipes of equal length. If in one pipe we apply pressure, the beam in it is at once retarded, but more than would correspond to the increase of density, since the pipe becomes a little longer. This can be compensated (except for an easy final correction) by putting both pipes into a big water-filled pipe without pressure. The principle of compensation is important too as regards engineering and practical science, where certain conditions must be kept constant, for example the length of a time-measuring pendulum.

13. Substitution, and especially compensation, when refined leads to the so-called null method. When small A-dependent changes of B are to be investigated, the greatest sensitivity is obtained by masking B through compensation, so that it does not appear until A is changed. Let A be temperature and B electric resistance which depends on it. We now compensate B by means of an equal resistance until the attendant galvanometer deflection is annulled (Wheatstone bridge). As B grows with temperature while the compensatory resistance remains unchanged, the galvanometer at once deflects (bolometer). If at two equipotential points of a current-carrying plate we connect the leads of a galvanometer, it will not deflect; but the slightest asymmetric change of equipotentials, say by magnetic changes in resistance, at once causes a deflection (Hall effect). The use of Soleil's double quartz for tests on the rotation of the plane of rotation is another example of the null method.

14. Processes that are too rapid for direct observation must of course be obtained indirectly. To this end one uses the method of composition. The unknown process to be investigated furnishes one component, which along with a second, known component produces an observable resultant. Vertical fall reveals itself in the parabola of a projectile combining fall with a known constant horizontal component, as in Morin's well-known device or in that of Lippich, if we use a horizontal component that oscillates harmonically, or most simply in a jet of water. A powerful impetus for developing this method came from Wheatstone when he used a rotating mirror to find the speed of propagation and the time span of electric discharges. Feddersen's refinement of this procedure led to our exact knowledge of electric oscillations. Another refinement led to Foucault's method for determining the speed of light. Besides, there are numerous acoustic applications.

Using an optical motion as known component commends itself because this does not affect the process to be examined. Fizeau's procedure for measuring the speed of light is a splendid example where this method is ingeniously applied. Other examples are rapidly rotating discs and cylinders to take on spark signals for time measurements that would be difficult otherwise, for example in experiments with projectiles, sound and discharges, the stroboscopic method, Wheatstone's kaleidophone, Lissajou's method of tuning, Helmholtz' vibration microscope, and so on.

Combining the outflow speed of an explosive gas with the speed of explosion enables us to determine the latter. The use of the speed of sound for measuring other velocities has become quite common, and there is no reason why the speed of light might not yet be similarly used for much finer time measurements. For the reason stated, using a motion as one component seems best, but there is no reason why on occasion it might not be useful to combine any two processes of which one is known and the other to be investigated, provided only that they are mutually independent or related in a known way.

15. Special interest belongs to experiments that not only yield correlated values of a pair of connected conditions, but provide a survey of a whole system of such values. An example is the combination of glasses by Hooke and Newton. When Newton takes this along with the spectrum and shows the contraction of rings from red to violet, he sets up a further such experiment. By spectrally resolving the refractions in a very short narrow vertical slit in the latter's direction (that is, at right angles to the direction of refraction),[10] we obtain the various monochromatic refractions one below the other. Further experiments of this kind are the axial images of chromatic polarization of crystal plates, the polarization apparatus devised by Spottiswoode and myself, Kundt's method of depositing red lead and sulphur powder on pyroelectric crystals, Chladni's sand patterns on sounding plates, the well-known magnetic curves; Herschel[11] calls them "collective instances", and W. S. Jevons[12] "collective experiments".

16. In order not to misinterpret an experiment, we must always pay attention to possible errors, especially if the effect to be expected is very slight. When Faraday was investigating the influence of a strong electromagnet on weak paramagnetic and diamagnetic substances he carefully tested the magnetic behaviour of the suspension, and of the paper and small glass vessels in which the materials to be tested were placed: only when the suspension gave no magnetic response did he trust the experiments on the substances themselves. This type of preparatory test is called a blind experiment. The same caution is required in doubling minute electric charges so that they can be observed more accurately: we must first make sure that the condenser electroscopes are free from residual charges from earlier experiments and that the doubling procedure does

not introduce extraneous charges. Before a chemist uses Marsh's apparatus to test a substance for arsenic content, he makes sure that there is no indication prior to introducing the sample; that is, he ascertains that the matter of the apparatus itself is free from arsenic.

17. The history of science shows that experiments with negative result must never be regarded as decisive. Hooke was unable to show the influence of the earth's distance on the weight of bodies by means of his own balances, but with the very much more sensitive ones of today this is easy. J. F. W. Herschel could not observe electric and magnetic rotations of the plane of polarization, but Faraday could. J. Kerr's experiments on double refraction in dielectrics had often been carried out before but with negative results. Bennet tried in vain to demonstrate the pressure of light on irradiated surfaces; Crookes succeeded with his radiometer, but A. Schuster showed that this pressure arose from forces internal to the aparatus and could not be explained by incident particles. Thus both the result and the interpretation of a negative experiment remain problematic.

18. The formative features of experiment here described have been abstracted from experiments actually carried out. The list is not complete, for ingenious enquirers go on adding new items to it; neither is it a classification, since different features do not in general exclude one another, so that several of them may be united in the experiment. For example, the methods of Fizeau and Foucault for measuring the speed of light contain the features of combining the known with the unknown to yield an observable result, the feature also of cumulative effects, as well as of that of stabilizing momentary phenomena: the respectively determining elements in the two methods are the maxima and minima of brightness and the displacements of images, both dependent on the speed.[13]

19. As regards the role of ideas in widening our knowledge through experiment, all ideas must have arisen from past experience and will be developed further by that yet to come. The thoughts that anticipate experience and the expectation prefigured by the experiment can concern only agreements and differeces between the new and the known. How far may we regard an experimental result as valid, and how far must we restrict it under changed conditions? These questions circumscribe the main

ideas of the enquirer about the experiment For the more special ideas, let us once more consider historical examples.

20. Suppose we know the result of an experiment and we now try to extend it as far as we can from a purely collective point of view. There are magnetic iron ores: are there other magnetic bodies? Is feldspar the only doubly refractive substance? What bodies can be electrified by rubbing? Which are conductors and which insulators? How far does phosphorescence reach?[14] Other cases are the search for all instances in which a phenomenon occurs, given that it was discovered by a single observation. Oerstedt tries to determine all the relative positions of magnetic needle and conductor of current and their behaviour, after observing a single instance of deflection, and so obtains complete knowledge of the magnetic field of the conductor.

21. Extending an investigation from a known to analogous cases is especially attractive. Analogies between heat, electricity, mechanical processes and diffusion and so on, have led to many experiments; witness Fick's researches on diffusion currents. Magnets influence each other, and so do currents and magnets: a current acts on a magnet like some other magnet; do currents behave to each other like magnets? Arago has pointed out that in using analogical metaphors we must be ready to find differences too. Magnets and soft iron attract each other; soft iron here behaves like a magnet, but two pieces of soft iron do not affect each other. Of course, currents and soft iron behave somewhat differently as regards magnets: the former shows polarity, the latter not.

22. Where phenomena occur in different degree we may conceive a contrast to be possible. Different strengths of magnetism suggest the idea of an opposite, diamagnetic type of behaviour. If we know one kind of double refraction, say the negative one, we try to find the contrasting positive one. Not everything that might thus have been found was in fact so found: discovery often came by chance, as for example Dufay's discovery of the opposite kind of electricity in contrast to the kind already known. Not everything that at first blush looks like a contrast must be one. Thus paramagnetism and diamagnetism are no longer regarded as a contrast but as a difference of degree with regard to an all-pervasive me-

dium, just as in air we no longer view rising bodies as having levity or negative gravity, but rather a smaller weight per unit volume than air. Similarly for hot and cold, positive and negative electricity and so on. Such changes belong to the field of theory.

23. Corresponding to continuous variation of conditions there is continuity of expectation as to experimental results. Unequal pressure in different directions causes double refraction in solids. The transition from solid to liquid as regards rigidity and viscosity being gradual, we may expect that plastic solids and viscous fluids too can be made doubly refractive by means of appropriate compression or tension, and this has indeed been observed. Since no liquid is quite devoid of rigidity and viscosity, it will depend only on the magnitude of the forces and speed of the deformations whether double refraction will become noticeable. Between gases and vapours, too, properties vary continuously, whence the idea of liquefying gases under appropriate temperature and pressure arose quite naturally. There are rigid and liquid bodies that rotate the plane of polarization, so that one may surmise that the same happens in gases and vapours. Magnetic rotation has been demonstrated for every state of aggregation, most recently for gases, by Kundt and Lippich independently, in 1879. Is there a fourth state of aggregation? (Crookes).

24. When phenomena vary with conditions we come to ask for the form of the former for extreme values of the latter. Thus we examine the behaviour of bodies, at the highest and lowest temperatures attainable, as regards hardness, elasticity, electric resistance and so on. Melting, freezing and evaporating bodies are subjected to the highest pressures. We investigate the properties of the perfect vacuum, try to obtain the highest electric tensions and currents, and examine the shortest and longest light-waves. Such attempts are always likely to yield much that is new.

25. Just as experience is enhanced by searching for the widest possible agreements, the same occurs through restricting, specializing and individualizing as circumstances may dictate in each case. If we know refraction as a universal phenomenon occurring at every transit between media, we still must determine the characteristic refractive index for each pair of media, or the speed of propagation in each medium. Such restrictions can

afford discoveries just as great as those resulting from generalization; witness Newton's discovery of dispersion by assigning special refractive indices for the various colours and the classification of colours in terms of lengths of period, and all quantitative determinations of the constants characteristic of individual substances, such as densities, specific heats, coefficients of expansion and elastic moduli, electric resistance, dielectric constants, specific magnetization and so on.

26. A fruitful guiding feature is that of combined and opposite action. More precisely, if condition A determines the appearance of condition $+B$, then condition $+B$ determines the appearance of $-A$, the opposite of A. Examples are pressure and reaction in mechanics; heated gas expands, gas expanding against pressure cools down; a current moves a magnetic pole and vice versa, but in opposite directions; current flow heats a resistance, heating the resistance reduces the current flow. Direct current makes iron magnetic, a magnet that is brought near or one of growing intensity generates a current while the change lasts and in a sense tending to remove or weaken the approaching magnet. Seebeck's thermal current, flowing through a heated junction from M to N, will cool that junction, as Peltier[15] has shown. Again not all phenomena that could have been discovered by this method were found in this way. As counterpart to the excitation of an electromagnet by a current, Faraday tried to excite a current by introducing a magnetic core into the coil, but he found only a momentary "induced" current, when actually introducing or removing the magnet. Peltier, too, was looking for a counter-phenomenon to the Seebeck effect, thinking of an influence on the thermal conductivity of metals: by warming the metals of the thermopile by means of a current he found that the soldered joins were unequally heated according to the direction of the current. By enclosing two thick[16] rods, one of bismuth the other of antimony, in the vessel of an air thermometer, we find that a positive current from antimony to bismuth causes heating, whereas with the current flowing the other way there is an unexpected cooling. If to a given phenomenon we seek the counter-phenomenon, the above rule can give us a hint, but cannot guide us on its own. A direct current can indeed produce a magnet, but a stationary magnet cannot produce a current, which would indeed represent work without expenditure of energy. Not until we combine the principle of energy with the

law of induction do we obtain a complete and coherent system of effects and countereffects. Thus the above rule has to be completed by means of special findings, because in observing phenomena we are seldom dealing with simple, pure and immediate connections. Of two bodies in direct mutual interaction, the one can gain momentum, heat, charge and so on, only at the expense of the other. If all relations were as simple as this, the rule would be a sure guide: when the mutual connections are mediate, things are not so simple, and direct inversion is not permissible.[17]

NOTES

[1] G. A. Colozza, *L'Immaginazione nella scienza*, Turin 1900, p. 156.
[2] *P* 3, 1903, pp. 287f.
[3] Claude Bernard's advice is to ignore all theory when experimenting. Duhem rightly objects that in physics, where experiment without theory is unintelligible, this is impossible. I think it is no different in physiology. In fact all one can do is to recommend careful assessment whether the experimental result is compatible with the prior implicit theory at all. Cf. Duhem, *La théorie physique*, pp. 297f.
[4] *P*, p. 263f.
[5] *A* 4, 1903, p. 209.
[6] Tissandier, *La physique sans appareils*, Paris, 7th ed.
[7] J. F. W. Herschel, *A preliminary discourse on the study of natural philosophy*, London 1831, pp. 151f.
[8] Hirn, *Théorie mécanique de la chaleur*, Paris 1865, pp. 26–34.
[9] Joule, 'On the Changes of Temperature Produced by the Rarefaction and Condensation of Air', *Phil. Mag.* 1845.
[10] Fraunhofer, *Gesammelte Schriften*, Munich 1888, p. 71.
[11] Herschel, *l.c.* p. 185.
[12] W. S. Jevons, *The principles of science*, London 1892, p. 447.
[13] Foucault, *Recueil des travaux scientifiques*, Paris 1878, p. 197. He describes his method as 'l'observation d'une image fixe d'une image mobile', which seems to capture the essential point.
[14] J. P. Heinrich, *Die Phosphoreszenz der Körper*, Nuremberg 1820. A. E. Becquerel, 'Sur la phosphorescence par insolation', *Ann. chim. phys.* **22**, 1848.
[15] *L'Institut* 1834. 21 [actually 26] Apr. and 11 [actually 16] Aug.
[16] Because this clearly brings out the Peltier effect at the soldered joins, as against the Joule heating.
[17] Cf. *A*, pp. 69–76.

CHAPTER XIII

SIMILARITY AND ANALOGY AS A LEADING FEATURE OF ENQUIRY[1]

1. Similarity is partial identity: the characteristics of similar objects are in part identical and in part different. Not a single observable mark of one object need coincide with a mark of the other, and yet the marks of the one may be interrelated in exactly the same way as those of the other. Jevons[2] calls analogy a more deep-seated similarity; one might say, an abstract similarity. Analogy may in some circumstances remain quite concealed from direct sense observations and reveal itself only through comparison of conceptual interconnections between the marks of one object with the corresponding connections in the other. Maxwell[3] not only defines analogy but also underlines those features of it that are most important for scientific enquiry, when he describes analogy as that partial similarity between the laws in one field and those in another, so that each illustrates the other. However, we shall see that Maxwell's approach is not different from ours. Hoppe[4] regards the concept of analogy as superfluous, since as with similarity in general it is merely a matter of conceptual agreement of certain marks in the objects between which analogy is found. Although this is correct, there are good grounds for taking analogy as a special case of similarity and distinguishing it from the general concept. Above all, it is the enquirer into nature who is driven to this view, since taking notice of analogies greatly furthers his work. Furthermore it is clear that analogy or sameness of relation between elements may occur in objects whose similarity is immediately observable by the senses, and this may seem so obvious that the analogy is overlooked.

2. Similarity perceived by the senses already determines an unconscious and unpremeditated similarity in behaviour and motor reactions towards similar objects; when the intellect becomes conscious it will do likewise, as Stern[5] has shown at length as regards popular thought. Besides, Tylor's works[6] offer ample evidence. As conceptual thought becomes stronger, deliberate and purposive endeavour to free oneself from practical or

intellectual unease will likewise be guided by similarities and soon by more deep-seated analogies as well.

3. I have elsewhere[7] defined analogy as a relation between systems of concepts, in which we become clearly conscious both that corresponding elements are different and that corresponding connections between elements are the same.

It seems that mathematics, where things are indeed simplest, was the first field in which analogy first clearly revealed its clarifying, simplifying and heuristic role. At any rate Aristotle, insofar as he speaks of it, relates analogy to quantitative relations of proportionality. Some simple analogies must have struck enquirers even in antiquity. Thus Euclid (Book 7, definition 16) calls the product of two numbers "area" and the factors "sides", a product of three factors "body" with the factors as "sides" (definition 17), and a product of two or three equal factors "square" and "cube" respectively (definitions 18, 19).[8] Where Plato touches on geometric notions he uses similar terms. The invention of algebra rests on seeing the analogy of arithmetic operations whatever the different numbers: what is conceptually identical here will then be dealt with at one go and once and for all. Where magnitudes enter analogously into a calculation, then from one result we obtain all the others by simple analogous interchange of symbols. Descartes' geometry makes extensive use of the analogy between algebra and geometry, Grassmann's mechanics or vector theory of that between lines and forces, areas and torques and so on. Every physical application of mathematics rests on taking note of analogies between facts and mathematical operations.

4. The great value of analogy for cognition was clearly recognized by Kepler[9]. In dealing with conic sections as to their optical properties, he says: "The one focus of a circle is A, namely at the centre; in the ellipse there are two foci, A and B, equidistant from the centre of the figure in the more pointed part. In a parabola, one focus D is inside the section, the other must be set either outside or inside on the axis infinitely distant from the first, so that the line HG or IG drawn from this blind focus to any point G of the section is parallel to the axis. In the hyperbola, the external focus F is the nearer to the internal one E, the more obtuse the curve; and the one outside either of the opposite branches is inside the

other and conversely. It follows by analogy that in a straight line either focus (we speak of straight lines not so much in the ordinary way but rather to complete analogy) falls on the line: there is but one, as in the circle. In the circle, the focus is thus at the centre, as far from the circumference as possible, in the ellipse already less and in the parabola much less, and finally in the straight line it is at minimal distance, that is, it falls on the line. Thus, in the extreme cases of the circle and the straight line the foci coalesce, in the former as far as can be from the curve, in the latter falling fairly on it. In intermediate cases, the two foci are infinitely far apart in the parabola and finitely distant from the sides of a hyperbola and ellipse; in the ellipse the second is inside, in the hyperbola outside, whence the ratios are of opposite sign.... For the geometrical voice of analogy ought to be at our service; indeed I very dearly love analogies, my faithful masters, aware of all nature's secrets: we must particularly look up to them in geometry, with infinitely many cases placed in between extreme ones, when however much we infer the intermediate case by absurd phrases, the analogies put any object's total essence clearly before our eyes."

5. Here Kepler not only emphasizes the value of analogy but also and rightly the principle of continuity which alone could lead him to the degree of abstraction that allowed him to grasp such deep-seated analogies. As to the workshop of ancient science we know rather little, we barely have their most important results of enquiry. However, the form of presentation is often such as to conceal the actual paths of enquiry, as is strikingly shown by the example of Euclid. Unfortunately the ancient example has often been copied in more recent times, for the sake of a rigour that is overrated and to the detriment of the true interests of science. A thought is most completely and rigorously founded when all the paths and motives that led to and confirmed it are clearly set out. Of this foundation the logical link with previous more familiar and uncontested thoughts is only one part. A thought, the motives for whose generation are completely clarified, can never be lost so long as those motives hold, while it can be given up immediately when they are recognized as decrepit.

6. The study of the classics from the renaissance of natural science is so enjoyable and rich in permanent and irreplaceable lessons precisely because those great and naive men tell us all the details of what they

found and how, in the friendly joy of search and discovery, without any professional and learned secrecy-mongering.

Copernicus, Stevin, Galileo, Gilbert, Kepler furnish examples of the greatest successes of scientific research that teach us without pomp what were the leading motives of enquiry: thus the methods of physical and thought experiments,[10] the principles of simplicity and continuity and so on become familiar to us in the simplest way.

7. Besides this cosmopolitan trait of openness the science of that period is distinguished by an unusual growth in abstraction. It is from individual experience that science develops, and it is at the level of the individual that ancient science generally sticks. However, if one starts by inheriting a rich fund, one is in a more favourable position to cast frequent, varied and rapid glances at the whole range of special findings with a view to comparison, thereby discovering common features even in what is far apart, where the original finder or novice was still put off by differences. It is especially when a change takes place continuously or in small steps, that the affinity of far distant members of a series becomes noticeable, making one aware of what, in spite of change, has remained the same. Thus, a pair of intersecting lines may appear as a hyperbola, one straight line as two collapsed hyperbolic branches, a segment of a line as an ellipse and so on. Parallels and intersecting lines differ for Kepler only by the distance of their point of intersection. To his younger contemporary Desargues[11] a line is a circle whose centre lies at infinity, a tangent a secant with coincident points of intersection, an asymptote a tangent at an infinitely distant point and so on. All these steps that are by now obvious offered insuperable difficulties to ancient geometers. With the help of the principle of continuity, we reach a higher level of abstraction, and so of ability to grasp analogies. In our geometrical intuition, analogies of continuously variable magnitudes lead to infinitesimal calculus both in its Newtonian and Leibnizian forms, comparing algebraic symbolism with ordinary language gives Leibniz the idea of a universal characteristic or conceptual script and so leads to logical discoveries that are only just coming back to life.[12] Lagrange, at a high level of abstraction, was able to see the analogy between small increments as due to changes of independent variables and small increments as due to variation of functional form, which leads to the creation of the calculus of variation.

8. If an object *M* has marks *a, b, c, d, e* and another *N* agrees with it as regards *a, b, c*, one is inclined to expect that it will agree in *d, e* as well, an expectation not logically justified. For logic only guarantees agreement with what has been fixed, which cannot be contradicted as long as it is retained. However, our expectations depend on our physiological and psychological organization. Inferences from similarity and analogy are not strictly matters of logic, at least not of formal logic, but only of psychology. If *a, b, c, d, e* above are directly observable, we speak of similarity, if they are conceptual relations between marks, analogy is closer to normal usage. If the object *M* is familiar, then looking at *N* will by association remind us of *d, e* alongside the observed *a, b, c*, which ends the process provided *d, e* are indifferent. Not so when *d, e* have strong biological interest because of some useful or noxious property, or are specially valuable for some applied or purely scientific and intellectual purpose. In that case we feel compelled to search for *d, e*, the outcome being awaited with close attention. This will be reached either by simple sense observation or by means of complex technical or scientific conceptual reactions. Whatever the outcome, we shall still have extended our knowledge of *N*, by obtaining a new agreement or difference with regard to *M*. Both cases are equally important and contain a discovery, but with agreement we have the further significant feature of an economic extension of a uniform conception to a larger field; that is why we are especially keen on looking for agreement. This amounts to a simple biological and epistemological account of why we value inferences from similarity and analogy.

9. Considerations of similarity and analogy are in several respects fruitful motives for extending knowledge. A still rather unfamiliar area of facts *N* may show some analogy with another *M* that is more familiar and more accessible to direct intuition: we feel at once impelled to look in thought, observation and experiment for what in *N* corresponds to the known features or relations between features of *M*, and usually this will reveal hitherto unknown facts about *N* which are thus discovered. Yet even if our expectations are disappointed and we find unexpected differences between *N* and *M*, we have not worked in vain: we have come to know *N* better and thereby enriched our conceptual grasp of it. What initiates our use of hypotheses is just this attraction of similarity and analogy:

a hypothesis enlivens our intuition and phantasy and therefore stimulates physical reactions. Besides, the function of hypotheses is partly to be reinforced and sharpened and partly to be destroyed, which in either case amplifies our knowledge.[13]

10. Several equally known areas *M*, *N*, *O*, *P* may enter into analogy, in groups of two or more. Of course these areas have differences as well as agreements, otherwise they would be identical. Hence in analogizing we may prefer to start from now the one and now the other, and this will reveal different analogies each justified in its setting: clearly, this process will show what is adventitious and arbitrary in our conceptions, and which of them are most widely and uniformly applicable and therefore most conformable with the ideal of science.

11. There is no lack of examples for the importance of analogy; it can indeed hardly be overrated in natural science. Even in ancient times water-waves directly visible clarified and made intelligible the process of sound propagation.[14] As to light, the appropriate ideas were developed from the case of sound.[15] Galileo's discovery of the moons of Jupiter was more powerful than any other arguments in supporting the Copernican system: we have here a small scale model of the solar system. Huygens greatly valued this support.

12. The rotation of the plane of polarization of light by electric currents, demonstrated by Faraday in 1845, is one of the most remarkable examples of a great discovery through analogy. Twenty years earlier, J. F. W. Herschel had already surmised this relation between light and electricity, and was guided in his experiments by the right idea, even though the results were negative because the forces applied were too small (letter to Faraday, 9 November 1845).[16] The rotation of the plane of polarization as rays proceed in certain rigid and liquid media gave Herschel the impression of a screw. He therefore looked for a helicoidal dissymmetry in quartz, and indeed discovered it: optical helicoidal dissymmetry thus corresponds to the same feature in the medium. Now clearly a straight electric current shows this screw structure in its associated magnetic field, so that Herschel expected it to influence polarized light rather as quartz does: to test this he passed a ray of light first through a current-carrying

coil (as Faraday later did), and then, in another experiment, parallel to and between two parallel wires with opposite currents, without positive result.

13. Another example, where analogy operates between known areas, is Fourier's theory of heat currents, apparently developed by analogy with water currents. In turn, from his theory of heat conduction other theories were developed by analogy, for instance that of electric currents and diffusion currents. Independently and alongside these, there developed a corresponding theory of attraction by forces at a distance. Comparing these theories that give a comprehensive account of vast areas of fact, many analogies emerge. W. Thomson[17] (Lord Kelvin) compared the theories of heat conduction and attraction and found that formulae of the first go over into those of the second if we replace the concepts of temperature and temperature gradient by those of potential and force respectively, a striking relationship seeing that the original fields seem so different since heat conduction was taken as based on contiguous action and attraction on action at a distance. These thoughts must have inspired Maxwell, since in this same way he recognized that Faraday's theories of electricity and magnetism, based on contiguous action, were just as valid as the theories of action at a distance, then alone recognized by mathematical physicists; thus he came to turn his attention to the great advantages of the former.[18] In similar vein, by recognizing the analogy between the equations of light propagation and electro-magnetic oscillations, he came to found the electro-magnetic theory of light,[19] with subsequent experimental confirmation by Hertz.[20]

14. Maxwell[21] consciously developed the use of analogy into a very perspicuous physical method: he thinks that phenomena are too much lost from sight if we represent our results only in mathematical formulae. If, however, we use a hypothesis, we look through coloured spectacles, and explaining things from a one-sided point of view makes us blind to facts. To him, the phenomena of electrostatics, magnetism, electric currents and so on reveal common traits reminiscent of the flow of a fluid. To complete the analogy, he idealizes the fluid: it has no inertia (no mass) and is incompressible and taken to flow through a resistive medium whose resistance is proportional to the speed of the current. The picture is

imaginary and based on analogy but nevertheless intuitive: we do not take it as real and we know precisely how it coincides conceptually with the facts to be represented. The pressure of the fluid corresponds to the various potentials, the direction of current to that of force and current, the pressure gradient to forces and so on. Without giving up intuition, Maxwell thus succeeds in preserving his open mind and conceptual purity, combining the advantages of hypothesis and mathematical formulation.[22] His picture is such that its mental consequences are pictures of the consequences of facts, to adapt a phrase of Hertz. Thus Maxwell comes close to an ideal method of scientific enquiry: hence his uncommon success.

15. To conclude, we emphasize again that incomplete analogies too can promote enquiry, by revealing differences in the fields being compared. Thus, a theory of energy that merely rested on the conformity of energies would have remained within the bounds of the first law of thermodynamics whereas it is precisely by taking notice of differences that we reach the important recognition of dissipation.[23] A very instructive and historically important example of prematurely dropping a fruitful analogy occurs in Newton's *Optics*, where (question 28) he is discussing Descartes' pressure theory and Huygens' wave theory[24]: having dismissed the first, he pronounces against the second because he finds no refraction of light into a shadow; although he knows that water-waves are refracted more strongly than sound, his experiments were such that the even slighter refraction of light into shadows could easily be missed, while only the opposite kind was noticed, and this last he prefers to attribute to a deviating force emanating from the body grazed. This assumption precludes any understanding of Huygens' work, and so Newton clings to his corpuscular projectile theory, trying to explain everything from the innate and immutable properties of rays, a task he regards as quite difficult enough as it is.

NOTES

[1] Reproduced from Ostwald, *Annalen der Naturphilosophie*, Vol. I, with amplifications.
[2] Jevons, *The principles of science*, London 1892, p. 627.
[3] Maxwell, *Transact. of the Cambridge Philos. Soc.*, Vol. X, 1855, p. 27. (Ostwalds Klassiker No. 69).
[4] Hoppe, *Die Analogie*, Berlin 1873.
[5] W. Stern, *Die Analogie im volkstümlichen Denken*, Berlin 1893.
[6] Tylor, *Die Anfänge der Kultur*, German translation Leipzig 1873.

[7] *P* 3, 1903, p. 277.
[8] Euclid's *Elements*, (quoted in the German from the edition by J. F. Lorenz, Halle 1798.)
[9] Kepler, *Opera*, ed. Frisch, Vol. II, p. 186. The relevant diagrams will be obvious and are omitted.
[10] Cf. Ch. XI.
[11] *Oeuvres de Desargues*, ed. Poudra, Paris 1864.
[12] Cf. Couturat, *La logique de Leibniz*, Paris 1901.
[13] Mach, 'Bemerkungen über die historische Entwicklung der Optik', *Poskes Zeitsch. f. physik. u. chem. Unterricht* **XI**, 1898.
[14] Vitruvius, *De architectura* V, Cap. III, 6.
[15] Huygens, *Traité de la lumière*, Leiden 1690.
[16] Bence Jones, *The life of Faraday*, London 1870, Vol. II, p. 205.
[17] W. Thomson, *Cambridge math. Journal* **III**, Feb. 1842.
[18] Maxwell, *A treatise on Electricity and Magnetism*, Oxford 1873, Vol. I, p. 99.
[19] Maxwell, 'Dynamical Theory of the electromagn. field', *London Phil. Trans.* 1865.
[20] Hertz, *Untersuchungen über die Ausbreitung der elektrischen Kraft*, Leipzig 1892.
[21] Maxwell, *Trans. Cambr. Phil. Soc.* **X**, p. 27, 1855. When I myself mentioned these analogies in a similar way, in the Prague periodical *Lotos* (Feb. 1871) and in *Erhaltung der Arbeit* (Prague 1872), the work of Thomson and Maxwell was still unknown and inaccessible to me. It seems that S. Carnot was the first consciously to have adopted this mode of thought.
[22] Cf. Mach's article mentioned in Ch. XI, note 1.
[23] Cf. *W* 2, 1900.
[24] Newton, *Optice*, ed. Clarke, London 1719, p. 366.

CHAPTER XIV

HYPOTHESIS

1. Isolated facts exist only because of our limited sense and intellectual equipment. Instinctively and of itself, thought spins observation further and completes a fact as regards its parts, consequences and conditions. A hunter finds a feather and his phantasy immediately produces the image of the whole bird that has lost it. A sea current carries exotic plants, animal carcases, finely carved wooden objects, and Columbus visualizes the far-off and as yet unknown land from which these objects originate. Herodotus (II, 19–27) observes the regular floods of the Nile and imagines the strangest causes for these events. Even higher animals are accustomed to draw out observed facts in this way, though in very primitive form. A cat looking for its image behind a mirror has formulated a hypothesis, albeit instinctive and unconscious, as regards its bodily character, and thus begins to test it; but while at this point the cat stops, it is precisely here that man in analogous cases begins to wonder and reflect.

2. Indeed, the formation of scientific hypotheses is merely a further degree of development of instinctive and primitive thought, and all the transitions between them can be demonstrated.[1] In a well-known area of facts, only obvious and customary surmises will present themselves, whose hypothetical nature will hardly be noticeable, although there cannot be said to be a qualitative difference. This is the case in the examples above. Whether Columbus surmises a western continent or Leverrier a perturbing planet exerting a pull in a given direction, in both cases an observation is simply completed in the customary manner according to the observer's every day experience. The newer, stranger and more unusual the initial observation, so also the surmise. However, here too the surmises must be derived from the stuff of experience, however strangely they may be combined. A stroke of lightning and the even rarer fall of a meteorite produces the thought of a thunderbolt hurled by Titans. The remains of mammoths found in Siberia led the inhabitants to the surmise that these were giant rats burrowing subterraneously that died at the first

contact with air. The horn of rhinoceroses in desert country rich in gold were taken for the claws of birds, and this gave rise to the idea of the griffin protecting gold. Deposits of shells at high altitudes suggest the idea of a great flood.[2]

3. Scientific views arise directly out of popular ones, from which they are at first inseparable and then gradually develop away. For physiological reasons the sky appears as a sphere of a certain moderate radius: this is the popular view and also the first scientific one. The nocturnal appearance of it leads us to ascribe rotation to this sphere, with the stars fixed to it to prevent them from falling. Looking more closely we now observe the irregular motions of planets, moon and sun, and this leads to the view of several transparent spheres rotating differently inside each other. Thus gradually emerges the epicyclic theory, the ptolemaïc system, the ancient heliocentric view and the system of Copernicus. That the moon is related to the tides does not escape even the vulgar: so long as enquirers know only pressure and impulse as causes of motion, they believe in an air pressure wave that the moon drives along below itself, but when they become familiar with action at a distance, the pressure is replaced by a pull.

4. The effect of provisional completion of fact in thought is in the first place to extend experience more quickly. The sailor in whose phantasy objects swept up on the coast vividly provoke the picture of a far-off land will go to look for it. Whether he finds it or not, whether its location and character correspond with his idea or not, if instead of the surmised Indian or Chinese coast he discovers a new one, in any case he has widened his experience. Somebody who pursues a mirror image expecting it to be corporeal and fails to find it from now on knows a new kind of visual objects that lack body but whose occurrences requires the presence of other bodily objects. Even in those cases where completion by thought cannot trigger off new experiences, it does at least put the old ones into clearer perspective. Take the case of the mammoth: that it is found in the ground, its flesh still fresh, though never alive, all this follows from the idea one has formed of it. Similarly with the astronomical example. If the completion occurs with vivid sensible intuition and one is convinced that what has been added in thought can be discovered, then this process is particularly suitable for sparking off the requisite activities that will

extend experience: completion in thought is an intellectual experience that drives us on to physical experience.

5. Looking now more closely at natural science, we notice first that anything not as yet directly ascertainable by observation can become the object of completion in thought, of surmise, assumption or hypothesis. We may assume that parts not directly observed in fact exist, as geologists and palaeontologists often have occasion to do. We can make assumptions about the consequences of a fact when they do not at once supervene or elude direct observation. The forms of factual laws are often assumed since it would actually take infinitely many observations, with all interference excluded, to furnish the law. These assumptions or hypotheses relate to the conditions that make a fact intelligible, that is they are explanatory. 'Hypothesis' traditionally means the sum of the conditions under which a mathematical proposition or thesis is valid and deducible or provable from them. The hypothesis is the given, tied to no conditions other than mathematical and logical ones, while the thesis is the inferred. In natural science we have to do the reverse, starting from the given, secure facts and inferring back to the indefinite conditions. This offers many possibilities, the more so the less complete our present experience, which is here an even more important extra-logical factor than in mathematics. A provisional and tentative assumption that cannot yet be established but helps us to understand a range of facts, we call a hypothesis.[3] As regards its provisional character, this may last a mere instant (as in the example of the mirror image) or a century or a thousand years (as in the corpuscular theory of light and the ptolemaïc system respectively): the duration does not alter the psychological and logical nature of a hypothesis.

6. Newton was decidedly averse from hypotheses. His first philosophic rule for enquiry states that to explain nature we must admit only causes that are actual and such as are sufficient to account for the phenomena[4] – a clear hint not to invent explanations if what is in fact known provides adequate understanding. Somewhat later he adds characteristically that he had been unable to deduce from the phenomena the reason for gravitation, and would not make up any hypothesis: what cannot be deduced from the phenomena must be called a hypothesis and these have no place

in experimental philosophy whether they be metaphysical, physical, of occult qualities or mechanical; in that philosophy propositions are deduced from phenomena and generalized by induction.[5] The often-quoted "hypotheses non fingo" may here properly be applied to gravitation: from the phenomena Newton had derived the inverse square law, which is actually operative and this is not an hypothesis. What causes this feature of gravity, however, he does not know and cannot derive from the phenomena, and he declines to offer any invented explanation. That this is his view is quite clear from two letters to Bentley (17 January 1692/93, 25 February 1692(93), where indeed he declares unmediated action at a distance to be an absurdity and suggests that gravitation must be caused by some agent, material or immaterial, but he offers no view on this.[6] 'You sometimes speak of gravity as essential and inherent to matter. Pray do not ascribe that notion to me; for the cause of gravity is what I do not pretend to know, and therefore would take more time to consider of it.' (Jan. 17, 1692–1693.)

'It is inconceivable, that *inanimate brute matter* should, without the mediation of something else, which is not material, operate upon, and affect other matter without mutual contact; as it must do, if gravitation, in the sense of Epicurus, be essential and inherent in it. And this is one reason, why I desired you would not ascribe innate gravity to me. That gravity should be innate, inherent and essential to matter, so that one body may act upon another at a distance through a vacuum, without the mediation of any thing else, by and through which their action and force may be conveyed from one to another, is to me so great an absurdity, that I believe no man who has in philosophical matters a competent faculty of thinking, can ever fall into it. Gravity must be caused by an agent acting constantly according to certain laws; but whether this agent be material or immaterial, I have left to the consideration of my readers.' (Febr. 25. 1692–1693.)

7. Newton's attitude and way of proceeding thus seem clear: led to the assumption that masses act on each other from afar (as the earth does on bodies falling on to it) according to the inverse square law, he found by his mathematical investigations that this explained all motion on earth and in the planetary system, so that the assumption ceased for him to be a hypothesis: it became a result of the analysis of phenomena. This he

distinguished sharply from the question whether action at a distance itself could in turn be explained by being reduced to something simpler: that last point only remained the object of speculation or 'hypothesis'. It would certainly have greatly impaired scientific progress to confuse these two matters as equivalent or to refrain from uttering the assumption of action at a distance because it was really or apparently inexplicable. However we cannot maintain the view that Newton's rejection of hypotheses was confined to mechanics and gravitation, for in optics too he speaks slightingly of hypotheses,[7] even if he himself develops many of them (though always marking them as such and keeping them carefully apart from fact). His 'analytic' method is to conduct experiments and observe phenomena, generalizing from them by induction, and disregarding all hypotheses.[8]

8. Much effort has been spent on trying to reconcile Newton's preaching with his practice. Even if this were not entirely possible, it would not matter. In a euphoric mood, even great men may say or write things beyond their power to abide by. Several such cases occur in Newton and many in Descartes. However, what he said and did as an enquirer is quite understandable. If one took the "hypotheses non fingo" unconditionally, it would mean "I surmise nothing beyond what I see and never reflect beyond observation". This view is refuted on every page of Newton's writings: what sets him apart from others is precisely a wealth of surmises, among which he is quick to sort out by experiment those that are useless and do not stand up to test. What cannot be derived from the phenomena he calls a hypothesis. Therefore what can so be derived cannot in his sense be a hypothesis, but a result of analytic investigation, to use his mode of thought. He does indeed use images to make his thought more intuitive but he does not attach any special value to them. If we could ask him what he regarded as essential in his idea of the polarization of light, he might reply the different sides of the ray, because they were a result of analytic investigation, while the corpuscles with magnetic properties were an unimportant intuitive picture that might just as well be replaced by some other. This sharp distinction in principle between, and very different estimate of, definitely established knowledge and mere surmise or intuitive representation, is found everywhere in Newton. Errors in detail are unimportant as against this general tendency.

9. Different authors have tried to specify the conditions that a good hypothesis in natural science must satisfy. J. S. Mill[9] has discussed the matter at length. His demand that a hypothesis must rest on the assumption of an explanatory cause that is already known to be present, a true cause in Newton's sense, has been exhaustively shown to be untenable by F. Hillebrand[10]: one cannot follow Mill's principles without constantly contradicting them, for as soon as conscious research began it would permanently explain the current lack of knowledge and no further essential discovery would be possible, at least not by means of thought.[11] Jevons, who strikes the enquirer into nature as someone fully conversant with the topic, thinks it enough if a hypothesis agrees with the facts.[12] Examples illustrate the point better than abstract explanations.

10. The essential function of a hypothesis is that it leads to new observations and experiments, which confirm, refute or modify our surmise and so widen experience. Priestley, in his history of optics, offers very sound views on this.[13]

The very imperfect views and conclusions of the philosophers of this period exhibit an amusing and instructive prospect; as they demonstrate that it is by no means necessary to have just views, and a true hypothesis, *a priori*, in order to make real discoveries. Very lame and imperfect theories are sufficient to suggest useful experiments, which serve to correct those theories, and give birth to others more perfect. These then occasion farther experiments, which bring us still nearer to the truth, and in this method of approximation, we must be content to proceed, and we ought to think ourselves happy, if, in this slow method, we make any real progress.

The point is perhaps best explained by a procedure for approximating to the roots of an equation. Suppose we wish to solve $f(x)=x^4+ax^3+$ $+bx^2+cx+d=0$ and we first try the value x_1 which gives $f(x_1)=+m_1$ instead of 0. Another substitution x_2 now gives, say, $f(x_2)=-m_2$. Then we can look for a root between x_1 and x_2. If we have found a value x' such that $f(x')=\mu$ is small, we can take $x-x'$ as proportional to μ and so approximate indefinitely to the true value x of the root.[14]

11. Consider the caloric hypothesis, which contains an intuitive idea that is an imaginary mental feature added by association to the sensible characteristics of the heat of a body. By observing fire and the way a body is warmed by another warmer than it and at the latter's expense, we

develop the idea of a substance or fluid in quite naïve, natural and involuntary fashion. This idea represents in vivid pictures the facts that have provoked it, and moreover, by meeting observation halfway, it facilitates the discovery of new facts, such as Richmann's rule of mixtures, the difference between specific heats, the latent heats of vaporization and liquefaction. In a very similar way we form the ideas of electric fluids guided by the facts as to the transmission of electric states, spark formation and so on. The idea of a fluid, mobile in a conductor and held in an insulator, producing forces of attraction and repulsion, not only gives an intuitive account of known facts, but promotes the finding of new ones; such as the charges on conductors residing on the surface, charge distribution as a function of curvature, induced charges, even the quantitative laws of Coulomb. How largely such ideas continue to be valuable as indirect descriptions[15] long after being superseded and no longer taken seriously can be seen for example from the fact that we still think of the production of a certain quantity of electricity, according to Faraday's laws of electrolysis, as being proportional to the quantity of matter dissociated.

12. The emission theory of light is another stuff hypothesis. Observing rays, their concentration and dilution with increase and decrease of brightness, we are quite automatically led to regard a ray as a jet of liquid, dust or corpuscles, and it is only the transience of light that eventually comes to stand in the way of this hypothesis. The great adaptability of hypotheses to fact shows itself in that the stuff hypothesis of light that seems so clumsy today did not prevent Malus from finding the so-called squared cosine law of the division of a polarized ray into two components at right angles: what Fresnel deduced from the conservation of the kinetic energy of light, Malus probably obtained by following the tacit idea that in decomposition the quantity of light stuff must remain constant, which again simply requires that law. Jevons[16] is wrong in separating such stuff hypotheses from properly explanatory ones as merely descriptive. Every hypothesis must begin by representing the fact for which it has been framed: this much follows even from the one requirement that Jevons himself lays down. It is a matter of luck how much or how little the hypothesis reaches beyond the fact that gave rise to it and how fertile it is as regards new discoveries.

13. In framing a hypothesis one tries to do justice to the properties of a factual situation under the special restricted conditions just revealed by observation, without knowing beforehand whether these properties will continue to hold under other more general conditions; that is, without knowing whether the hypothesis will continue to hold and how far it will hold in general. The matter or elements for hypothetical ideas we can borrow only from our current sensible surroundings, by noticing cases that offer similarity or analogy with prevailing ones. Yet similarity is not identity, but partial sameness and partial difference. That alone ensures that an hypothesis based on analogy will with wider experience apply in some cases and certainly not in others. Thus, it lies in the nature of hypotheses to be changed in the course of enquiry, becoming adapted to new experience or even dropped and replaced by a new one or simply by complete knowledge of the facts. Enquirers who keep this in mind will not be too timid in framing hypotheses: on the contrary, a measure of daring is quite beneficial. Huygens' wave hypothesis was not a perfect fit and its justification left much to be desired, causing not a little trouble even to much later followers; but had he dropped it, much of the ground would have been unprepared for Young and Fresnel who would probably have had to confine themselves to the preliminary run up.

14. The hypothesis of emission in optics gradually adapts itself to the growing field of experience. Grimaldi is no longer satisfied with a uniform jet. His refraction strips lead him to the view of a wavelike flowing away of the luminous fluid, probably in analogy with pressure waves. Newton no longer thinks in terms of one simple jet but of many qualitatively different ones merged with each other; he even manages to account for the periodicity of light, even if inadequately and clumsily and partly on the basis of false factual premisses. In the end wave hypothesis openly replaces the emission theory. At first, in Huygens' form, it ignores periodicity and polarization. Hooke does indeed introduce periodicity, but fails to relate this in the proper way with colour, not to mention other flaws in his account. At last, Young and Fresnel combine in their hypotheses the advantages of Huygens and Hooke; Fresnel particularly manages to remove the flaws of both and to bring in new properties regarding polarization. Thus experience constantly works at transforming and perfecting our ideas.[17]

15. However, the ideas which we have formed, in turn exert their influence on the course of experience. Grimaldi's strips lead us to ascribe periodicity to each single ray, although we cannot observe this directly but only in combinations of rays under specially favourable conditions. This idea becomes very lively and intuitively clear by means of the wave hypothesis. By maintaining the idea of periodicity found in one particular instance in all cases where there are light rays, we enrich every optical fact by this addition. To every case we add in thought more than can be seen in it, namely the content of Grimaldi's idea. A physicist thus prepared will henceforth behave differently towards individual cases, just as anyone with enhanced experience would in ordinary life. He will expect more and other things and organize his experiments differently. Thus it becomes intelligible that Fresnel who is constantly aware of Grimaldi's experience, thinks and experiments differently from Newton, Huygens and Malus as regards refraction, the colour of thin plates, reflexion and polarization.

16. Apart from the elements essential for representing the facts from which a hypothesis has been derived, the latter always or at least usually contains other elements that are not. For the hypothesis is framed by analogy, whose points of identity and difference are incompletely known, since otherwise there would be no need for enquiry here. For example, the theory of light speaks of waves, whereas only periodicity is needed to understand it. These further elements beyond the necessary are precisely the ones that are subject to change in the mutual reactions of thought and experience, until they are gradually eliminated in favour of necessary ones. Thus nothing remains of the idea of emission save the high velocity of propagation of many different kinds of light of different periodicity within the one ray. This idea coincides in essential points with the wave hypothesis that came to replace it, although that in its turn eventually had to drop the accessory elements of longitudinal vibrations derived from acoustic analogies.

17. The ideas we have formed on the basis of our observations arouse expectations and urge us towards new observations and experiments. This strengthens the tenable elements and casts out untenable ones, modifying or even replacing them by new ones, special importance attaches to those experiments that force us to decide between two ideas or groups

of ideas that both represent the facts. The question whether colours arise through refraction or exist already beforehand and become visible because of their different refractive indices was settled by Newton in a "crucial experiment". The term was introduced by Bacon, and adopted by Newton, for experiments deciding between two such views. An important example is Foucault's experiment for showing that the speed of light is smaller in water than in air, which makes the emission theory untenable and decides in favour of the vibration hypothesis. Galileo's discovery of the phases of Venus decided in favour of the Copernican system from which these phenomena were deducible; similarly for Hooke's observation of the expected deviation of free fall from the vertical and Foucault's pendulum experiment.

18. A hypothesis can be problematic in very different ways and degress. To explain suction in pumps, the well-known hypothesis that nature abhors a vacuum was excogitated. If we nowhere under any circumstances met a vacuum this view might be maintained. Another hypothesis bases the same phenomena on the pressure due to the weight of air. Although the weight of air had by then been demonstrated, this explanation nevertheless remained a hypothesis until Torricelli's experiment and the work of Pascal, especially the test on the Puy-de-Dôme, showed that all the phenomena in question could be explained without exception and that there is neither call nor scope for another parallel explanation. Although one explanation is, bluntly put, a free invention and the other operates only with factual elements, both are hypothetical when first put forward. Another example is the explanation of cosmic motions by means of gravity. The idea of the factually given gravitational acceleration suitably generalized is introduced into astronomy. I cannot agree with F. Hillebrand[18] that hypothesis played no role in Newton's theory of gravitation. It is indeed true that in the finished account everything is reduced to the appropriate description of cosmic motions in terms of accelerations, and the acceleration of a particle near the earth's surface neatly goes over into that of terrestrial gravity as a special case, so that we need no hypothesis. It is logically conceivable that somebody who analyses the kinematics of Kepler's motions should adopt the notion of describing them by centripetal accelerations inversely proportional to the square of and along the solar radius, but to me this seems psychologically unthink-

able. How is somebody without guiding physical ideas going to stumble precisely on to accelerations, why not on the first or third differential coefficients? How is anybody going to select among the infinitely many ways of resolving the motion into two directions the very one that yields so simple a result? Even parabolic projectile motion is difficult to analyse without the guiding idea of gravitational acceleration, which could be gained only from a much simpler case and yet is applied here.

19. Science in its development moves among surmises and parables, there is no denying it; but the more it approaches perfection, the more it goes over into description of fact only. Analogies between one fact and others helps us to look for new properties. Whether in this we find new agreements or differences, in either case our experience is enhanced: both indicate in equal measure new conceptual determinations of the properties of facts. That an enquirer should start where his predecessors left off, thus avoiding any loss of previously acquired experience, is here just as important as the change of enquiring individuals, peoples and races that guarantees a many-sided and unprejudiced outlook.

20. Thus in its self-destroying function, a hypothesis in the end leads to the conceptual expression of facts. Let us recall the long sequence of assumptions and corrections that led to the view of light as a transverse vibration, which was at first quite adventurous and without analogy and therefore regarded as doubtful. Yet the recognition that the periodic properties of rays behave like summable segments in a two-dimensional space (the plane at right angles to the ray's direction) is merely a conceptual expression of the facts. In the same way the properties of the aether, or light-propagating space, behaving partly like a fluid and partly like a rigid body, were gradually determined conceptually. The resulting views are then no longer hypotheses, but presuppositions of the intelligibility of facts and results of analytic investigation. These we can retain as certain, even if we can find no analogy for them and never encounter any transverse vibrations or fluids that could sustain them. Had Young and Fresnel kept quiet about assuming transverse waves because they are difficult to explain, science would have suffered as serious a loss as if Newton's law of gravitation had been suppressed because of analogous doubts. We must not shy away from unusual views if they are soundly

based. For the possibility of finding fundamentally new facts existed not only in earlier periods of enquiry, it goes on existing and has never ceased for a day. Mill's rules for restricting hypotheses imply a great overestimate of what has already been found as against what remains to be investigated.

21. If our thinking were abstract enough, we should ascribe to a fact only those conceptual marks that it must have. We should then never have to eliminate anything, but by the same token lack inspiration for new experiments through intuitive analogies. Such a purely conceptual representation can be used for completed parts of science, where there is no room for hypotheses that have a beneficial role only in growing areas. The use of pictures deliberately employed is here not only not excluded but highly appropriate. There are facts that we perceive directly through the senses, surveying them at a glance as it were. Other facts do not appear until we apply a complicated system of observations, concepts and reactions. That light is periodic is not seen immediately; indeed the extreme shortness of the periods makes it difficult to come to grips with the fact. Polarization, likewise, is not immediately recognized. Since we are much more familiar with intuitive ideas of sense and are more adept at operating with them than with abstract concepts that are in any event based ultimately on intuitive ideas, instinct alone tells us to imagine a light ray as a wave of an intuitively sizable wave length with a definite plane of vibration related to the reflecting surface of a polarizing mirror, such that under analogous tests the wave would behave like that light ray. By means of such ideas we obtain a much quicker conspectus of optical phenomena than by abstract concepts. These ideas are pictures of facts whose mental consequences are pictures of the factual consequences, to adapt a phrase of Hertz. Once we have accurately determined wherein a picture conceptually coincides with the facts, it combines the advantages of intuitive clarity with conceptual purity. It now lends itself to taking on without reluctance such further determinations as may be required by new facts, say of electrodynamics or chemistry.

22. Although there is a widespread opinion that in mathematics hypotheses have no role to play, let us emphasize that on the contrary they do, as in any growing field of science. What gives rise to this false view is the fact that mathematicians more than others tend to eliminate all trace of

development as soon as they present their findings. The perfectly clear recognition of mathematical propositions is by no means attained all at once, but is preceded and prepared by incidental observations, surmises, thought-experiments and physical experiments with counters and geometrical constructions, as mentioned earlier and to be further discussed.[19]

NOTES

[1] Cf. *P* 3, p. 256.

[2] Cf. Tylor, *Urgeschichte*, pp. 398–403.

[3] This is a slight modification of the formulation given by P. Biedermann, *Die Bedeutung der Hypothese*, Dresden 1894, p. 10. In this excellent treatise it is shown that what in scientific thought is called hypothesis and in ordinary thought conjecture are closely related. In any case we can speak of supplementing the facts in imagination or thought; though if this done deliberately and consciously, the expression 'conjecturing' or 'assuming' is more appropriate.

[4] *Philosophiae naturalis Principia mathematica*, Lib. III Regulae philosophandi, reg. 1.

[5] *Ibid.*, Lib. III, Sect. V.

[6] *Newtoni Opera*, ed. Horseley London 1782, Tom. IV, pp. 437–438. In his correspondence with Bentley, Newton's aim is to gain a proof of the rule of divine wisdom from the ordering of the universe. His expression 'inanimate brute matter' shows clearly that Newton regarded animate matter as something quite different, thinking it more versatile than the former. This dualism, so ingrained from our savage ancestors down, is not overcome even today. W. Thomson too, in his work *On the dynamical theory of heat* (1852), felt bound to say "it is impossible, by means of inanimate material agency, to derive mechanical effect from any portion of matter by cooling it below the temperature of the coldest of the surrounding objects". Even H. Hertz (*Die Prinzipien der Mechanik*, 1894), who assumes that all physics can be put on a mechanistic-atomist basis, regards it as necessary, 200 years after Newton, explicitly to confine this view to inanimate nature (p. 165). Finally, Boltzmann (1897) discusses the question of 'the objective existence of processes in inanimate nature'. I freely admit that 'inanimate' matter seems no less puzzling to me than animate, and that I regard the contrary view as a residue of some old superstition. As long as it is believed that all physics can be exhaustively treated by mechanics, regarding the latter in turn exhausted by the simple doctrines discovered up to the present, life must indeed appear to be something hyperphysical. However, I reject both these views.

[7] If Newton's opposition to hypotheses seems exaggerated, it may be more readily understood in the light of their misuse in research in Descartes' time.

[8] *Newtoni Optice*, London 1719, pp. 412, 413.

[9] J. S. Mill, *Induktive Logik*, ed. Gomperz, 1885, **II**, pp. 208–225.

[10] Hillebrand, 'Zur Lehre von der Hypothesenbildung', *SB. d. Wiener Akademie, philos.-histor. Cl.* **134**, 1896.

[11] Cf. also A. Stöhr, *Leitfaden der Logik*, pp. 172f.

[12] Jevons, *The principles of science*, 1892, p. 510.

[13] Priestley, *History and present state of discoveries relating to vision, light and colours*, London 1772, Vol. I, p. 181.

[14] A review of the present volume by Prof. G. Vailati in *Leonardo* has drawn my attention to three small dissertations by G. L. Le Sage, 'Sur la méthode d'hypothèse', and

two supplements on analogy and exclusion, reprinted in P. Prevost in Vol. II of his *Essai de Philosophie*, Genève, An. **XIII**, pp. 253–335. Le Sage indeed explains the logical aspect of the use of hypotheses very well in mathematical examples. However, the psychological importance of hypotheses seems not so well acknowledged. What is moreover interesting for a German reader is the circumspect character of Prevost's philosophy which never loses touch with positive science at a time when in Germany the unrestrained demon of speculation conquered all academic chairs. I am obliged to Prof. Th. Flournoy in Geneva for letting me see this now almost unobtainable book.

[15] Cf. *P* 3, pp. 267f.

[16] Jevons, *l.c.*, pp. 522f.

[17] Duhem (*La Théorie physique*, pp. 364f.) argues that enquirers do not so much choose arbitrary hypotheses at will, but that in the course of historical development, under the impact of gradually discovered facts, hypotheses obtrude themselves on enquirers. Such an hypothesis usually consists of a whole complex of ideas. If now a result supervenes that is incompatible with the hypothesis, for example by a crucial experiment, we can at first regard this result only as contradicting the whole complex. On this latter point cf. Duhem, *l.c.*, pp. 311f.

[18] Hillebrand, *l.c.*

[19] For detailed accounts of hypotheses closely linked with special sciences and their degree of development, see E. Naville, *La logique de l'hypothèse*, 2nd ed., Paris 1895.

CHAPTER XV

PROBLEMS

1. When the results of partial mental adaptations fall into such oppsition that thought is driven in different directions and disquiet mounts to the point that we consciously and deliberately seek for a thread to lead us through this confusion, then a problem has arisen. A stable and customary area of experience to which thoughts have adapted themselves rarely gives rise to problems; at least it would require great mental sensitivity to differences if here too problems were to arise. However, if the area of experience becomes wider through certain circumstances and thoughts come into contact with hitherto unknown facts to which they are inadequately adapted and if the thoughts modified by novel adaptation react on the results of earlier adaptations, then a host of new problems develops, as the history of civilization in general and of science in particular shows. Problems arise when thought and fact, or thought and thought no longer agree. We have not the power to adduce hitherto unknown facts that depend in unknown ways on circumstances within our sphere; they meet us against our will, without or against our expectations, and although they lie outside the scope of our work or investigation, they arise from chance, that is through circumstances that may not be without rule but beyond our ken and influence. Moreover, it is mental chance that brings thoughts together that may long have lived in an individual without ever coming into mutual contact, and so failed to come near enough to react and thereby create a problem. In most cases, chance unveils the remaining incongruities between thought and fact or thought and thought, thus promoting further adaptation by making these flaws felt.[1] Forming and solving problems thus involves chance not in a minor role but as a central aspect in the nature of the case.

2. Once the incongruity is clearly recognized and the problem posed, we must seek the solution. The intellectual activity of a man who, with definite aim and purpose, is looking for a solution of which he knows only certain properties while ignorant of others, is like that of recalling some-

thing forgotten, as William James[2] has aptly remarked. What is forgotten was once known, and, upon recall, is at once correctly recognized. By contrast, the solution looked for is new, and it requires a special test to show that it is correct: that is the difference between the two cases. If one is trying to recall a forgotten solution, say a mathematical substitution, the second case changes into the first and easier one. Suppose I want to recall a quotation important to me here and now, when I have forgotten the exact words or source: I think of the time and occasion when I first learnt it, of the matters that occupied me at the time and the related writings that I might have read, of the authors whose mode of thought might correspond to the quotation, of the place where I studied and the means and inspirations afforded by my surroundings and so on. Just so I should behave if I were looking for a long unused instrument that I had mislaid. The more numerous and stronger the available associations that lead to the forgotten, the easier it will be to use one or several of them to drag the latter into the light of consciousness.[3]

3. Fairly close to this is the case of rediscoveries of an invention following news of its existence, as a remarkable historical example will illustrate. In Venice, Galileo learnt that the Dutch had invented an optical instrument that showed distant objects as nearer, larger and more distinct.[4] The night after his return to Padua he succeeded in improvising a telescope using a leaden organ pipe and two lenses, of which he sent instant news to his friends in Venice, with whom he had discussed the matter the day before. Six days later he was able to exhibit a much more perfected instrument in Venice. Galileo admits that without the news from Holland he might never have thought of such a construction, but contests the objection that merely knowing that the Dutch instrument existed did very much to detract from his own invention, as his opponent Sarsi would have people believe: let them try to re-invent Archytas' flying dove, or Archimedes' burning mirrors and so on. He appeals to public opinion by describing the line of thought that had led him to the reconstruction: the instrument might consist of one or more pieces of glass; a plane piece is ineffective, a concave one reduces, a convex one, while magnifying, gives hazy images; therefore one piece is insufficient, and proceeding to two, leaving aside the plane piece, he succeeded by trying a combination of the remaining two types. The last step he seems to have taken in quite a

groping fashion, as was natural at the time. Kepler[5] had indeed found the correct theory of the eye as early as 1604, but a more complete account of dioptrics, in particular as regards a systematic survey of the properties of lenses he could not provide until 1611, two years after Galileo's invention and probably with its help.[6] For the rest, Galileo's line of thought was not free from subjective chance elements, it might well have come out otherwise and particularly in a more general and comprehensive manner. Suppose we know only the real images of convex lenses, the empirical properties of reading and magnifying glasses and convex and concave spectacles, all of which were then known. These are sufficient for the following reflection: one convex lens of large focal length whose real image can be clearly seen from a distance less than this, already constitutes a (Kepler) telescope, whose eyepiece is replaced by the eye itself. If we further approach the image and prevent it from becoming hazy by using a magnifying glass in front of the eye, we have an actual Kepler telescope. If we go past the image near the object lens, a concave glass before one's eye can restore clear vision, and we have the Dutch telescope. If therefore we regard magnitude and clarity of image as the aim of the construction, we reach all possible solutions. Galileo's path remained restricted probably through his eager haste in rediscovery; his lucky and of course chance discovery precisely of the Dutch form became immensely valuable through his ingenious idea of applying it to the observation of celestial bodies.

4. That we here put invention and scientific problem solving on a level need cause no surprise: indeed the only difference between them, and that not always easily maintained, is the practical as against the theoretical aim. There are many instances in the history of science and technology where information about the success of predecessors has occasioned further identical or different solutions of the same problem. They would be even better known if the rediscoverers were less secretive, no doubt because of the suspicions they meet with. Nor is the multiple solution of a problem superfluous; on the contrary, it is very beneficial, since it usually illuminates the same question from different angles. Thus the accidental discovery of the Dutchman Lippershey inspires the more scientific one of Galileo and the quite different approach of Kepler. Whether the second or third inventor has an easier time of it depends

entirely on the scientific vision, intellectual tools and experience that he happens to have.[7] Even a multiple posing of the same problem from different sides without solution is not indifferent to science, especially if at the time when the problem arises it is as yet regarded as unsolvable or even absurd. In such cases competitors encourage each other, by no means the least prerequisite for success.[8]

5. Before considering further special examples, let us look at the methods of problem solving in general. These universally applicable methods, invented and developed by the philosophers of ancient Greece in connection with the simple and perspicuous subject of geometry, remain an important part of the method of scientific enquiry. Proclus, in his commentary on Euclid, ascribes the main merit to Plato. The three methods mentioned are the analytic (starting from the result and working back to admitted premisses), the synthetic (starting from admitted premisses and working forward to the result) and the indirect method or reductio ad absurdum (showing that the contradictory of the result is impossible).[9] We must not suppose that Plato invented all these single-handed, since they were partly used before his time, but Diogenes Laertius explicitly credits him[10] with having introduced the analytic method and passed it on to the geometer Laodamus of Thasos. These three methods can be used in enquiry as well as in demonstrating what is already known. Moreover,

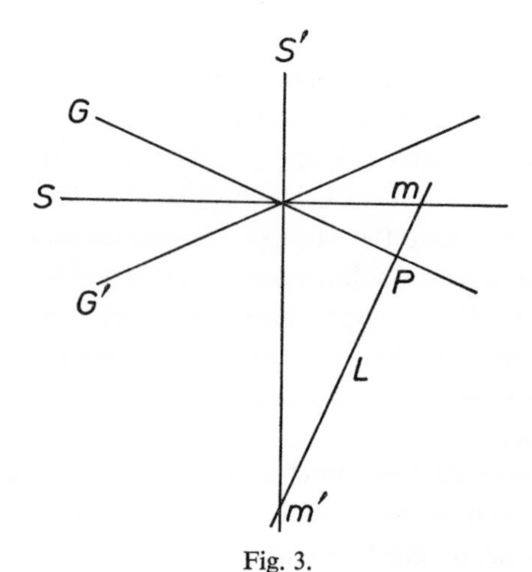

Fig. 3.

while the analytic and synthetic methods exclude each other, each can be used directly or indirectly.

6. A simple example will illustrate the synthetic method: construct a circle touching two intersecting coplanar lines G, G', one of them at the point P (Figure 3). Because of symmetry, the centre of circles touching two such lines must lie on one or other of the bisectors S, S'. Since P is one of the points of contact, the centres must lie on the line L normal to G at P, which determines the only two centres m, m', intersections of L with S, S'. The respective radii are mP, $m'P$. The example shows how the various conditions that the solution must obey are separated to draw from each the requisite consequence for the solution. Moreover, we see that a scientific procedure differs from mere trial and error which might solve the problem at least approximately, in that we go ahead in a planned way, carefully using what is already known or established once and for all. We look only in families of circles that already satisfy the separate conditions. Finally, we notice that the scientific procedure is not essentially different from ordinary puzzle-solving, except that in the latter case the field is usually wider and less well known or previously explored, so that planned searching is more difficult. Any problem of geometrical construction can easily be presented in the garb of a puzzle, as was well known to those Indian mathematicians who went so far as to pronounce their problems in verse.

7. Suppose we had to solve this same problem without prior knowledge of the theorems used. According to ancient practice, amplified by some hints from Newton,[11] we should then proceed by the analytic method, regarding the problem as solved and starting by drawing an arbitrary circle with two tangents G, G' and marking the point of contact with G as P. By examining the connections of the centre m and the radius mP with the tangents and points of contact, we are led to those theorems that give us the reverse procedure from G, G' to m and mP and so the construction.

To illustrate the value of the analytic method, consider a somewhat more difficult example: construct a circle touching the lines G, G' and passing through an arbitrary point P (Figure 4).[12] Suppose the circle touching G is given, its centre C is thus on the bisector S, and the line CP must be equal to the perpendicular CH on to G, that is equal to the radius r.

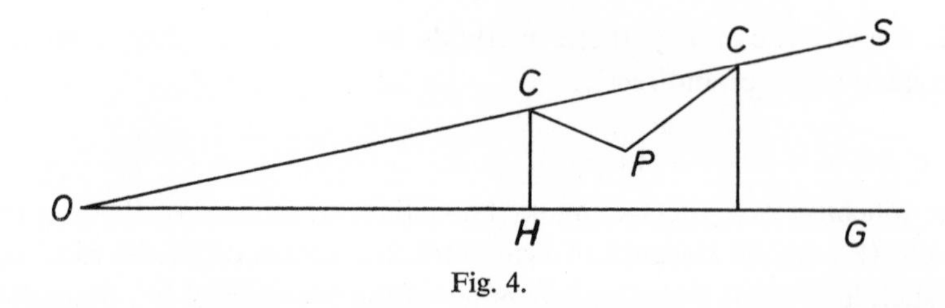

Fig. 4.

If from this we can find C, H, or r, the problem will be solved. By moving CH past P, we see that there are two solutions. Let us express the condition as an equation, using G as the axis of abscissae, putting tan $SOG=a$, and denoting the coordinates of C by x and $y=ax$, and of P by m and n. Then

$$a^2x^2 = (x - m)^2 + (ax - n)^2$$

or

$$x = (m + an) \pm \{(m + an)^2 - (m^2 + n^2)\}^{1/2},$$

which gives the construction of $x=OH$. Without calculation and using the ancient method of drawing, we can find the solution by taking into account the point P', symmetrical to P with regard to S and drawing the line $P'PQ$ (Figure 5), and then constructing the point of contact H

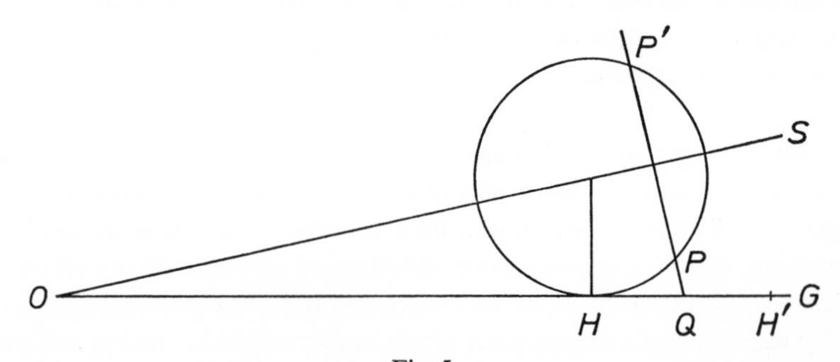

Fig. 5.

according to the theorem $\overline{QH}^2=QP\cdot QP'$. The second solution is obtained by taking $QH'=QH$. However, the simplest and most elegant solution derives from the simple observation that there are infinitely many constructions that with regard to O are similarly situated to the required one. If therefore, we draw the line OP (Figure 6) and any circle K with its

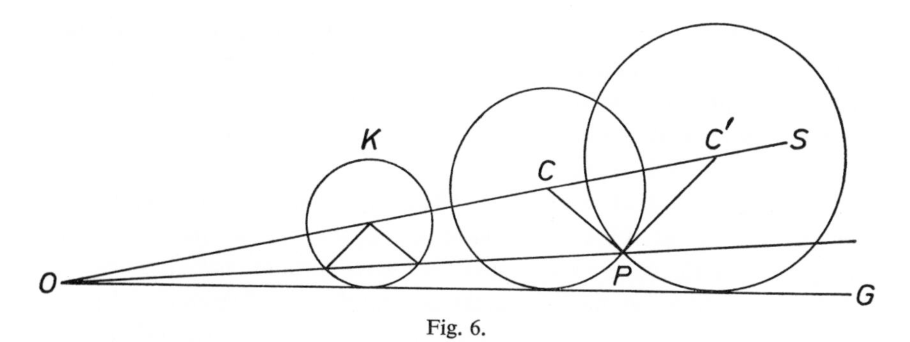

Fig. 6.

centre on *S* and touching *G, G′*, its intersections with *OP* may be regarded as points homologous to *P*. The parallels through *P* to the two radii thus lead to the required centres *C, C′*.

8. It must be a happy mental instinct peculiar to ingenious intellects that led Plato to discover the analytic method. One knows only what one has accidentally experienced before through the senses or in thought. In a field where one has no experience one cannot solve problems. To reduce the unknown to a minimum there is no better method than to imagine the sought for and the known combined in a case that is already familiar, and then in construction to retrace the path, now more readily visible, from the latter to the former. This holds good not only of geometry. If you want a tree trunk laid across a stream in order to walk over, you imagine the problem solved: by considering that the trunk must be dragged into place, but first the tree must be felled and so on, you tread the path from the sought for to the given, which in actual construction of the bridge he has to traverse in the opposite direction, reversing the sequence of operations.[13] This is a case of very ordinary practical thinking. Most great engineering inventions seem to rest on this process, insofar as they were not gradually provided by chance but rapidly called into existence by spontaneous effort. Fulton imagines a fast moving ship, with continuously rotating paddle wheels (by analogy with land vehicles) rather than rhythmically acting oars, a steam engine driving the wheels and so on. One can show likewise that the greatest and most important scientific discoveries owe their origins to the analytic method, although we cannot quite exclude the intervention of synthetic procedures. Once again the intellectual

activity of the enquirer and inventor shows itself as not essentially different from that of the common man. What the latter does by instinct, the enquirer erects into a method. However, this method became conscious already in the most ancient and simple exact natural science, namely geometry.

9. Before proceeding to examples of analogous methods of enquiry in natural science, let us take a further look at geometry. The first geometric insights, even the more complicated ones, were certainly not obtained by deduction, which belongs to a more advanced level of science and presupposes a solid body of knowledge and a demand for simplification, order and system. Rather, such insights were obtained, as in natural science, through practical needs of exact observation, by measuring, counting, weighing, estimating; through intuition and only later through deduction from prior knowledge, by speculation or thought experiment under the guiding principles of comparison, induction, similarity and analogy. Here the writings of Archimedes,[14] a relatively late ancient enquirer, are very instructive. He tells us that he and others knew various theorems before they found exact forms and proofs. For example, the quadrature of the parabola was approximately obtained by covering the drawing with thin sheets cut out and weighed. From the results Archimedes guessed the exact law and later succeeded in proving it. Even in modern times such problems are first empirically found and solved by approximation, and later treated exactly. In 1615 Mersenne drew the attention of mathematicians to the method of generating cycloids. Galileo was able only to use weighing to show that the area of the curve is approximately three times that of the generating circle, and in 1634 Robeval proved that this was exactly so.

10. If we form a surmise concerning the existence of a certain proposition *C*, we can try to derive it by forward synthesis from known propositions, but this requires fairly secure information about the foundations. Otherwise we are likely to try working back analytically to the adjacent condition *B* of *C*, and then to the condition *A* of *B*. If *A* be known or self-evident, then we have found the deduction "*A* entails *B* and *B* entails *C*". If not-*C* follows from *B*, and *B* from *A* and *A* turns out to be impossible, once again *C* would be proved. This last result is unconditional. On the other hand, if the analysis is undertaken for the sake of a direct proof

we have to make sure that the propositions "*C* is conditioned by *B*", "*B* is conditioned by *A*", and so on, are all convertible, for only then can the reverse sequence be regarded as a proper proof of *C*. Not all propositions are convertible: from *N* conditions *M* it does not follow that *M* conditions *N*. Take for example: in a square (*M*) the diagonals are equal (*N*). The converse: two equal diagonals (*N*) define a square (*M*), is obviously false. To obtain a converse we should either have to widen the concept *M*, replacing it by *M′* which comprises all the many quadrilaterals with equal diagonals, for which so far no name has been invented, or we might specialize *N* to some *N′*. This last procedure would lead to the convertible proposition: in a square (*M*) the two equal and mutually perpendicular diagonals intersect at their midpoint (*N′*). Congruent figures are similar, but similar figures must be equal in area to be congruent. Two equal sides in a triangle lie opposite equal angles and conversely. These examples will suffice to show that care is needed in applying theoretical or problematic analysis.

11. It has often been justly regretted that the enquirers of antiquity have told us so little about their methods of invention and investigation and have indeed concealed the pathways of research by synthetic exposition. Against this, Ofterdinger has emphasized that synthetic representation has certain advantages as regards system. Careful scrutiny of Euclid's proof of Pythagoras' theorem for example shows us that its elements allow us to construct all the explanations and theorems in the order in which, as Book I, they precede it. Hankel's, Ofterdinger's and Mann's remarks[15] on Geometrical methods are well worth reading.

12. We may prepare the solution of a problem in natural science by eliminating prejudices that stand in the way and lead to blind alleys. An example of such a case is the prejudice handed down from antiquity that colours arise from diluting white light by mixing it with darkness. Boyle opposed this view and so prepared the way for Newton's correct solution of the problem of colour. Eliminating the view of heat as a substance of constant quantity, enabled one correctly to solve thermodynamic problems. Hering's solution of the problem of three-dimensional vision required prior removal of many old prejudices: physiological space had to be distinguished from geometrical space, the theory of directive lines

removed, and visual sensations recognized as different from other mental formations. Johannes Muller, Panum and Hering himself prepared the ground.[16]

13. Moreover, the solution of problems is essentially promoted through the emergence of attendant paradoxes that will not let the intellect rest until they have been eliminated. A historical investigation of the rise of paradoxes or a study of all that followed from conflicting views to the last consequence, leads one way or another to the point whose removal dissolves the paradox and usually solves or at least clarifies a problem. Thus if we go to the origin of the controversy between Descartes and Leibniz as to the measure of force by mv or mv^2, we recognize that we have here a mere convention, in that we can measure the force of a moving body either in terms of time or in terms of the distance travelled against another force, whichever we prefer.[17] The paradoxical cyclic process of W. Thomson and J. Thomson using water at freezing point, if viewed in all its aspects and consequences, leads to the discovery that the freezing point is lowered by pressure.[18]

14. Not all problems that arise in the course of scientific development are solved; many are in fact dropped as empty. The dismissal of problems that rest on a wrong question, by showing that they are insoluble because incapable of a sensible or even any answer, represents an essential progress of science. Thus relieved of useless and noxious burdens, science gains deeper and clearer vision that can be turned on new and fruitful tasks. We all readily see that a circle cannot be made to pass through four arbitrary points, since three of them already determine it; but if we can prove that the circle can be only approximately squared,[19] that there is no algebraic solution of equations of the 5th degree,[20] that various problems vainly pursued by many generations are insoluble or senseless, these are achievements that cannot be overrated. For example, a most valuable proof is that which shows perpetual motions machines to be impossible, and so is the uncovering of contradictions between our best established physical knowledge and the assumption of such a device. This abolition of a problem resulted in the discovery of the principle of conversation of energy, a most powerful source of further more specific discoveries. In every field we find abandoned problems, or such as have been so radically

modified as hardly to resemble the originals any longer. Cosmogonies in the old sense are no longer put forward. Nobody asks for the origin of language in the sense in which this was still done a hundred years ago. Soon nobody will be tempted any longer to reduce mental phenomena to the motions of atoms, or to explain consciousness by a special substance, quality or form of energy.

15. A proposition in natural science, like any in geometry, is of the form "if *M* exists, then *N* exists", where *M* and *N* are more or less complicated groups of characteristics of phenomena; one group determining the other. Such a proposition may result directly from observation or indirectly through reflection and mental comparison of already known observations. If the proposition seems not to agree with observations or their attendant thoughts, then it presents a problem that can be solved in two ways. The proposition "if *M* exists then *N* exists" may be derived or explained from propositions expressing facts already known, by means of a sequence of intermediate propositions. In that case our thoughts were already more adapted to facts, and to each other, than we had assumed or known; they corresponded to the new proposition too, except that we could not immediately see it. This solution consists in a deductive, synthetic, geometrical derivation of a new propostion from principles already known. All easier, secondary problems belong here. We are bound always to try this way first for luck. Whether we succeed in solving the problem depends of course on what we know already. Thus Galileo explains the floating of very heavy dust in water and air in terms of the low speed of fall because of the large resistance produced by the fine distribution. Huygens completely derives the motion of pendulums from Galileo's mechanical principles. Similarly, Segner, Euler, d'Alembert and others manage to give a mechanical account of the clear and striking phenomena of spinning tops. The upward flow of water in the shorter arm of a hydraulic jack is intelligible in the same way as the sliding down of a chain over the edge of a glass because of the greater weight of the overhanging part, except that the links of the chain are automatically connected whereas water is kept in contact through air pressure, or, as was previously supposed, by horror of the void. In the same way, the colour phenomena observed by Brewster in two parallel glass plates of the same thickness, surprising as they are, can be derived from already known principles of optics. Arago's rota-

tional magnetism was explained by Faraday's law of induction. However, closer reflection reveals that at an earlier stage of science these and similar problems could not be solved in this manner, and some indeed were not. This leads us quite naturally to the second way.

16. Suppose we can find no known principles with which the observed facts or their consequences agree. In that case we simply have to look for new principles, by means of fresh adaptations of thought.[21] The new approach may relate directly to the facts in question, or we may proceed analytically. We look for the nearest condition of the fact, then for the condition of the condition and so on. A new way of taking one or other of these conditions will usually make the strange or seemingly too complicated fact intelligible. Although geometry is a well-known and much investigated field, the analytic method still leads to new conceptions that allow us to derive new theorems and solve problems much more easily than was possible from older points of view; witness for example similar and similarly situated figures, and the wealth of projective relations as such. The field of natural science in general is incomparably richer and wider than geometry, it is all but inexhaustible and almost unexplored. We may therefore expect that the analytic method will yet produce fundamentally new principles. If we pay attention to what makes up this new adaptation or conception to which we are led, we find that its peculiarity resides in noticing conditions or characteristics previously unnoticed. Consider some examples, an easy one to start with: we see that bodies exert pressure from the top downwards and fall. This direction and the sense from top to bottom is physiologically determined for us humans who are geotropically organized. For a person at one place this becomes a physical orientation (sky above, earth below) that we regard as absolute and valid for the whole world. When astronomic and geographical enquiry reveals that the Earth is a sphere inhabited on all sides, we fail at first to understand why antipodean movable objects do not fall off. As children we have all behaved in this way, and very few of us have consciously gone through the stupendous and historically vital change that consists in taking gravity as determined by the direction of the Earth's centre and not by our local sky and ground. Most of us have dreamt our way from one view to the other under the influence of what we were taught at school.

We soon become familiar with the motion of isolated heavy bodies, but if a lighter body is raised by a heavier one as on a pulley we learn to notice the relations between several bodies and their weights. If we add findings from levers with unequal arms or from other machines, we are driven to consider not only the weights but their corresponding displacements in the direction of gravity and to the products of their measures, that is, work done. If we see immersed bodies sink, remain suspended or float, our desire for clear and secure ordering of these facts leads us to take notice of the weights of equal volumes. The raising of water under a piston against gravity yields the ingenious thought of horror of the void. This conception provides a principle that at first makes everything intelligible, especially the surprising paralysis of gravity; but then we find cases in which the principle fails. Torricelli measures the horror vacui by means of various columns of liquids and finds one definite fluid pressure to be sufficient to make all cases intelligible. He and Pascal thus take the analytic process one step further back to a more remote condition. Heavy bodies when projected may rise or fall, two cases that Aristotelian physics treated as different. Galileo takes notice of the acceleration of the motion, which makes all these cases similar and equally intelligible. Thus chance constantly reveals inadequate adaptions that drive us to new analytic steps for noticing new conditions, conceptions and adaptions doing justice to ever wider areas of experience. Nature offers us propositions analogous to geometrical ones but without derivations, or solved problems without solutions, leaving us to look for the principles of derivation and solution, which is immensely difficult given the incomparable complication of the whole world as against mere space.[22]

17. These few examples already show that the greatest and most important discoveries are found by way of analysis. For further evidence, take the previously mentioned discovery by Newton of the principles of general mechanics, celestial mechanics and optics. The analytic search for the premisses of what is given is a much less determinate task than deduction from given premisses, and therefore often succeeds only in tentative steps with the help of hypotheses, which may combine correctly guessed items with false or indifferent ones. Hence the line of thought adopted by different enquirers is here much influenced by accidental features. The similarity of behaviour of light and water waves or sound waves leads Huygens[23] to his theory of light, while its similarity with projectiles and

the faulty observation of refraction that suggested the absence of this phenomenon led Newton[24] to his theory of emission. Hooke[25] notices the periodicity of light which Huygens leaves completely out of account while Newton interprets it differently. Nevertheless each of these enquirers has earned great credit in this question. Chance led each of them in different directions and all three now combine into one complete analysis.

18. The function of hypotheses is further clarified in the light of what Plato and Newton have said about analytic method. Suppose we wish to find out the unknown conditions of a fact. About the unknown, however, we cannot form thoughts of sufficient clarity. Therefore we provisionally imagine intuitive conditions of a kind we know and tentatively regard the problem awaiting solution as solved. The path from the assumed conditions to the fact now becomes fairly easy to survey. Next we modify the assumptions until the path leads exactly to the given fact. By inverting the course of ideas we find the path from fact to conditions. After eliminating all superfluous and imaginary features from the assumptions we have completed the analysis. As to method, analysis is the same in natural science and geometry, both using hypotheses as a means. However, in the wider, less explored and less completely known field of natural science, the choice of hypotheses is less methodically restricted, so that is it left more to whim, chance and luck, and is thus more exposed to error.

19. Considering in particular Newton's analysis of light, we see that it was initiated by the then inadequate quantitative agreement with the assumed law of prismatic refraction. The divergence of the emergent coloured rays in the direction of dispersion was about five times as large as could be expected given the elevation of the sun, while at right angles to this direction the spread agreed with theory. Already Marcus Marci had noticed the increase in divergence when rays pass through a prism, but given his inadequate knowledge as to the law of refraction he could not draw the appropriate conclusions from this fact. To make this discrepancy intelligible, Newton assumed rays of different refractive indices: taking red as having the smallest and violet the biggest, each a constant for all refractions in the same material, makes all the phenomena intelli-

gible. Besides, one need not assume colour to arise from refraction, or from a mixture of light and dark (this had already been doubted by Boyle and Grimaldi). Newton was able to pronounce on it thus: colours are invariable independent constituents of white light, they are substances or 'stuff'. He was strengthened in this assumption by the characteristic periodic length that revealed itself in the analysis of the colour of thin plates. It remains established today that coloured lights are independent, invariable and constant components of white light, only the view of them as 'stuff' in the physico-chemical sense was arbitrary and one-sided. Indeed, it meant that Newton, though recognizing the principle of superposition of rays, failed to recognize the principle of superposition of phases, that results from the approach of Hooke and Huygens. In order fully to appreciate the import of Newton's analysis, we must remember the constancy of pigments as against the evanescent colours of rainbow, soap bubbles, mother of pearl, and reflect how differently and under what different conditions all these appeared. After Newton all this could be viewed from a uniform point of view, and the most remote members of this series of phenomena were related through the principle of selective absorption.

20. Let us try to reconstruct the line of thought that revealed the impossibility of the perpetual motion machine. Stevin already knew this, and derived from it many difficult theorems of statics and hydrostatics. The evidence moreover puts beyond doubt that Stevin took over from his forerunners many special cases of propositions in statics; while his representation of systems of pulleys shows that his aim was to bring under one expression everything that was common to all these cases. In this connection he expresses the theorem of virtual displacements for simple conditions. Suppose, then, he had asked himself what was common to all cases of statics, what principle would have to hold to embrace all these various phenomena. Given the then familiar method of measuring force by weight he would no doubt have recognized that a disturbance of equilibrium or start of motion occurs only if an excess of heavy mass moves downwards. A motion in which the mass distribution remains the same cannot occur. Stevin now derives special intances of the laws of equilibrium by showing that their non-existence would lead to the absurdity of infinite motion without change in the distribution of equilibrium. Special investigations

thus lead him to the general condition of equilibrium. Once this had been recognized, it served in turn as a prop to other special researches which now constituted a kind of test for the calculation. In this he provides a paradigm for all great enquirers. That our assumption concerning Stevin's line of thought is correct seems to be confirmed by the fact that Galileo thought almost in the same manner when dealing with inclined planes. A general principle such as Stevin's has the advantage over the derivable propositions that its contradictory stands in very strong contrast with all our instinctive experiences. When Galileo came to establish the dynamics of heavy bodies, he found by various reflections and trials that the speed of fall reached depended on the distance fallen, so that increase and decrease in velocity meant respectively lower and higher position. A remarkable pendulum experiment especially led him to recognize the general condition of all these special features. On whatever path a heavy body might move, the speed it reached in falling from rest at a certain level enabled it at most to regain that height. Huygens extended this notion to a system of heavy bodies; he obtained a special case that was later called the principle of conservation of vis viva, whose contradictory again stands in strong contrast with our instinctive experience. This principle states (like Galileo's principle) in Huygens' explicit words that heavy bodies do not rise of themselves. Confidently applying it, Huygens solves the difficult problem of the centre of oscillations, just as Galileo had solved special problems by means of his conception. In the more precise terms of Huygens, Stevin's principle would read thus: heavy masses can become accelerated only if their average height diminishes. By explicitly assuming that the mechanical principle of the conservation of vis viva cannot be violated by non-mechanical detours, S. Carnot first opened the way to the so-called principle of conservation of energy. This general point of view which once again is very close to our instinct has turned out to be immensely fruitful in the solution of special problems. As enquiry thus brings ever more details of experience into the light of conscious conceptual thinking, the most general principles forge ever closer and stronger links with the instinctive foundations of our mental life.[26]

NOTES

[1] *P* 3, p. 287.

[2] James, *Psychology*, Vol. I, pp. 585f.

[3] For individual examples, see *P*, pp. 303f.
[4] Galileo, *Sydereus nuncius*. To begin with the account of the story from Holland, the reconstruction, determination of magnification by binocular considerations and so on. *Opere di Galilei*, Padova 1744, **II**, pp. 4, 5. Once more, partly in more detail: *Il saggiatore, Opere* **II**, pp. 267, 268. The most important passage, on p. 268, reads thus: "My discourse was therefore as follows: The device requires either one piece of glass or more than one. It cannot be one alone, since its shape is either convex (thicker in the middle than at the edges) or concave (thinner in the middle), or else it lies between parallel planes. However, this last does not change visible objects at all, neither magnifying nor diminishing them. The concave piece does indeed diminish and the convex magnify them, but it shows them rather blurred and dazzling. Therefore one piece of glass alone is not enough to produce the effect. Going on, then, to two and given that the parallel plate does not change anything, as mentioned, I concluded that the effect could not be produced by using it with either of the other two. Hence I confined myself to finding out by experiment what would happen on combining the other two, that is, the convex and concave pieces, and I saw how that gave me what I was after. Thus came about the progress of my discovery, in which the conceptual notion that the conclusion was true was of no use to me."
[5] Kepler, *Ad Vitellionem paralipomena*, 1604.
[6] Kepler, *Dioptrice*, 1611.
[7] The first report of Edison's invention of the phonograph was told me as I was walking in the street, by a colleague who doubted the news. "Why not believe it?" I replied. "Imagine a barrel organ with a drum first formed by sound then reproducing it on being turned a second time." Before even reaching home I was almost certain that the phonograph was a slight modification of König's phono-autograph which, instead of the writing motion in the cylindrical surface of the drum, uses a motion at right angles to the drum. Nor was it difficult for me to guess this, for I had been occupied with acoustics and especially with König's instrument, often demonstrating the speechlike sounds heard when a fingernail glides over a ribbed book binding. I felt that the most difficult part of the construction was the choice of the drum material, which would have to be soft enough to take impressions yet resistant enough to reproduce them. Without special experience one cannot make this choice correctly. Gauss was the man not only to invent the electromagnetic telegraph but also to bring it to the highest degree of technical perfection, had he posed himself purely technical problems at all. When Wilhelm Weber in determining electrodynamic units used a current-carrying vibrating string to induce periodic currents in a second string, he could easily have invented the telephone, had he been technically minded. But how much more did these two men enhance the foundations of technology by turning to pure science! There simply are various different ways of progress and nothing is more regrettable than the narrow and one-sided arrogance of the theoretician towards the engineer and conversely.
[8] Thus I regard Fechner's greatest merit to be his way of posing the problem of psychophysics.
[9] Bretschneider, *Die Geometrie und die Geometer vor Euklid*, Leipzig 1870, p. 146.
[10] *Ibid.*, p. 147. For *analytic* and *synthetic*, see Euclid, *Elements*, **XIII** 1, in the edition by J. F. Lorenz, Halle 1798.
[11] Newton, *Arithmetica Universalis*, 1732, p. 87.
[12] In Figure 4 only the straight line *G* and one of the two angular bisectors are drawn.
[13] *P* 3, p. 296.

[14] Archimedes *Werke* (German translation by Nizze, Stralsund 1824). Cf. especially the piece on the quadrature of the parabola.
[15] Hankel, *Geschichte der Mathematik*, Leipzig 1874, especially pp. 137–156. Ofterdinger, *Beiträge zur Geschichte der griechischen Mathematik* (Programmabhandlung), Ulm 1860. Mann, *Abhandlungen aus dem Gebiet der Mathematik*, (Festschrift for Würzburg University's tercentenary) 1883; *Die logischen Grundoperationen der Mathematik*, Erlangen & Leipzig 1895.
[16] Cf. *A*, pp. 101f.
[17] Cf. *M* 5, p. 322.
[18] Cf. *W* 2, pp. 234f.
[19] F. Klein, *Ausgewählte Fragen der Elementargeometrie*, Leipzig 1895. F. Rudio, *Geschichte des Problems der Quadratur des Zirkels*, Leipzig 1892.
[20] Abel, 'Démonstration de l'impossibilité de la résolution algébrique des équations générales qui dépassent le quatrième degré', *Crelles Journal* **I**, 1826.
[21] One must of course be careful not to posit more principles than are necessary. Cf. Duhem, *La Théorie physique*, pp. 195f.
[22] *M* 5, 1904.
[23] Huygens, *Traité de la lumière*, 1690.
[24] Newton, *Optice*, 1719.
[25] Hooke, *Micrographia*, 1665.
[26] Cf. *M* and *W*.

CHAPTER XVI

PRESUPPOSITIONS OF ENQUIRY

1. A person who grew up and lives in a certain limited environment has time and again encountered bodies of fairly constant size and shape, colour, taste, gravity and so on. Under the influence of his environment and the power of association he has become accustomed to find the same sensations combined in one place and moment. Through habit and instinct, he presupposes this constant conjunction which becomes an important condition of his biological welfare. The constant conjunctions crowded into one place and time that must have served as foundation for the idea of absolute constancy or substance are not the only ones. An impelled body begins to move, impels another and starts it moving; the contents of an inclined vessel flow out of it; a released stone falls; salt dissolves in water; a burning body sets another alight, heats metal until it glows and melts, and so on. Here too we meet constant conjunctions, except that there is more scope for spatio-temporal variation.

2. The provisionally ultimate common constituents of our physical and mental experiences are what we have called elements. We observe their separate constancy, constancies of connections between them at one place and time, and more general constancies of these connections. Repeated and careful observation shows that separate elements are not in fact constant: if they seem so (say, colour under constant illumination, gravity in constant position with regard to the earth) this is merely because of the accidental constancy of other, connected elements. Even connection at one time and place is not absolutely constant, as the previous example illustrates, and as is shown daily by physics, chemistry and the physiology of sense perception in particular. We are thus left with the general constancy of connection, of which the other two are special cases. If we count the sensations of space and time amongst the elements, all constant connections consist in mutual dependences of elements.[1] Of course, biological needs direct us to begin by observing the simplest dependences immediately accessible to the senses, as amply shown already. Only later do we

succeed in deliberately ascertaining complicated and more general cases that require concepts for their representation, where the elements themselves hide in the concepts.

3. Just as we have learnt to grasp in the physical sense, by reflex and instinct, under the influence of our make-up, our biological needs and environment, and now exercise this skill consciously and deliberately in everyday life; so in enquiry, where it is a matter of grasping in the intellectual sense, in expectation of success as often experienced in the past, we learn consciously and deliberately to maintain the presuppositions that arorse instinctively from our mental make-up through association and from the influence of our environment.

4. The presupposition of mutual dependence of elements of experience need not be innate, on the contrary, we can observe its gradual development. In the life and linguistic development of peoples as well as individuals, concepts like 'because', 'since', 'consequently' must long remain at the level of spatio-temporal coincidence, before acquiring conditional or causal meaning.[2] Equally, it takes quite a time before the relation of mutual dependence between elements is viewed more comprehensively and correctly. This is indeed quite understandable: if everything ran quite regularly without interference as night follows day, we should adapt ourselves to this course quite unthinkingly.[3] Not until there is a change from rule to lack of rule are we forced, for the sake of immediate or mediate biological interests, to ask why events are now thus, now otherwise, which things are constantly connected and which only by accident. By means of this distinction, we attain the concepts of cause and effect. We call cause an event to which another (the effect) is constantly tied. To be sure, one finds that this relation is viewed mostly in a rather imperfect and superficial manner. Usually it is just two specially conspicuous marks of a process that are considered as cause and effect. Close analysis almost always reveals that the so-called cause is only one of a whole set of conditions that determine the so-called effect; so that according to which of these have been noticed or overlooked, the condition in question may differ greatly.

5. Once the presupposition of constant connection of elements has impressed itself on our thinking either as an instinctive habit or as a

conscious feature of method, we immediately look for a cause for any new and unexpected change. Why does the hitherto observed not go on existing? Has some neglected or unnoticed condition changed? Every alteration appears as upsetting stability, as a dissolution of what until then existed together, abolishing, the accustomed condition and so disturbing us and posing a problem, which drives us to look for a new connection and enquire into the cause.[4]

6. In the more highly developed natural sciences the concepts of cause and effect are constantly becoming rarer and more restricted in their use. There is a good reason for this: these concepts describe a state of affairs at best rather provisionally and imperfectly because they are insufficiently precise, as previously pointed out. As soon as we can characterise the elements of events by means of measurable quantities, as is possible immediately for space and time and by detours for elements of sense perception, the mutual dependence of elements is much more completely and precisely represented by the concept of function[5] than by those of cause and effect. This holds not only when more than two elements are in direct dependence (e.g. the gas laws $pv/T = \text{const.}$), but also and more importantly when the elements are in mediate dependence through several chains of elements. Physics with its equations makes this clearer than words can.

7. With two or more immediately dependent elements all of which are connected by an equation, each is a function of the others. In the old mode of expression we should have to say that in this case the concepts of cause and effect can be interchanged. If for example we have two isolated gravitating masses, or two heat conductors in contact, then the acceleration of the one is the cause of the other's and vice versa, and likewise for the temperature changes of the conductors. If a hot body *A* conveys heat by means of *B*, *C*... to *N*, it is no longer *A* alone that determines the change in *N*, but also all the intermediate bodies and their disposition. Nor can *N*'s change alone determine that of *A*: we can no longer reverse the relation. Even in the simple cases where all bodies can be regarded as points, we require as many simultaneous differential equations as there are bodies. Each equation generally contains the variables that relate to all the bodies. If we can obtain an equation containing the variable of only one body, then we can integrate it. This leads to the other integrals as well, in which the

constants are determined by the initial conditions. Working through even the simplest such example is enough to show how inadequate the ordinary concepts of cause and effect are, and indeed superfluous, given the concept of function.[6]

8. Looking carefully at physical processes, we can, it seems, regard all direct dependences as mutual and simultaneous. For the ordinary concepts of cause and effect the opposite holds, because they are applied to quite unanalysed cases of multiply mediate dependence. The effect 'follows' the cause, and the relation is 'not reversible'. Take the explosion of powder in a gun and the impact of the shell, or a shining object and the sensation of light: in both cases there are chains of mediate dependence with immense numbers of links. The target that is hit does not restore the work done by the powder, nor the sensitive retina the light; both are merely links in the chain of dependences, which continue differently from the way they began. The target may yield flying fragments, the perceiving person may grasp for the shining object. The process as a whole need not be instantaneous and reversible because it is based on a multiple chain of simultaneous and reversible dependences, of which more later.[7]

The conception of cause has thus not always been the same, but has changed in the course of history and may do so in future The problem of Hume and Kant having been discussed by me elsewhere,[8] a few minor additions may here suffice. Mental individuality develops through mutual relations between subject and environment. The organism certainly contributes innate features, perhaps many more than Kant had supposed, above all as regards the fact that reflexes can be stimulated. The innate comprises not only the sensations of space and time, but also the specific energies of all the senses with the systems of possible sensations that they contain.[9] Still, it has become clear that physiological space and time unaided by physical experience cannot yield foundations for a science of geometry or mathematics. Kant's question, how is pure mathematics (*a priori*) possible therefore undoubtedly contained a vital seminal idea of enquiry, but it would have been more important still not to presuppose that mathematical knowledge is gained *a priori:* only positive psycho-physiological research can determine what is innate, not philosophical decrees. As to the causal view, the innate at most comprises the foundations that make association possible, namely orgainc connection, for associations themselves can be shown to be acquired individually (see p. 24 above). The idea of an innate view of causality has misled so eminent an enquirer as Whewell to the strangest contortions, although we must really call him a rather free Kantian. Fries and his school, especially Apelt, to whom we owe much as to the foundations of a rational methodology for natural science, greatly strain at the Kantian fetters without quite breaking loose (see the examples given on pp. 100–101 above). Beneke was the first of the Germans to make essential progress. He says explicitly: "In what precedes we have carried through the principle that all concepts without exception, Kantian categories included, arise by association of ideas, so that in this particular we cannot adopt Whewell's view"[10]... "The most general division of the sciences from this point of view is into those that refer to what is conceived by external impressions, and those that refer to what is internally predetermined. The latter do indeed contain, as it were, knowledge of what is given within us prior to all experience. Yet attempts at defining this relation more closely have hitherto failed in that they presuppose that the forms standing out in the fully fashioned soul are given even before experience, or rather before the development of the soul, as innate. This is wrong: the forms at first available for knowledge arose only in the course of the soul's development, before which they were merely predetermined in innate dispositions and conditions that evince quite different forms."[11] I can add nothing significant to these excellent general remarks.

10. It is thus a matter of natural development, that what the subject interacting with this environment has gradually formed in the way of instinctive expectations of constancies, he will in the end carry over into enquiry as a postulate, as a deliberately conscious methodological presupposition often tested by past success and promising more of it. Indeed we cannot decide to enquire into a field unless we assume that it is investigable,[12] which presupposes constancies; for if not, what is there to investigate? These constancies are mutual dependences of given elements, functional relations or equations between them. If an equation is satisfied, this amounts to a widened and generalized substantial view, but also a more developed, sharpened and calrified causal one. In general it matters little whether we regard the equations of physics as expressing substances,

laws, or, in special cases, forces: at all events they express functional dependences. Consider the simple and readily intelligible example of the law of energy, capable of various interpretations, which we thus cannot really take to be so very different as they often seem.[13]

11. There is no way of proving the correctness of the position of 'determinism', or 'indeterminism'. Only if science were complete or demonstrably impossible could we decide such questions. These are presuppositions that we bring to the consideration of things, depending on whether we give greater subjective weight to past successes or to past failures of enquiry. However, during enquiry every thinker is necessarily a theoretical determinist, even if he is concerned with mere probabilities. Jacob Bernoulli's law of big numbers[14] can be derived only on the basis of determinist presuppositions. If so convinced a determinist as Laplace with his cosmic formula could occasionally be led to remark that the combination of chance events can yield the most amazing regularities,[15] we must not take this to mean that for example statistical phenomena are compatible with a will exempt from all law. The propositions of the calculus of probability hold only when chance events are regularities masked by complications.[16] Only then can the mean values obtained for certain time spans make any sense.[17]

12. In assuming constancy in general we do not exclude the possibility of failure in individual instances. On the contrary, the enquirer must always be prepared for disappointments, since he never knows whether he has taken into account all the dependences applicable in a particular case. His experience is limited in space and time and offers him only a small segment of the totality of events. No facts of experience repeats itself with absolute accuracy, each new discovery uncovers flaws of insight and reveals a so far unnoticed residue of dependences. Therefore even the extreme theoretical determinist must in practice remain an indeterminist, especially if he does not wish to render highly important discoveries impossible by speculation.

13. Science is itself a fact, but it is not possible without a certain, though most imperfect, stability of facts and corresponding stability of thoughts (through adaptation): from the latter we can infer the former, which must

be presupposed; indeed stability of thought is a part of the stability of fact. Perhaps there is no perfect stability, but what there is of it suffices to furnish a beneficial ideal of science.[18]

14. Once we have gone so far as to notice and deliberately look for the mutual dependences of elements, the method of search results automatically. Things that depend on each other usually vary together: the method of concomitant variation is the universal guide, and on it rest Aristotle's scant hints to enquirers as well as Bacon's more extensive account. J. F. W. Herschel, considering the indissoluble link of cause and effect and their sequence in that order, and noting that increase, disappearance or reversal of the former conditions the same changes in the latter, established the guiding rules of enquiry.[19] The many provisos he found himself forced to make show clearly that as an experienced enquirer he strongly felt the inadequacy of the two concepts. No experimenter can fail to know that parallel variation, which generally[20] applies in simple cases of dependence, cannot be unconditionally presupposed for more complicated mediated dependences. The most detailed account of the instructions for enquiry in schematic form is due to J. S. Mill.[21] If cause and effect are taken to be measureable and capable of all values, then Mill's methods are special cases of the method of concomitant variation. If in *ABCD*, *A* is the cause of *D*, then *D* is present in all complexes that contain *A* (method of agreement). If *A* vanishes, we get the complex *BC*, in which *D* too vanishes (method of difference). Other specializations give the remaining methods. The guiding thoughts and the difficulties and complications are essentially the same in Herschel and Mill. Whewell[22] has aptly criticized Mill's account and examples. Schematizing the enquirer's thought processes and so making him consciously aware of their form is certainly not useless but we must not expect this to make enquiry much easier in any special case. The difficulty is rather in the finding of the important elements of the complex, and not in the forms of inference. However, if with or without the help of Mill's schemata we have found that one element depends on another, this is a mere preliminary step, as every scientific enquirer knows, for the main task is only just beginning: namely, to find how *D* depends on *A*. In most cases, Mill's schemata make sense only if both *A* and *D* are regarded as a whole complex of elements. Given the aim and purpose of the enquiry, the enquirer will try as far as possible to examine complexes such

that they mutually determine each other without ambiguity. For only if he knows such complexes can he complete in thought what is only partly given in fact; or predict if the completion relates to the future. In this, he will hardly be helped by Mill's instructions.

15. Equipped with the concept of function and the method of variation, the enquirer sets out on his journey. Whatever else he might need, he must learn from special acquaintance with his field. For this no special rules can be set up. The method of variation is at the base of both quantitative and qualitative investigation; it is used in the same way in observation and experiment and equally guides thought experiments that lead to theory.

NOTES

1 *Erhaltung der Arbeit*, Prague 1872, pp. 35f. *A* 4, p. 258.

2 Geiger, *Ursprung und Entwicklung der menschlichen Sprache und Vernunft*, Stuttgart 1868.

3 J. F. W. Herschel, *The study of natural philosophy*, London 1831, p. 35.

4 *A* 4, p. 249.

5 *Ibid.*, pp. 74–78; *Erhaltung der Arbeit*, pp. 35f.

6 I have read somewhere that I am leading a 'bitter struggle' against the concept of cause. Not so, for I am no founder of religions. For my own needs and goals I have replaced this concept by that of function. If somebody does not find this more precise, liberated and enlightened, he can simply retain the old concepts. I neither can, nor wish to, convert everybody to my views. On learning that somebody had been indicted for not believing in the resurrection, Frederick II is said to have decreed: "If on Judgement day he does not want to rise with the rest, let him stay put for all I care". This mixture of humour and tolerance is on the whole to be recommended. Our successors will one day be amazed at the things we quarrel about and even more at how excited we grew in doing so.

7 A trifling though instructive experience led me to this last explanation. A man who was obviously no scientist but highly gifted in philosophy and poetry came to the view that just as an image on the retina must provoke a sensation so conversely a vivid visual idea must produce an image on the retina, which must be capable of being demonstrated in some way; and he asked me to carry out this hopeless experiment. The concept of function could hardly have misled him so badly as that of cause had here.

8 *W* 2, pp. 432f.

9 Cf. F. J. Schmidt, *Grundzüge der konstitutiven Erfahrungsphilosophie*, Berlin 1901.

10 Beneke, *System der Logik als Kunstlehre des Denkens*, Berlin 1842, p. 23.

11 *Ibid.* II, p. 282.

12 Cf. Oelzelt-Newin, *Kleinere philosophische Schriften*, Vienna 1901, 'Naturnotwendigkeit u. Gleichförmigkeit des Naturgeschehens als Postulate', pp. 28–42. His explanations are very close to my own views.

13 *W*, pp. 423f.

14 Jac. Bernouilli, *Ars conjectandi*, Basle 1713.

[15] Laplace, *Essai philosophique sur les probabilités*, 6th ed. Paris 1840.
[16] *A* 4, p. 65.
[17] Fries, *Kritik der Prinzipien der Wahrscheinlichkeitsrechnung*, Brunswick 1842.
[18] Cf. *Erhaltung der Arbeit*, p. 46. Also: Petzoldt, 'Das Gesetz der Eindeutigkeit', *Viertelj. f. wisswrsch. Philos.* **XIX**, pp. 146f; *A*, p. 274.
[19] *Preliminary Discourse*, pp. 151f.
[20] If one uses the concept of function instead of that of cause, it is at once clear that two functionally related variables need not vanish together, indeed one may change without the other. Take for example temperature and electromotive force of a point of contact between two metals: as the temperature rises, the emf first increases, then declines through zero to become finally negative.
[21] J. S. Mill, *System der deduktiven und induktiven Logik*, German translation by Th. Gomperz, Leipzig 1884.
[22] Whewell, *On the Philosophy of Discovery*, London 1860, pp. 238–291.

CHAPTER XVII

PATHWAYS OF ENQUIRY

1. A brief and generally applicable description of the endeavours, the activities and the goal that satisfies the enquirer into nature would amount to this: he wants to bring his thoughts into the best possible agreement with the facts and with each other. The same idea is expressed with minor variations by "complete and simplest description" (Kirchhoff 1874), "economic representation of the facts" (Mach 1872), "agreement of thought with being and thought processes amongst themselves" (Grassmann 1844). Conveying to others the adaptation of thought to fact turns it into description, and if this is complete and as simple as possible, into economic representation. Every avoidable incongruity or incompleteness, logical differentiation or superfluity of the describing thoughts involves a loss and is uneconomic. General and indeterminate as this characterization of enquiry may seem, it is likely to contribute more to an understanding of the enquirer's activity than more specialized and therefore more one-sided accounts of it, as some examples will show.

2. Scientific astronomic ideas developed from naïve and vulgar views. The rotation of the celestial vault or sphere of fixed stars round the earth is an immediate expression of observation. The motions of sun, moon and planets differ from that of this sphere. Hipparchus[1] first tries to represent solar and lunar motion by means of epicycles and thereby succeeds in deriving the irregularities of motion from a much simpler geometrical idea. Ptolemy[2] extended the method of epicycles to planetary motion. Philolaus[3], Archytas[4] and Aristarchus[5] prepared the heliocentric view that finally broke through for good with Copernicus[6]. This makes the eleven motions of the geocentric system superfluous, as Kepler[7] showed in 1596. Starting from the presupposition that the planetary system must be ruled by mystic numerical and geometrical relations Kepler endeavoured to explore them by means of highly phantastic constructions using the five regular solids.[8] However, after 22 years these speculations led him to the discovery that the cube of the distance divided by the square of the orbital

period was the same number for all planets (his third law). He illustrates this in the case of Earth and Saturn.[9] A study of the motion of Mars based on Tycho's observations yielded the law of areas[10] as a physical hypothesis that proved true in retrospect (1609). For he conceived the 'motive winds' that drive the celestial bodies round the central body as diminished with increase of central distance. This idea presumably led him to the third and second laws (law of areas)[11]. After many fruitless attempts he hit upon the elliptical planetary orbit[12] with the sun in one focus. These three laws he then extended to the other planets.[13] Newton's achievement consisted in making these still numerous individual descriptions derivable from the assumption of planets being accelerated inversely as the square of their distance from the sun. These accelerations he regards as special cases of mutual accelerations of masses of which free fall of heavy bodies near the Earth is the best known special case. Thus Newton makes astronomical motions a problem of general physical mechanics. But this step, too, was prepared already in the views of Copernicus[14] and especially Kepler,[15] who regarded gravity as a universal mutual attraction of masses: Kepler not only used motive winds to account for circular impulsion, but further mentioned that the moon would fall to the Earth if it were not held back by some wind-like force or by some other equal weight somewhere in its orbit.[16] Both still lacked the new dynamical insights of Galileo and Huygens to bring this step too into play.

3. This development unmistakeably shows the increasingly accurate mental reconstruction of astronomic facts. First the apparent motions of celestial bodies on the sphere of fixed stars must be viewed in rough terms, then irregularities attract our attention and finally the variable distances from Earth as well. Today the sphere of fixed stars can no longer be regarded either as a sphere or as fixed. The process is not finished and may well never be.[17] At the same time we see the mental reconstruction or description becoming ever simpler and more economical, so that in the end it is no longer confined to the facts for which it was originally made, but holds over a much wider field. That the steps leading to simplification do not rest on inferences of the moment obtainable by means of some formula, is seen from the required lapse of time. Kepler's *Astronomia nova* is especially instructive here, because of his admissions and candid accounts of his erroneous paths. It took 22 years of work before he suc-

ceeded. Of Newton, too, we know that years lay between the birth of his idea and its execution. A luxuriant phantasy engenders all kinds of abortive ideas, before one or the other is recognized as the right means towards simplification and is confirmed by experiment. Planned searching can be of little use if one has not yet found the idea that brings deliverance, which must be guessed before it reveals itself as such to the surprised enquirer. Here it is much better to dig amongst the productions of phantasy, while keeping one's eye fixed on the goal. Kepler's *Mysterium cosmographicum* (1596) and *Harmonice mundi* (1619) are very instructive on this. Astronomy, whose development has spun its web through millennia in the most varied heads, shows very vividly that science is not a personal matter but viable only as a social one.

4. The need for clarifying and simplifying thoughts must of course spring from the field under investigation, but the ideas themselves may well come from a different field. Epicycles are readily handled by any experienced geometer or practical mechanic.[18] Copernicus was evidently helped by everyday experiences concerning apparent motion and displacements of perspective, which in Kepler were accompanied by mystical and animistic thoughts. Finally Newton, as the physicist and pre-eminent geometer, added his own work and eliminates what is now superfluous. In the contest for the solution of such questions, breadth of intellectual vision is perhaps equally essential for victory as is sharpness of critical judgement about the economy of the ideas that happen to have been chosen and put to the test. Of course, the path chosen must be psychologically possible, even with the greatest genius – how else could normal average men follow him? Dynamics must be prepared and already to hand if it is to be applied to astronomy. How great the influence of individual mental development still remains is shown by the following careful consideration. Huygens the astronomer and physicist did himself develop all the tools required to explain the planetary system, but still did not solve the problem, nor could he summon up any proper appreciation for the finished solution. Anyone thinking of gravity as the determining factor for astronomic motions must indeed soon find the heart of the matter: for gravity could not be independent of distance, since otherwise not even stones near the Earth would fall to the ground and Kepler's third law could not exist. Thus one had to look for some other dependence of the acceleration

of fall on distance, and the third law clearly points to the inverse square. In fact Hooke, though not in the same class as Huygens as a mathematician, was carried further by thoughts of gravitational radiation and in this way grasped the vital point, gaining priority even over Newton. However, Newton was the only one to master the whole mathematical problem.

5. Consider another example. Electric and magnetic phenomena known since antiquity were viewed very superficially and often confused, until Gilbert[19] sharply emphasized the difference, and Guericke[20] initiated a more precise study of electricity. The discovery of two different electric states by Dufay[21] the recognition of the difference between conductors and insulators and the wealth of gradually emerging phenomena, enabled Coulomb[22] to found a more complete dualistic mathematical theory in contrast to an older unitary one by Aepinus[23]. As for magnetic phenomena, Coulomb was able to deal with them in very similar fashion. Both theories were further developed by Poisson[24] and the analogy between electricity and magnetism came out once again. This mere analogy sufficed to suggest a connection between the two fields, a surmise reinforced by chance observations such as magnetization of steel pins by electric discharges though this still did not lead to a tangible result. When Volta[25] built his pile he gave a new impulse to the study of electricity and there were further unsuccessful attempts to track down this connection. Oerstedt was finally lucky enough to find one: probably by accident, he noticed during a lecture that a magnetic needle was disturbed by closing the circuit of a voltaic pile. Suddenly he held the thread that he and others had been looking for all this time, and now it was a matter of not losing it. He put the needle into all possible positions with regard to the closing wire and was able to give a comprehensive description of all the relevant phenomena, an account that is quite correct though not very attractive to modern readers because of its awkwardness and unfamiliar terminology[26]. Ampère summarized the facts into the following rule: the north-pointing pole (north pole) of the needle turns to the left of an observer swimming with the positive current and facing the pole. The expression 'current' was first used by Ampère, while Oerstedt speaks of 'electric conflict'. Oerstedt recognizes that electric conflict does not determine any attraction, that it produces the same motions of the

needle through glass, wood, metal, water and so on, so that it does not exert any electrostatic attraction or repulsion and is not confined to the conducting wire but spreads afar in the space round the wire. He imagines that one electric substance whirls round the wire in one sense taking the north pole with it, while the other whirls in the opposite sense similarly taking the south pole. As we know, with a suitable arrangement a pole will whirl round the current carrier. These naïve ideas, which are much closer to those of today than were the official academic views of the mid-19th century, were further developed in this direction by Th. Seebeck[27] and Faraday.[28] Seebeck actually represents the circular lines of magnetic force and regards a current carrying circuit as a kind of circular magnet. This case in fact shows that a happy accident has here revealed something that was being looked for; although whether sought or not, it might have commended itself to an attentive observer, as for example did X-rays and many other discoveries. However, there were two circumstances that were unforeseeable and thus excluded any discovery according to plan: to start with, nobody could know that it required a dynamic electric state to determine a static magnetic one. Hence the many unsuccessful attempts, mentioned by Oerstedt, to find an effect of open circuits on magnets. How could those who knew only of static phenomena invent experiments involving dynamic states? Secondly, most phenomena in electrostatics[29] and magnetostatics are symmetrical with regard to positive and negative. Who could ever have expected that the north pole would give way unsymmetrically to one side from the plane determined by the needle and the current-carrying wire parallel to it? As regards discoveries according to a formula or rule, where we merely repeat intellectual situations that have occurred before, these are not really discoveries. Everybody who experienced Oerstedt's experiments in spirit, must have been mightily shaken, for he suddenly gained a glimpse of a new and hitherto unsuspected world. What was this strange physical agency that here disturbed symmetry when seemingly it was otherwise perfect?

6. Oerstedt's find greatly stimulated the phantasy and eagerness of enquirers grown tired through lack of success, and there quickly followed important discoveries that further revealed the connection between electricity and magnetism. One could expect, and Oerstedt had shown, that

movable current-carrying conductors, by way of mechanical reaction, could be set in motion by magnets. Ampère suspected that currents react on each other, because of the magnet-like behaviour of currents. He himself thought this surmise daring, since pieces of soft iron behave magnetically towards magnets but are indifferent to each other, but experiments showed him right. Although his mathematical theory[30], strongly influenced by Newton's idea of elementary forces at a distance, cannot stand up to critical examination today, he nevertheless showed that as far as their effects are concerned all currents may be thought of as replaced by magnets and vice-versa. In the shortest time and in most brilliant manner, he created for the physics of his day an excellent tool for further enquiry.

7. If towards magnets currents behave like magnets, we may expect that they behave likewise towards iron and steel. However, it seems that it was not this alone, but also a chance observation that led Arago[31] to the discovery of electro-magnetism. A current-carrying wire dipped in iron filings enveloped itself with these to the thickness of a quill, dropping them again when the current was stopped. This led him to magnetize thin rods of iron and steel needles, the former temporarily and the latter permanently, by holding the current-carrying wire over and across them. At Ampère's suggestion, Arago[32] then laid the rods into current-carrying coils. He owed a further discovery to the chance observation of the strong damping of a magnetic needle oscillating over a copper disc. Assuming the reaction, he was led to set the disc into rapid rotation, which caused the magnetic needle to rotate as well, so that the copper seemed to exhibit 'rotational magnetism'. The problem of using a current to turn a piece of soft iron into a magnet was solved. Faraday[33] long tried in vain to use magnets to produce currents, until a happy accident helped him to find the track. While inserting and removing the magnetic core of a coil he observed that a galvanometer connected with the coil showed a momentary deflection. This secured the discovery of induction, and soon he knew all its forms and rules. Now he could easily show that there were currents in Arago's disc which naturally had magnetic effects. Nobody had tried this out before, although it seemed an obvious thing to do given Ampère's principle of equivalence of currents and magnets. This illustrates that by no means all possible or even obvious lines of thought

are followed up; but the more enquirers there are, the more their differences as individuals will ensure that all psychological possible paths will be followed and the more rapidly will science progress. Of course, had Arago's rotating disc been thoroughly investigated in all respects, induction must have been discovered seven years earlier than it was. Moreover, induction is curious in another respect, since we here almost repeat the intellectual situation of Oerstedt, as is easily seen in retrospect. An *A* is indifferent to a *B*, but not to a change in that *B*. In the one case *B* is the static state, in the other the stationary current. A genius like Faraday is of course even less likely to think according to such a formula, which is always easy to abstract afterwards.

Space allows no more than briefly mentioning that the equations of Maxwell and Hertz[34] contain merely a more complete clarification of the relation between electricity and magnetism, which now form an inseparable whole and are in the process of absorbing the field of optics; a second example of a scientific development that reaches from ancient to modern times.

8. The peculiar smell emanating during the action of an electrification machine, especially when electricity flows out through probes, was observed by Van Marum[35]. In 1839, Schönbein several times observed this smell in lightning accompanied by the formation of a bluish haze, and later in the oxygen provided by electrolysis of water. The active and complementing phantasy of the chemist related this smell to a gas-like substance, for only that could affect the sense of smell. This was all the easier since gold or platinum dipped into the smelling substance quickly became negatively polarized while silver and other metals were rapidly oxidized by it, thus revealing chemical properties that disappeared again on heating. Equally naturally, Schönbein regarded this substance, mized with oxygen, as a compound; he called it ozone. The observation that slowly burning phosphorus in air produces the same characteristic smell, led to chemical experiments to isolate ozone, which provoked many controversies. In 1845, De la Rive proved that ozone is an allotropic form of oxygen, as Marignac had surmised. The example clearly shows how important, in the course of inventions, is the role of phantasy, by the way it compares and adapts sensations with experience (memories) gained under different conditions.[36] A closer study of the ozone problem further

reveals how differently the same matter is mirrored in different heads and how important and beneficial it is for individuals of different intellectual cast to take part in the treatment of the one question.[37] Here is a typical example of how a chance observation touching the curiosity of an individual can open up new pathways of enquiry.

9. When Daguerre tried to illuminate iodized silver plate in a camera obscura to produce pictures, he failed in spite of many attempts. He then stored the plates in a cupboard, and when he took them out again weeks later, he found the most beautiful pictures on them, but could not explain how they might have arisen. Removing apparatus and reagents from the cupboards changed nothing: exposed plates stored in it always showed pictures after a few hours. At last it became clear that a mercury bath that had remained was the cause of the miracle, since mercury vapours had settled on the exposed parts, somewhat in the way of Moser's breath images. He succeeded in fixing the erasable pictures by means of a gilding process.[38] Thus accident led to an invention that had been looked for and to a discovery that had not. It lies in the nature of the method of variation that it makes no difference whether the decisive concomitant conditions of the process are found by physical variation or, if thoughts are sufficiently adapted, by thought experiment. To realize in how many ways physical and psychological accidents intervene in discovery and invention, one merely has to enumerate some famous names like Bradley, Fraunhofer, Foucault, Galvani, Grimaldi, Hertz, Hooke, Kirchhoff, Malus, J. R. Mayer, Roemer, Röntgen and others. Almost any enquirer has experienced the influence of chance.

10. The stems of plants on the whole grow upwards against gravity while the roots grow downwards with it. Given the constant conjunction of these two facts the thought naturally occurs that gravity is a condition of the direction of growth. Moreover, Du Hamel[39] conducted special experiments showing that a growing plant always compensates any change of direction forced on it, by curving back and so growing in the normal direction. Knight[40] has added some very important experiments. On the axle of a small vertical waterwheel he fixed a second wheel of 11 in. diameter which turned at 150 revolutions a minute. On it beans were allowed to germinate and grow in the most varied position. The direc-

tion of gravity with regard to the plants varied so fast and regularly that it could no longer have a decisive influence: instead the plants now aligned with the direction of centrifugal acceleration, the roots outwards and the stems towards the axle, beyond it and then turning round towards it.[41] On a horizontal wheel of the same diameter and at 250 revolutions a minute centrifugal and gravitational acceleration combined into a resultant whose direction now determined growth.[42] The clinostat of Sachs[43] which is small and counterbalances gravity at very low speeds of rotation without any noticeable centrifugal acceleration, allows the plants fixed to it grow in any direction. However I think he is wrong to place little value on such experiments.[44] To the unprejudiced observer it may be highly probable that gravity determines the direction of growth, and yet this may be caused by quite different circumstances so far overlooked. Not until Knight's experiments, with their variation of size and direction of mass acceleration, was it clearly shown that the latter is the determining factor. Besides, it took experiments to enable us to separate the influence of other factors (light, air, soil humidity) from gravity. Mill has well shown that the method of agreement can never be so sure a guide as that of difference or concomitant variation. Even though gravity was now known as the determinant of the direction of growth, the nature of this effect remained a mystery for almost another century. Noll[45] was the first to surmise that gravity stimulated geotropic adaptation in plants in a similar way as statholiths do in animals. The investigations of Haberlandt and Nemec (1904) have shown that in plants the role of statholiths is taken over by grains of starch, which determine geotropic adaptation through special organs or perception and release.[46]

11. One of the strangest questions that has exercised men from immemorial is that of the genesis of organic life. Aristotle believed in original generation of the organic from the inorganic, and the late Middle Ages still agreed with him. Van Helmont (1577–1644) still gives instructions on how to produce mice. The thought of making a homunculus in a retort may not then have seemed so adventurous. Redi (1626–1697), a member of the Accademia del Cimento, showed that in putrefying meat no 'worms' appear if egg-laying flies are kept off by a piece of gauze. When next the microscope helped to discover a host of minute organisms that were hard to pursue in detail, such questions once more become difficult to decide.

Needham[47] was the first to hit upon the idea of heating organic substances in glass vessels in order to kill all germs and then sealing the vessels hermetically. Nevertheless after some time the enclosed fluids appeared to be animated by new infusoria. Spallanzani[48] thought that by this experiments he could prove the opposite, while Needham objected that Spallanzani in his procedure had spoiled the air required for animal life. Although Appert successfully applied Spallanzani's method in order to make conserves, and although other enquirers took part in the investigation (men like Gay-Lussac, Schwann, Schroeder, Dusch and others), the question still remained undecided because the source of error in these difficult experiments had not been completely uncovered. Pasteur[49] was led to the question of the origin of life through his study of fermentation, in which he thought he recognized definitely organic beings. By aspirating large quantities of air through a pipe whose far end was barred by a pellet of cotton wool, he caught the dust which he then obtained by dissolving it out of the pellet with ether and alcohol. Microscopic examination showed a content of organic germs that differed in kind and quantity depending on whether it was town, country or mountain air. If water containing sugar and albumen is boiled for some minutes in a retort and on cooling air is admitted through a glowing platinum pipe whereupon the retort is hermetically closed, it can remain for several months at 25–30°C without any organisms developing in the fluid. If now we carefully introduce the dust-carrying pellet, making sure that only air that has gone through the glow-tube is admitted during the operation, then, on resealing the retort, organic formations regularly appear after 24–48 hours. Asbestos that has been brought to glowing point will produce organic formations in the retort only if it has been first aspirated with dust. In open retorts with several bends in the thin neck the boiled liquid remains unchanged even after cooling, since dust is trapped in the wet curved pipes; but if one tries to seal off the liquid by inverting the open retort and dipping it into mercury, the germs on the surface and in the interior of the mercury soon come to life.

12. These experiments, which are valuable also as uncovering the sources of error, prove conclusively that the organisms we know develop only from organic germs. The general question of the origin of life however goes too far and too deep to be decided by a simple physical experiment.

One might agree with Fechner[50] that it is the organic rather than the inorganic that is primary, and that the former can go over into the latter as its final and most stable state, but not the reverse. Nature is not bound to start with what is simpler to our understanding. On this view there arises the difficulty of understanding the beginning of the organic on the Earth, which used to be at a much higher temperature. Even if organic germs had been transferred to Earth by means of meteoric fragments of other cosmic bodies, a live transfer can be conceived only for the lowest organism. Only a highly developed theory of descent could solve this difficulty. What, then, forces us to assume so abrupt a difference between organic and inorganic and to believe that the transition from the former to the latter is absolutely irreversible? Perhaps there is no sharp dividing line. Chemistry and physics are indeed still far from understanding the organic, but something has already been achieved and more is added daily. Pasteur still thought that all fermentation was organic. Today we know that similar processes analogous to the catalytic acceleration of possible chemical reactions (Ostwald) are to be found in the inorganic field too. Imagine a state of civilization in which we are as yet fairly ignorant of the nature of fire, able to extinguish but not to produce it, and therefore forced to rely on using naturally occurring fire. In that case we should rightly say that fire can descend only from fire. Yet today we know better.[51] How people could conceive the notion that the question as to the origin of life was connected with the principle of conservation of energy is quite incomprehensible.

13. The above scientific developments mostly begin with very primitive ideas in the depths of pre-history, but are by no means concluded today. Instead of the problems that have been solved or recognized as sham, more numerous and usually more difficult new problems have cropped up. Knowledge is gained on very tortuous paths and the single steps, though conditioned by prior ones, are partly determined by purely accidental physical and mental circumstances as well. Modern astronomy must carry on where ancient astronomy leaves off. The latter borrows from geometry, the former gains help from physics and particularly dynamics, which happen, quite independently, to have developed, as have technical and theoretical optics, two further aids to modern astronomy. Later, we even find chemistry in mutually beneficial relation with astronomy. How

could our modern theory of electricity exist without the help of glass and metal technology, without the air pump and without chemistry? Yet how much has been contributed by the great historical chance thoughts, and by gravitation, from which potential theory began! Schematizing the cognitive stages may perhaps benefit further enquiry when similar situations recur, but there can be no widely effective instructions for enquiry by formula. Nevertheless, it remains always correct that we aim at adapting thoughts to facts and to each other. What corresponds to this in the case of organic development is adaptation of organisms to environment and parts of organisms to each other.

NOTES

[1] Born c. 160 B.C.
[2] Observed c. 125–150 A.D.
[3] C. 410 A.D.
[4] C. 400 A.D.
[5] C. 310–250 B.C.
[6] Copernicus, *De revolutionibus orbium coelestium*, 1543.
[7] Kepler, *Mysterium cosmographicum*, 1596, Ch. I.
[8] *Ibid.*
[9] *Harmonice Mundi*, 1619, Lib. V, pp. 189, 190.
[10] *Astronomia Nova, De Motibus stellae Martis*, 1609, p. 194.
[11] *Mysterium cosmographicum*, 2nd ed., Ch. II p. 75.
[12] *Ibid.*, pp. 285f.
[13] *Epitome astronomiae Copernicanae*, 1619.
[14] Loc. cit. Lib. I ch. 9, where gravity is already attributes to all celestial bodies.
[15] *Astronomia nova*, especially Introductio p. 5, where he speaks of the mutual gravity of Earth and Moon, states that the Moon would attract terrestrial water if the latter had no gravity towards the Earth, and so on.
[16] As in note 2 above.
[17] Since we know that the sphere of fixed stars is variable and the stars are at an incomparably large distance, the original Copernican co-ordinate system is once again subject to uncertainty; but even a purely terrestrial system could hardly be fixed precisely enough.
[18] Any mathematician will note that the representation of an arbitrary periodic motion by means of epicycles rests on the same principle as the application of Fourier series. Thus modern mathematical physics has points of contact with ancient astronomy.
[19] Gilbert, *De Magnete*, 1600.
[20] Guericke, *Experimenta Magdeburgica*, 1672, pp. 136, 147.
[21] *Mém. de l'Académie de Paris*, 1733.
[22] Coulomb, *Mém. de Paris*, 1788.
[23] Aepinus, *Tentamen theoriae Electricitatis et Magnetismi*, 1759.
[24] *Mém. de Paris*, 1811.
[25] *Philos. Transact.* 1800.
[26] Oerstedt, *Gilberts Annalen* 1820.

[27] Th. Seebeck, 'Über den Magnetismus der galvanischen Kette', read in the Berlin Academy, 1820–1821.
[28] Faraday, 'Electro-magnetic Rotation Apparatus' (*Experimental Researches in Electricity*, Vol. II, p. 147); 'On the Physical Character of Lines of Magnetic Force' (*Ibid.*, Vol. III, p. 418, n. 3265). Electromagnetic rotations were important because it was from them that Ampère recognized that ponderomotive actions at a distance of currents could not be reduced to electrostatic action, but that something radically new was involved. Cf. Duhem, *La Théorie physique*, pp. 203f.
[29] If we ignore one-sided discharges, Lichtenberg figures and so on.
[30] Ampère, *Théorie des Phén. électrodynamiques*, Paris 1826.
[31] *Ann. de chimie et de physique*, 1820, T. XV, p. 94.
[32] *Ibid.*, 1825, T. XXVIII, p. 325.
[33] *Philos. Transact.* 1832.
[34] Hertz, *Werke*, Leipzig 1895, I, p. 295; II, pp. 208–286.
[35] Van Marum, *Déscription d'une très grande machine électrique*, 1785.
[36] Cf. the detailed account in Kahlbaum & Schaer, *Ch. F. Schönbein, Ein Blatt zur Geschichte des 19. Jahrhunderts*, 1901.
[37] In the same work it is shown how much Schönbein was at a disadvantage compared to other scientists, because he spurned the help of atomistic ideas.
[38] Abbreviated from Liebig, 'Induktion und Deduktion', *Reden und Abhandlungen*, 1874, pp. 304–306.
[39] Du Hamel, *La physique des arbres*, Paris 1758, Vol. II, p. 137.
[40] *Philos. Transact.* 1806.
[41] The centrifugal acceleration at constant period of revolution is proportional to the axial distance. Inversion thus occurs where the mass acceleration for the plant reaches the threshold value.
[42] To judge by the dimensions and periods of revolution of the wheel ($\phi = 4\pi^2 r/t^2$) Knight used centrifugal accelerations which at the outer rim were equal to, three and a half times and ten times as large as gravity. The ratio varies as the axial distance if the period is constant.
[43] Sachs, *Vorlesungen über Pflanzen-Physiologie*, 1887, pp. 721f.
[44] *Ibid.*, p. 719.
[45] Noll, 'Über Geotropismus', *Jahrb. f. wissensch. Botanik* **XXXIV**, 1900.
[46] Haberlandt, *Physiologische Pflanzenanatomie*, 1904, p. 523–534.
[47] Needham, *New microscopical discoveries*, London 1745.
[48] Spallanzani, *Opuscules de Physique animale et végétale*, 1777.
[49] Pasteur, *Ann. de chimie et de physique*, 3rd series, **LXIV**, 1862.
[50] For a comparison of the views of Fechner and Boltzmann on the Second Law of thermodynamics, see *W*, p. 381.
[51] How old and instinctively obvious the relation between life and burning really is we can see from a report of Herodotus (bk. III, Ch. 16) following a misdeed of Cambyses: "The Egyptians regard fire as a live animal that devours everything it can reach and dies along with what it consumes". Cf. in Ostwald (*Vorlesungen über Naturphilosophie*, 1902, pp. 312f.) a detailed parallel between the self-preservation of life and of a flame. Cf. also W. Roux (*Vorträge und Aufsätze über Entwicklungsmechanik*, 1905), which has particularly attractive accounts of initial generation and of the comparison of a flame with an organic being, pp. 108f.

CHAPTER XVIII

DEDUCTION AND INDUCTION PSYCHOLOGICALLY VIEWED

1. According to Aristotelian doctrine, there are two kinds of inference or modes of deriving one judgement from others without contradiction: from a more general to a more particular judgement, by syllogism; and from particular judgements to the general one that comprises them, by what is now called induction. Judgements that form a science or system are perfectly adapted to each other without contradiction, if they can be derived from each other according to these modes. This alone shows that the opening up of new sources of knowledge cannot be the task of the rules of logic, which rather serve only to examine whether findings drawn from other sources agree or disagree, and, if the latter, to point to the need to secure full agreement.

2. Take first the syllogism represented graphically in Figure 7, using the conventional example: all men are mortal (general major premiss), Caius is a man (special minor premiss), therefore Caius is mortal (conclusion).

J. S. Mill[1] has pointed out that syllogism cannot yield insights that one did not have already, since one cannot utter the major premiss in general unless one is also certain of the special case of the conclusion. Mortality cannot be asserted of all men unless it holds of Caius. In order to establish the major premiss, the mere logician must wait for the death of any future

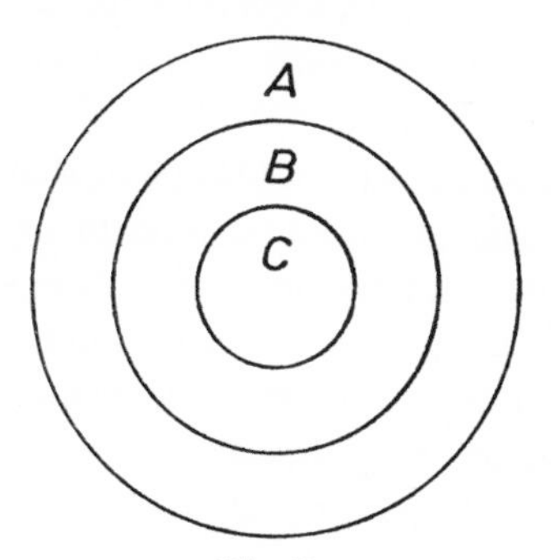

Fig. 7.

Caius and no Caius relying on the syllogism can experience the certainty of his own mortality. To be sure, few will ever have believed in the creation of knowledge by the power of logic alone, but Mill's criticism has had a useful clarifying effect, as the attendant discussions show.[2] Indeed, Kant had long since recognized that sciences like arithmetic and geometry are not built up from mere logical derivations, but require sources of knowledge.[3] Pure intuition a priori, however, turns out not to have been such a source. Beneke[4], too, is quite clear that syllogism in no way goes beyond what is given. They merely make us more clearly aware of the way judgements depend on each other. For the careless observer of mental processes, syllogism may of course produce the appearance of wider insights. Let us for example start from the proposition that the external angle u of a triangle is equal to the sum of the two opposite internal angles $a+b$. If now the two sides meeting at the vertex of the external angle are equal then $u=2a$. If with this vertex as centre we draw a circle through the other two vertices, then the new construction shows the central angle u to be twice the peripheral angle, that is $2a$. If, however, we carefully remove all ideas of additional construction and specialization not introduced by syllogism, we find no more than the original proposition about the external angle.

3. Enquiring into the ultimate source of this proposition, we find that it is a fact of experience[5], according to which the angular sum of any plane triangle we can measure does not demonstrably deviate from two right angles. In longer derivations the semblance of novelty emerges even more strongly. Take Euclid's proof of Pythagoras' theorem. The square on *ab* equals twice the area of *acf* which is congruent with the triangle *aeb*; but twice that triangle equals the rectangle *agde* formed by the perpendicular from *b* to *ac*. Dealing likewise with the right-hand part, not shown in Figure 8, completes the theorem. Here we have used simple theorems of congruence (determining the size and shape of triangles in terms of sides and angles) and theorems about equal areas of figures. The strange and unexpected relation between the squares on the sides that results will surprise any beginner, but the novelty again depends only on construction and not on the form of derivation. Remembering that the theorems used rest on facts about figures being displaceable[6] without change of size and shape, that, except for the constructions, is all we see in Pythagoras' theorem.

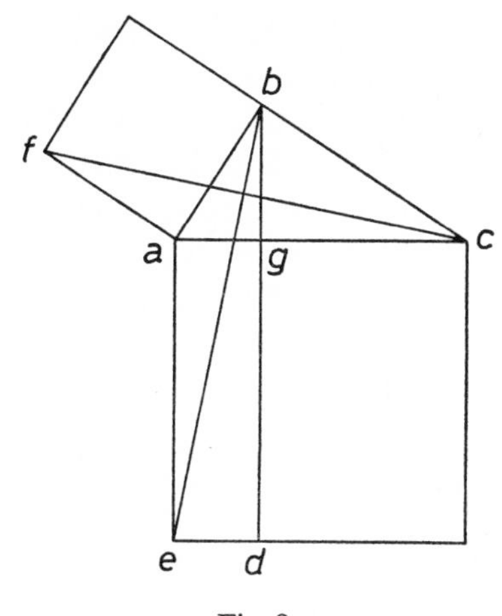

Fig. 8.

A beginner will learn a proposition about parallelograms perhaps from a skew-angled figure and will apply it to a rectangle of which he may never have thought in connection with that proposition. However, if he is surprised by the result, he cannot have considered the parallelism of the opposite sides, without regard to the angles of the adjacent sides, in a sufficiently abstract way. To abstract and concentrate attention on essentials while ignoring irrelevancies requires practice, without which attention is derailed now this way now that, as every student will have experienced. Repeated reflection, for example when making a deduction, gives occasion to notice and correct these derailments and thus to improve abstraction. Somebody who has the practice will see for example that the mutual bisection of the diagonals in a square is common to all parallelograms, equal diagonals to all rectangles and intersection at right angles to all rhombi and certain other quadrilaterals.

Syllogistic deduction, starting from more general propositions (rarely imagined explicitly in their specialized forms) and progressing to more special propositions, by means of changing and combining various points of view through several intermediate links, can here deceive us into seeing new insights not contained in the premisses. However, the same propositions could have been seen directly, even if it is easier to grasp them by establishing the separate elements. It is in this that the proper value of deduction lies, and not in creating new knowledge.

4. The weakness of abstraction[7] is greatly remedied if a successful case is linguistically fixed in definitions and propositions, to be stored in memory. This relieves thinking and preserves it from fatigue, since it will not have to face the same efforts every time. Even though the basic knowledge with which syllogism operates must be obtained from elsewhere, the logical operation is not useless. It makes us clearly aware of the way various insights depend on each other and saves us from having to seek for a special grounding for a proposition that is contained already in some other. Even if the propositions from which we logically begin are not absolutely secure, they still remain logically usable. Suppose we had a major premiss *B* is *A* that had not been established, it would still be the case that, if *B* is *A* and *C* is *B*, then *C* is *A*. It is in this sense that we really should take all propositions of contemporary science and even of mathematics when applied to real objects, whether natural or artificial, since they never correspond completely to the abstract ideals.[8]

5. Let us now look at induction, the counterpart of syllogism. Let C_1, C_2, C_3 ... be the members of the class whose concept is *B* (Figure 7). We observe that each falls under the concept of *A*. If the C_1, C_2, C_3 ... exhaust the extension of *B* and all come under *A*, then *B* as a whole does likewise, and the induction is called complete. If we cannot show that all the *C*'s fall under *A* and we still infer *B* is *A*, without having exhausted the extension of *B*, then the induction is incomplete. In that case the inference is not logically justified[9]. However, through the power of association and custom we can find ourselves mentally in the mood to expect that all *C*, and therefore *B*, will turn out to be *A*.[10] For the sake of intellectual advantage and scientific practical success we can desire that it be so, and tentatively assume that it is so, whether by instinct or by way of deliberate methodological stipulation in expectation of possible or likely success.

6. Complete induction affords no more widening of experience than does syllogism. By collecting individual judgements into a class judgement we merely make our knowledge more concise and express it more compendiously. On the other hand, incomplete induction anticipates a widening of knowledge, but thereby includes the danger of error and is from the outset designed to be tested, corrected or even rejected. The vast majority of our more easily obtained general judgements are gained by incomplete

induction and very few by complete induction. Forming a general judgement in this way is not the affair of a moment nor takes place in a single individual alone. All contemporaries, all classes, indeed whole generations and peoples collaborate in the consolidation or correction of such inductions. The more widely experience spreads in space and time, the sharper and more comprehensive the control of inductions becomes. We may recall the great events in world history, the crusades, the voyages of discovery, intensified international communication, the development of technology and the way in which this leads to changes of views and opinions. The inductions that resist correction longest are those false ones that reach into subjective areas where testing is difficult or impossible. We may remember the view of comets as harbingers of misfortune, astrology, belief in witches, spiritualism and other forms of official and private belief and superstition. Besides this direct testing of inductions by experience there is another indirect variety that is no less important: induction meets induction and they show themselves compatible or incompatible, either immediately or mediately through derived consequences. How, for example, does free will in the sense of indeterminism measure up to the results of statistics? How totally different in value is the induction contained in the mortality tables of insurance companies, from that in the proposition "all men are mortal".

7. The major premiss of a syllogism may have been obtained in various ways, just as the individual judgements on which induction rests. These in turn may be the result of inductions, or direct findings or even of deductions. The propositions from which the geometers of ancient Greece may have started will no doubt have been the result of immediate induction. Thus it seems that the proposition "the straight line is the shortest distance between two points" was obtained directly from observation of stretched strings. It remains a basic principle with Archimedes. However, we may equally start from propositions whose exact test by experience is difficult but whose consequences everywhere agree with it. It is from such propositions, which should really be called hypotheses, that Newton's mechanics begins.

8. In deriving mathematical propositions, in geometry for example, complete induction often plays a mediating role. In Euclid's proof of the theorem relating central and peripheral angles, three cases are distin-

guished in which the derivation runs differently. Only after showing that the proposition holds in all three cases, does he pronounce it in general. Besides there is here a tacit induction, or at least one that is not explicitly mentioned. For if we consider one of the cases in particular we observe that the vertex of the peripheral angle can be shifted within certain limits without changing the mode of inference applied. Finally we may regard the central angle as continuously running through all values without having to alter our approach. In short, one is using a complete induction as a means of proof. Similarly with other derivations: one must always gain a complete survey of all possible cases, a process speeded up by experience and practice. A deficiency in this, when a derivation from a special case is taken as valid in general, has on occasion led to serious mathematical errors. Where mathematics is applied to physics, chemistry or other sciences, this tacit induction is automatically included, for in mathematics a complete survey of all possible cases is reached with relative ease because of the uniformity and continuity of its objects; moreover, we are there concerned with our own familiar ordering activity often tested in practice.

9. Even incomplete induction has often been used in mathematics for heuristic ends. Wallis[11] used it to derive the general term and sum of series formed according to a certain law. These investigations might be regarded as an arithmetization of Cavalieri's[12] thoughts about quadrature and cubature, that is, as the beginnings of integral calculus. Jacob Bernoulli[13] next found the beautiful method of turning such incomplete inductions into complete ones. He first explains it by means of a very simple example. Suppose we are to form the sum of the natural integers up to n, zero included, and a simple induction yields $\frac{1}{2}n(n+1)$. To show that this is true in general for any n, we add an extra integer, and find

$$\tfrac{1}{2}n(n+1)+n+1=\tfrac{1}{2}(n+1)(n+2).$$

Hence the same formula for the sum remains valid, so that since this process can be repeated indefinitely, the formula is valid in general.

10. This example is so simple, intuitive and perspicuous that it hardly requires special proof.[14] Next, Bernoulli mentions that this procedure can be used to evaluate the sum of squares, triangular numbers and so on.

For the former we obtain for example

$$\sum_{1}^{n} n^2 = \frac{n^3}{3} + \frac{n^2}{2} + \frac{n}{6}$$

by simple induction, which by means of Bernoulli's procedure turns out to hold also for $n+1$ and therefore for any n.[15] The more general schema of the procedure is this: if $f(n)$ is the general term of a series, $F(n)$ the sum obtained by induction, then if $F(n)+f(n+1)=F(n+1)$, the formula for the sum is valid in general.

11. Jacob Bernoulli's method is important for enquiry into nature too; for it tells us that a property A found by incomplete induction on the members C_1, C_2, C_3... of a concept B, can be ascribed to B only if it is recognized as tied to B's characteristics and therefore independent of the variations of B's members.

12. Thus syllogism and induction do not create new knowledge, but merely make sure that there is no contradiction between our various insights and show clearly how these are connected, and lead our attention to different sides of some particular insight, teaching us to recognize it in different forms. Obviously, then, the genuine source from which the enquirer gains knowledge must lie elsewhere. In view of this it is rather strange that most enquirers who have dealt with the methods of enquiry nevertheless denote induction as the principal means of enquiry, as though the natural sciences had nothing to do but directly classify individual facts that lie openly about. Not that we wish to deny that to be important too, but it does not exhaust the enquirer's task: he must above all find the relevant characteristics and their connections, which is much harder than classifying what is known already. The name 'inductive sciences' for the natural sciences is therefore not justified.

13. The name is intelligible only from a long outdated tradition and convention that is still maintained. Looking at Bacon's tables of favourable or unfavourable instances ar regards an assumption, or at Mill's schemata of agreement and difference, we see that comparison can make us aware of a hitherto unnoticed connection, even if it is not noticeable enough to attract immediate attention. If we concentrate on the elements

that depend on each other, while being distracted form the less important ones, this is called abstraction.[16] Thus we bring about a situation that may lead to a discovery, but also to error if the attention is misdirected. This operation has nothing to do with induction. However, if we consider that observing or enumerating many cases that agree in spite of variation in some characteristics leads more readily to an abstract view of the stable characteristics than does looking at a single case, then we are indeed reminded of the similarity of this procedure with induction. That, perhaps, is why the name has survived so long.

14. As to the views of different exponents of the methodology of science on what should be called induction, there is great variety both in general and in special applications. For Mill[17], induction is inference from particular to particular on the basis of agreement in certain characteristics. For Whewell[18], on the other hand, only those inferences are inductive that yield new general propositions containing more than lies in the particular case. Inferences by analogy from particular to particular, such as animals too may draw and such as guide every practical activity, he will not allow to be inductive, in contrast with Mill. It seems difficult to draw a sharp psychological dividing line here. Kepler's discovery of the elliptic motion of Mars is to Mill a mere description, an achievement somewhat like sailing round an island so as to determine its shape; while Whewell regards the discovery as an induction just as Newton's discoveries, adding that different theories may indeed be viewed as different descriptions[19] of the same facts: what is essential in induction is the introduction of a new concept, such as Kepler's ellipse, Descartes' vortices, Newton's inverse square law of attraction. According to Apelt[20], Kepler's discovery is based on a genuine induction, since it amounts to finding all positions of Mars to lie on one ellipse. Galileo's law of free fall, however, Apelt regards as the result of a deduction. For myself, the only difference I can see between the finds of Kepler and Galileo is that the former guessed the helpful concept only after the observation, while the latter did so before. Whewell thinks that there is something mysterious in induction[21], something that is hard to put into words, a point to which we shall return. From these differences of view we gather at least that there is some lack of precision in terminology. Since the term induction has gained a fixed meaning in formal logic, and is used in the methodology of natural science for many

very different activities as just adumbrated, we shall henceforth no longer use the name.

15. Let us now try to analyse the procedure of enquiry without letting any nomenclature distract us. Logic provides no new knowledge. Whence, therefore, does it come? Always from observation, which can be 'external' through the senses or 'internal' through ideas. Depending which way our attention is attuned, it will emphasize now one connection of elements now another, fixing it in concepts: if this turns out to be tenable and stands the test vis-à-vis other findings it constitutes knowledge; if not, error[22]. The basis of all knowledge is thus intuition[23], which may relate both to sense perception and to intuitive ideas, as well as to what is potentially intuitive and conceptual. Logical knowledge is only a special case, concerned purely with the finding of agreement or contradiction, inoperative unless there are sense perceptions or ideas from findings previously fixed. Whether it is pure physical or mental chance or planned extension of experience through thought experiments that leads us to a new factual finding in sense perception or idea, it is nevertheless only this finding that gives rise to new knowledge. Once our interest in a new finding is aroused, because biologically it may be directly or indirectly important or because it agrees or disagrees with other findings, the mental mechanism of association alone will concentrate our attention on two or several elements connected in this finding: there is involuntary abstraction and neglect of seemingly unimportant elements, giving the individual case the character of a general one that represents many similar special ones. If several similar findings accumulate, this psychological situation is the more readily produced, but if our interest is lively, a single finding may be enough. The experienced enquirer may however make a tentative abstraction, deliberately and fully aware of the daring involved, neglecting subsidiary conditions in prospect of possible success. The more general thought must then stand the test of observation and experiment in order to become tenable. Still, in thus tentatively extending the idea of the special case into a more general thought, we have some scope for arbitrariness in the process of provisional completion. For part of the extension one or several special cases may provide starting points. Thus it is obvious to Kepler that Mars is moving in a closed oval orbit, to Galileo that the speed and distance of fall increase, to Newton

that a hot body cools the faster the colder its surroundings; however, the rest must be added by their own efforts from their own store of thoughts. Thus the tentatively assumed ellipse for Mars is Kepler's own construction, and likewise for Galileo's assumption that speed of fall is proportional to time and Newton's assumption that the speed of cooling is proportional to the temperature difference. Experience concerning their own conceptual activities, especially as to ordering, calculating and constructing, must help enquirers to formulate their general thoughts in concepts; observation alone cannot do this. Everything we have said about hypothesis, analogy and thought experiments is applicable here. Whether a thought so formed represents the facts accurately enough can now be tested in detail.

16. Merely to ascertain the facts accurately and represent them in thought requires more initiative than is commonly supposed. In order to be able to say that one element depends on another and how (according to what functional relation), an enquirer must contribute something of his own beyond what is immediately observed. It would be a mistake to think that one can belittle this by calling it description.

17. How far the finding of facts satisfies the enquirer depends entirely on his point of view and vision and on the level of the science of his time. Descartes was satisfied with vortices as a means for representing planetary motion. For Kepler, who had started from animistic ideas, the laws he finally found were a great simplification.[24] Newton, to start with, knew many relatively simple things from the mechanics of Galileo and Huygens which taught him how to determine the conditions of motion of bodies at any point in space and time. A motion that changes in speed and direction at every such point must have seemed something very complicated to him. In his urge towards completing things beyond what was observed, he surmised that there might here be simpler, perhaps already known, but overlapping facts. Practical mechanics tells us how to swing a body round at the end of a string, while theory tells us how to reduce this process to the simplest facts. This is the additional experience that Newton contributes. Following Plato's instruction, he takes the opposite path and imagines the problem solved and regards planetary motion as just such a whirling round. By analysis he finds what kind of tension in the string

would satisfy the conditions of the problem. In this last step resides the discovery of the simpler new fact, knowledge of which can replace all Kepler's descriptions. Yet to note this fact is itself no more than describing something factual, albeit much more elementary and general.

18. Similarly for other fields. The rectilinear propagation, reflection and refraction of light are noted in the same way as Kepler's laws. Huygens, supported by his experiences with sound and water waves, tentatively reduces these complicated and isolated facts to the few facts of wave motion, a step analogous to Newton's. The investigations of Newton on sound and water waves were continued in the 18th century and finally enabled Young and Fresnel to master the periodicity and polarization of light after Huygens' model. Here as everywhere findings gained by synthesis in one field are used for analysis of another. Plato's methods are always helpful in this, although they are less secure guides here and less easily applied than in the more familiar field of geometry. By gradually adducing ever wider fields of experience in order to explain the one currently being investigated, we find that in the end all fields become connected and enter into relations of mutual clarification, as is very evident even now in physics and chemistry.

19. If by tentative analytical procedures one has discovered a fundamental idea, that provides prospects for simpler easier and more complete views of a fact or set of facts, deduction of these facts with all their details from that basic idea serves as a test of the latter's value. If one could prove, as is possible only in the rarest cases, that that idea is the only possible assumption from which the facts can be deduced, this would amount to a full proof that the analysis is correct. Whewell points to the necessity of thus combining and mutually supporting deduction with what he calls induction. A general proposition that is a starting point for deduction is conversely the result of an inductive procedure; but whereas deduction proceeds methodically step by step, induction occurs in jumps that lie outside the reach of method. Hence inductive results must later be justified by deduction.[25]

20. From all this it emerges that the mental operation by which new insights are gained, which is usually called by the unsuitable name

'induction', is not a simple process but a rather complex one. Above all it is not a logical process, although logical processes may figure as auxiliary intermediate links. Abstraction and the activity of phantasy does the main work in the finding of new knowledge. The fact, emphasized by Whewell, that method can do little here, makes clear the mysterious feature that he thinks attaches to so-called 'inductive' findings. The enquirer seeks for a clarifying thought, but at first knows neither it nor the way in which it can be securely found. However, when the goal or the way to it has become revealed, he is as surprised by his find as somebody who was lost in a forest and suddenly, stepping out of a thicket, gains a free prospect and sees everything lying clearly before him. Not until the principle is found can method intervene in an ordering and adjusting capacity.

21. If one is guided by an interest in the connection between facts and lets the focus of attention roam repeatedly over the facts, whether present in sense perception or simply fixed in ideas, or varied and combined in thought experiments, then in a lucky moment one may perhaps espy the simplifying thought that furthers enquiry. That is all one can say in general. Here one will learn most by careful analysis of examples of successful reflection, by starting with problems where end and means are known and then turning to problems where one or the other is less sharply circumscribed, to finish with cases where indeterminacy, complication or paradox incite us to think. Since there is no adequate method to guide us towards scientific discovery, successful discoveries appear in the light of artistic achievement as was well known to Johannes Müller[26], Liebig[27] and others.

NOTES

[1] J. S. Mill, *System der deduktiven und induktiven Logik*, translated by Gomperz, 1884, I, pp. 209f.
[2] *Ibid.*, p. 235.
[3] Kant, *Prolegomena zu einer jeden künftigen Metaphysik*, Pt. 1.
[4] Beneke, *System der Logik als Kunstlehre des Denkens* I, pp. 225f.
[5] Cf. Ch. XXI of the present volume.
[6] *Ibid.*
[7] An expression often used by Schuppe in his epistemological writings.
[8] Cf. note 1 above.
[9] Apelt has already explained this well (*l.c.*, pp. 37f.), but he thinks that every incomplete induction is based on the a priori insight of a general causal law. Since however he himself admits that this knowledge tells us nothing about applicability in special

cases, it is of no help and can mislead as much as show the correct path. An arbitrary methodological presupposition would here serve just as well, indeed better since it is taken from experience and therefore has leading empirical features.

[10] A. Stöhr (*Leitfaden der Logik*) treats induction in the section 'Erwartungslogik' (=logic of expectation), pp. 94f., which seems to me to mark the proper point of view.

[11] Wallis, *Arithmetica infinitorum*, Oxford 1655.

[12] Cavalieri, *Geometria indivisibilibus continuorum nova quadam ratione promota*, Bologna 1635.

[13] Jac. Bernouilli, *Acta Eruditorum* 1686, pp. 360, 361.

[14] The same reflection is advanced by Galileo in discussing free fall geometrically.

[15] This example worked out by Kunze in Weimar is mentioned in Apelt, *Theorie der Induktion*, pp. 34, 35. We readily see how this leads to integral calculus. Take n very large and the lower powers vanish compared with the higher, and the expression differs only in notation from $\int x^2\, dx = x^3/3$. In the formula of the text dx is represented by 1.

[16] The importance of comparison was emphasized already by Whewell, that of abstraction especially by Apelt, but I feel that the significance of both as against induction is not sufficiently appreciated.

[17] J. S. Mill, *Logik* **I**, pp. 331–367.

[18] Whewell, *Philosophy of Discovery*, pp. 238–291.

[19] We therefore see that even then people were coming near to Kirchhoff's idea.

[20] Apelt, *Theorie der Induktion*, pp. 62f., 143f.

[21] Whewell, *l.c.*, p. 284.

[22] A single individual finding is always a fact and as such cannot be called either error or knowledge.

[23] Next to Kant I think Schopenhauer best appreciated the importance of intuition.

[24] Kepler liked to regard the Earth as animate, conceiving it to be an animal.

[25] Whewell, *The Philosophy of the inductive sciences* **II**, p. 92.

[26] J. Müller, *Phantastische Gesichtserscheinungen*, pp. 95f.

[27] Liebig, *Induktion und Deduktion*, 1874.

CHAPTER XIX

NUMBER AND MEASURE

1. Scientific knowledge arises from finding the connection between certain reactions or groups of reactions A and B in an object or a relatively stable complex of sensible elements. If for example we find that a species of plants systematically determined by a certain shape and position of leaves and flowers and so on (reaction A) shows in addition certain stimulated motions, geotropic and heliotropic phenomena (reaction B), this constitutes a finding in natural science. Fixing such knowledge in communicable form by a description that excludes misinterpretation remains an awkward business in spite of the development of a simplifying classificatory terminology. The same awkwardness repeats itself in the description of the behaviour of a closely related species of plant, which will again have many peculiarities that must be specially noted. What is even more difficult in view of these individual features, is to fix a wider group of insights in a comprehensive description. For a group of viviparous mammals it remains possible to demonstrate common physiological and anatomical reactions such as higher blood temperature, breathing through lungs, double circulatory systems and so on. However, if we consider the big anatomical and physiological differences of marsupials, or even monotremes, oviparous animals, platypus, anteater, with regard to 'mammals', to which in other respects they are quite close, we recognize how difficult it is to convey a wide group of zoological findings in comprehensive description. The aim to derive development and life cycle from the properties of cells and the dispositions of germ cells with regard to certain environmental conditions, can thus at best be regarded as an ideal hovering in the remotest distance.

2. In physics the picture seems to be in marked contrast to that just drawn. If two weights hang from the ends of a string over a pulley, we need only replace each by a number of smaller ones to be able to say that the weight consisting of the larger number of smaller ones will pull up the other. If weights are suspended from unequal arms of a lever, we divide

their lengths into smaller equal parts and form the product of the numbers for each arm and corresponding weight: the lever overbalances on the side of the greater product. The description of particular facts is thus easily obtained from counting the equal parts into which the constituent items can be divided. Thus all cases in the field (say, of levers) differ merely by the number of units in the relevant characteristics and are so similar that we can easily give a comprehensive description by indicating the relevant rule for deriving or calculating results from these numbers. For that reason, the comprehensive account will be possible for a rather wide factual region, for example for all machines, by means of the concept of work. Similarly, free fall or refraction can be described in most simple form by counting and recording results in a table, and a lucky glance at such tables may lead one to find the compendious rule of derivation that replaces them. The division of magnitudes of space, time and in tensity may be carried out by means of counting (measuring) in terms of arbitrarily small equal parts. This enables us to regard the measurable as made up of arbitrarily small elements ('infinitesimals') and to reduce their course to the behaviour of infinitely small elements in infinitely small time spans. For this we can set up general rules of calculation in the form of differential equations. A small number of such equations suffice to represent in principle all conceivable facts of mechanics, thermodynamics, electrodynamics and so on. The application of these equations can of course still present considerable difficulties in special cases. In the biological fields mentioned above the analogous step is as yet unattainable. Fields like chemistry that are so far only partially amenable to quantitative treatment, are half way between the two extremes.

3. If a quantitative reaction *abc* shows itself to be linked with another *klm*, this relation can at best be noted and fixed in language. Likewise for another pair *def* and *nop*. Even if the two facts are close, it will in general be difficult to bring them under a single expression. However, it will be the easier the more widely we can reduce qualitative differences to quantitative ones. Compare for example the facts of qualitative chemical analysis with the phase rule in physical chemistry. At closer range we notice that quantitative investigations are merely a special and rather simple case of qualitative ones. Physics has reached a higher level than for example physiology only because it deals with simpler and easier

problems and because these are all more of a kind so that their solutions are more readily given in comprehensive expression. Indeed, description by counting is the simplest imaginable and can be pushed to any degree of fine and accurate discrimination by means of a number system ready to hand without need for any new invention. The number system is a nomenclature of inexhaustible subtlety and extent and yet remains unsurpassed in clarity by any other nomenclature.[1] Moreover, by means of counting any number can be derived from any other, which is precisely what makes numbers so extremely suitable for representing dependences. The particular dependences differ from each other only by what is countable, and by considering this we similarly reach more general comprehensive rules of dependence. These obvious advantages of using quantitative features must incite us to look wherever possible for the quantitative that may be connected with qualitative aspects, in order gradually to reduce all investigations to quantitative ones. Thus the qualities of colour become refractive indices and wave lengths, sounds become frequencies and so on, all of them quantitative characteristics.

4. Quantitative investigations have moreover a special advantage over qualitative ones, where we wish to ascertain elements given in sensation in their dependence from each other, that is dependences external to our bodies and therefore belonging to physics in the widest sense. To keep these dependences pure we must as far as possible exclude the influence of the observer and of all elements internal to him. This is achieved by the fact that all measurement consists only in comparing what is qualitatively alike, noting what is equal and what unequal, thus taking out of play the quality of sense perception as such, which depends in part on the observing subject. Introspective psychology cannot at first eliminate qualitative features, and concepts of measure have hardly any meaning there as yet, but by building on physiology and indirectly on physics psychologists may change this state of affairs in the future.

5. Let us now try to clarify psychologically how the idea and concept of number arose from direct or indirect biological needs. Children of say between two or three, who have not yet acquired the concept of counting, notice at once if in an unobserved moment one removes something from a small group of like coins or toys, or adds something to it. Even animals

are no doubt driven by vital needs to distinguish small groups of the same fruit for example as to content, and to prefer the larger to the smaller. In the need to refine this ability to distinguish lies the origin of the number concept. The more members can be gathered into a group without losing the overview and the individuality of members, the more we shall value this ability. To start with, children manage groups of 2, 3 or 4. Contiguity in space and time may be helpful in forming the group, while difference of spatio-temporal position may condition the process of distinguishing the members. Thus arise the first idea of number, with or without name according to the influence of environment. These ideas develop through the senses of sight, touch and hearing, in the last case through attention to rhythm.[2] As we work with ideas of number while the objects change, we are led with the help of number names to the view of a uniform activity of reaction independent of the nature of the object, that is, to the concept of number.[3] To obtain clear ideas of the numbers of rather large groups, we arrange them into clearly ordered parts that are already familiar. This history of formations is embodied in the number symbols of the Assyrians, Egyptians, Mexicans, Romans and other peoples.[4] Our playing cards and dominoes too vouch for this history. We are thus right to take primary school children along the same road that all primitive peoples adopt spontaneously, by representing the groups of objects themselves in a clearly ordered and subdivided manner.[5] However, this device for maintaining a clear view of the numbers of members in a group does not take us far.

6. Apart from this device for ordering the members of a group another method commends itself: one assigns every member of the group to be surveyed to a member of a group of objects that is very familiar to us. Primitive peoples use the fingers of their hands, and sometimes the toes as well, as this second group[6]. As children we too have used this primitive means to strengthen our ideas of number by looking at these very familiar objects. If now during the coordinating process we call out the names of the fingers and quite undeliberately and from sheer habit we run through them always in the same order, these names of fingers, through frequent use and the loss of their original meaning, become number words[7]. The final name, because of the fixed order, determines the total content, the number of members of the coordinated, counted group[8]. That is the origin

of number words, as the history of human culture has shown. The need and the occasion for this development arose often enough, when one had to count the number of friends or enemies, or to divide the spoils of war or the bag of the hunt and so on.

7. By means of a small and obvious device, the method or coordination can be given unlimited scope of application, namely by counting groups of ten as members of a higher group, and ten of these as members of a yet higher group and so on indefinitely. Likewise any member can be regarded as a group of ten smaller equal members, an obvious procedure when counting (measuring) infinitely divisible quantities like length, but one that can be supposed operative anywhere.[9]

8. Let groups *A* and *B* each consist of members that are the same. Assign a member of *B* to every member of *A*: if this exhausts both groups we say that they have the same content, or more briefly that they are equal. If *B* is exhausted when *A* is not, then *A*'s content is greater than *B*'s. We call numbers those concepts by means of which we determine and distinguish from one another groups of like members with regard to content. Where number concepts replace ideas of number, there is no further need for immediate intuition but only for potential intuition. The number concept enables us to make the content of a group at least indirectly intuitive, wherever this may be necessary and we are prepared to make the effort. The learned quarrel whether cardinal or ordinal numbers are to be regarded as psychologically or logically primary need not concern us here. It is in any case impossible to select one of these systems set up after the event as being exclusively decisive for cultural developments. Names for small numbers may doubtless arise without an ordering principle. However, where numbers go beyond the directly intuitive, such a principle is essential to the formation of the concept of number, even if it is not explicitly mentioned. If we count objects that are, or are to us to pass for, identical, we affix the number words as signs of difference to objects that are otherwise hardly distinguishable; but we should soon lose control of them again, if these names were not also part of a simple and very familiar system of ordering signs. It is the ordering principle, by means of which every number potentially contains the idea of every preceding number and at the same time reveals clearly its position between two definite numbers

of the system, that makes for the great superiority of numbers as against ordinary names. Every alphabetic register, the page numbers of a book, every inventory that is numerically arranged and so on, clearly impress us with the value of order for rapid orientation.

9. Numbers are often called "free creations of the human spirit". The admiration for the human spirit expressed in this phrase is natural enough, given the imposing structure of arithmetic. However, an understanding of this creation is helped much more if we pursue its instinctive beginnings and consider the circumstances that produced the need for it. Perhaps this will lead us to the insight that the first formations in this field were unconsciously forced by biological and material conditions, and their value could not be appreciated until they were in existence and had often proved useful. It is only after the intellect was schooled by such rather simple formations that it could gradually rise to freer and conscious inventions that rapidly answered to the current needs.

10. Social intercourse and trade, buying and selling, demand the development of arithmetic. Primitive culture uses simple devices or calculating machines to facilitate ist calculations, for example the Roman abacus, or the Chinese tally board, which became known via Russia and has taken root in our primary schools. All these devices symbolizes the objects to be counted by means of small movable bodies, buttons, spheres or other markers; these latter, and not the heavy objects, are the items with which one operates. The groups of tens, hundreds and so on, are represented by special markers that have a special section assigned to them in the device.[10] If we take a somewhat freer and wider view of the concept of machine (or auxiliary device), we recognize that our Arabic or Indian numerals and their clear decimal notation, where a group that happens to be unrepresented is denoted by zero[11], is itself a calculating machine that can be constructed at any time by means of pencil and paper. This further relieves our attention, since the numerals save us the trouble of counting the members of each class of groups.

11. Various tasks now arise in social intercourse. For example, the need arises to combine two or more groups of like members into a single group and to give their number, that is, the problem of addition. The primitive

solution no doubt consisted in counting through the combined group regardless of the single groups having been counted or not. Indeed children still do this with small numbers, thus acquiring counting experience that they apply in adding larger decimally written numbers by adding units, tens and so on and carrying resulting units of higher order. This simple example is enough to show that calculating consists in saving direct counting by replcaing it as simply as possible by counting operations previously carried out, using the experience gained with such operations. To calculate is to count indirectly or mediately. Suppose we had to add 4 or 5 multi-figure numbers, and we proceeded first by direct counting and then by the usual rules, we recognize the enormous saving in time and effort by the latter. Equally readily the tasks of practical life give occasion to problems of subtraction, multiplication, division and so on, and again one could show that these are cases of simplified and shortened counting using prior experience; we shall refrain from further detail.[12]

12. The material environment is thus not nearly so innocent of the development of arithmetical concepts as is at times supposed. If physical experience did not tell us that a multiplicity of equivalent, immutable and permanent objects exists, nor biological needs impel us to gather these into groups, then counting would be without sense of purpose. Why count if the environment is totally impermanent and different at every moment as in a dream? If direct counting in order to determine larger numbers were not impossible in practice because of the time and effort required, the inventions of calculation or mediate counting would never have forced itself on us. By direct counting we take note only of what is given in direct sense perception. Since calculating is a form of indirect counting, it cannot tell us anything essentially new as regards the sphere of sense experience, indeed nothing that could not be learnt from direct counting. How, then, could mathematics prescribe a priori laws to nature? Indeed, it must confine itself to demonstrating agreement between results and starting points of calculations, using in this the experience concerning the mathematician's own ordering activity: having a good grasp of this activity still remains extremely valuable and continues to illuminate a fact from all angles.

13. The simple beginnings of arithmetic developed in the service of

practical life. Further progress results when arithmetic becomes a special calling for life. A person who frequently has to carry out similar calculations gains some special perspective and facility in this and will most readily consider how to simplify and shorten his procedures. Thus arises algebra, whose symbols stand not for special numbers but rather turn the attention to the form of the operations. Algebra settles all formally like operations once and for all, leaving only a residue of effort in calculating with special numbers. The theorems of algebra too, indeed of mathematics in general, always express equivalences of ordering activities. This holds for example of the two sides of the equation that expresses the binomial theorem. If alongside a quadratic equation we write the formula for the roots, we have fixed the equivalence of two operations, just as by placing together a differential equation and its integrals. Incidentally, the symbolic language of mathematics is again a kind of brain-relieving machine on which we easily and often perform symbolic operations that would otherwise tire us out. Besides, mathematical script is the most beautiful and perfect example of a successful pasigraphy, although it applies to a restricted field.

14. Looking at groups of equivalent objects leads us directly to the concept of integers. If objects are individuals not divisible into equivalent parts, then only integers can be properly used in counting them. However, division, as the analytic opposite of synthetic multiplication, leads us in special cases to divide the separate objects or units into fractions, which of course makes sense only for divisible units, or purely arithmetic operations like taking roots, as analytic opposite of the synthetic process of raising to a power, lead to the fiction of irrational numbers as not completely determinable by finite counting operations. Even the simplest operations such as addition and subtraction, provide inducement to the formation of new concepts. The operation $7+8$ is always performable, and so is $8-5$, but $5-8$ involves an impossible request if we are dealing with identical objects without opposite features. This last operation however becomes readily possible and acquires an intelligible sense, as soon as the units in question stand in contrast as credit and debit, forward and backward steps and so on. Thus we obtain the concept of the contrast of positive and negative numbers, which are denoted by means of the ordinary signs for addition and subtraction, the operation in which the need to

fix this contrast first revealed itself. Strictly speaking, we should have to use special symbols for this. The sign rule for the multiplication of signed numbers is given by noticing that the product $(a-b)(c-d)$ must agree with the one obtained by inserting the simple values m and n for the factors. For numbers without opposites such a rule has no sense. As it is, both negative and positive numbers have positive squares, which means that at first blush the square root of a negative number must seem impossible or imaginary. Indeed it was long regarded as impossible; like negative numbers; and must remain so as long as the only contrast is that between positive and negative. Wallis[13], guided by geometrical application of algebra, first viewed $\sqrt{-1}$ as the geometric mean between $+1$ and -1, that is $+1:i=i:-1$ and $i=\sqrt{-1}$. This view recurs several times more or less clearly until Argand[14] expounds it with precision in full generality. By relating proportionality not only to size but also to direction, he represents the expression $a+b\sqrt{-1}$ as a vector in a plane: from the origin we go a distance a along one direction and then b at right angles. The points of the plane can thus be represented by complex numbers.

15. The practice of arithmetic thus sometimes leads to analytic operations that at first sight seem to be impossible or to issue in results devoid of sense. However, on closer examination we find that a slight modification or extension of hitherto applicable arithmetical concepts dissolves the impossibility and the result allows an entirely clear interpretation, albeit in a wider field of application. When in this way mathematicians were forced against their intention to modify their concepts and had learnt the value and advantages of such procedures, it became rather more natural to let free invention meet needs more rapidly or even run ahead of them, witness the brilliant examples of the inventions of Grassmann, Hamilton and others as regards vector calculus, where number concepts are adapted to the needs of geometry, kinematics, mechanics, physics and so on.

16. Let us consider also a modern attempt to give sharper conceptual formulation not only to the infinitely increasing and decreasing but also to the actual infinite. Galileo in the first day of his dialogues (1638) mentions the paradox that the infinite set of integers seems much greater than the set of squares, whereas to every member of the former there

corresponds one of the latter so that the sets must be equal. He concludes that the categories of equal, greater and smaller do not apply to infinities. This reflection, traces of which go back to ancient times, lead to the investigations of G. Cantor on set theory. Galileo's example shows how one might reach the following definition: two sets have the same power if to each element of the one there is one and only one element in the other, and vice versa. Two such sets are called equivalent. A set is infinite if it is equivalent to a part of itself.[15] Cantor's investigations show that even in the field of the actually infinite, appopriate construction of ordering concepts enables one still to maintain a conspectus.

17. As regards the logico-mathematical representation of number theory, we may refer to the clear and attractively written book of L. Couturat[16]. Our point of view corresponds to the psychological and anthropological considerations that are at all events a necessary complement to the logical aspects. An intensive historical study of the subject's development may here have the same salutary effect as Felix Klein's[17] famous lectures.

18. Where from the outset we are dealing with objects that are discrete as far as our actual interest goes, the application of number theory is relatively simple. Many objects of enquiry, such as extension and duration, intensity of force and so on, do not immediately form groups of directly countable equivalent members. Of course there are many ways of dividing them into equivalent countable members, and likewise for each of these in turn and so on, but the limits of division must be made artificially perceptible and distinguishable while the degree of division at which we wish to stop and therefore the size of the last units is arbitrary and conventional. However, once we have prepared a continuum in this way, a portion of it which happens to be involved in carrying out an investigation can be determined by counting the parts, that is by measuring to any desired degree of accuracy. The artificially constructed numerical continuum is a means for pursuing the conditions of natural continua to any level of accuracy. Yet somewhere or other we must stop at a limit, because the senses even when artificially supported are imperfect. For we cannot observe with limitless accuracy that a measuring scale covers the object to be measured, or that their ends coincide. This same uncertainty likewise infects the number that indicates the result of measuring the relation

between the object to be measured and the scale of measurement. The same flaw indeed attaches also to practical applications of arithmetic to discretely countable objects, since their presupposed perfect equivalence is never actually satisfied.

19. If we have to reduce continuously variable physical conditions or magnitudes to a measure, we must first choose an object of comparison or unit of measurement and lay down how another object is to be judged equal to the standard. We regard objects as equal in a certain respect if under unchanged conditions they can stand for each other without loss of result. Two weights are equal if they cause the same deflection when singly and separately placed in the same pan of the same balance; two electric currents are equal if when successively passed through the same unchanged galvanometer they determine the same deflection of the needle; similarly for magnetic poles, degree and quantity of heat and so on. If n weights equal to the unit are put in the same pan, n units of current are passed through the same galvanometer wire (or through closely adjacent wires) and so on, then if the units are entirely interchangeable, the result is determined only by the numerical measure n.[18]

20. If the decisive conditions of a series of similar physical cases have been determined by means of numerical measures, we can often represent the way they depend on each other by means of a simple rule of derivation with an accuracy sufficient for the representation of the facts. This is illustrated by such examples as the law of refraction, Boyle's gas law, Biot-Savart's law. When such laws are once known, they can often facilitate indirect measurement where it is difficult or impossible to measure directly. For example, it is difficult continuously to vary the intensity of a light source, but easy to judge by eye that two light sources are equal, by the equal brightness of illumination of two contiguous equal areas equally distant from the source and illuminated at right angles. If now we can show that an area illuminated at right angles by one source appears to be just as bright as an equal illuminated area by 4, 9, 16... closely gathered lights equal to the first but at 2, 3, 4... times the distance, then the measurement of any state of illumination can be reduced to ascertaining the relations of distance at equal brightness, even though the eye is restricted to judging equal and unequal brightness.

21. When putting together a physical quantity out of similar parts, we must alwasy take care whether this fitting together corresponds to a real addition. Whereas for example a more intense light can without reservation be put together from similar, independent (incoherent) lights and the total intensity is the sum of the partial ones, it is well known that under certain conditions we can no longer do this with the lights of the one small light source. Likewise, the intensity of sound of several tuning forks of the same note is not in general the sum of the individual intensities, but only in the case where the phases coincide. Other such caveats are mentioned in *W*2, pp. 39–57.

NOTES

[1] Cf. also Natorp, *Die logischen Grundlagen der exakten Wissenschaften*, Leipzig 1910.

[2] Sighted and blind, hearing and deaf all alike learn to count. The deaf and dumb Massieu himself said that he knew numbers before receiving instruction, his fingers having taught him. (Tylor, *Einleit. i. d. Studium d. Anthropologie*, p. 372; cf. also *Anfänge d. Kultur* **I**, p. 241 f.).

[3] Number concepts are acquired only by carrying out counting operations in different cases. Cf. also Ch. VIII, note 6.

[4] Look at Table I in M. Cantor, *Mathem. Beiträge zum Kulturleben der Völker*, 1863.

[5] G. Schneider, *Die Zahl im grundlegenden Rechenunterricht*, Berlin 1900.

[6] Details in Tylor, *Anthropologie*, pp. 372f. The Tamanacas on the Orinoco river say 'whole hand' for five, 'both hands' for ten, 'whole man' for twenty. Traces of this primitive mode of counting survive in highly advanced cultures, for example 'quatre-vingts' for eighty.

[7] Tylor, *Anfänge* **I**, pp. 248f.; *Anthropologie*, p. 373.

[8] A. Lanner, *Die wissenschaftlichen Grundlagen des ersten Rechenunterrichts*, Vienna & Leipzig 1905, which contains very good psychological observations on children's learning to count, the first number concepts and so on. The concept of 'unit' can emerge only as specialised abstraction from the general number concept. The problem 1×2 of 1×1 can be understood only if the problems 2×2 or 3×2 are, and likewise for a^1 after a^2, a^n and so on. Similarly Ribot, *L'évolution des idées générales*, Paris 1897, p. 160.

[9] Our decadic system, in analogy to which any other system could be devised, owes its natural origin to our ten fingers.

[10] The mechanical calculating machines of Pascal, Leibniz, Babbage, Thomas and others, which by turning a handle and gear transmissions carry out arithmetical operations, as well as modern integrators are a natural development of primitive calculating machines.

[11] The important invention of zero is ascribed to the Indians.

[12] My account of these questions in 1882 (*P* 3, p. 224) comes very close to the views of Helmholtz and Kronecker reported in the *Festschrift* for Zeller (1887). Other points I have tried to elucidate in *W* 2, pp. 65f. Cf. also the splendid and detailed treatment in M. Fack, 'Zählen und Rechnen' (*Zeitschr. f. Philos. u. Pädagogik*, Flügel & Rein **2**,

196f.). Also Czuber, 'Zum Zahl- und Grössenbegriff' (*Zeitschr. f. d. Realschulwesen* **29**, 257).

[13] Wallis, *Algebra*, 1673, Chs. 66–69.

[14] R. Argand, *Essai sur la manière de représenter les quantités imaginaires*, Paris 1806. His view becomes clear by the following examples. Draw a vector r from an origin, and a vector nr from the same origin at an angle ϕ to the first, and likewise n^2r at 2ϕ in the same sense. Then the second vector is for him the mean proportional between the first and third. His essay is a model example for representing a new idea.

[15] G. Cantor, *Grundlagen einer allgemeinen Mannigfaltigkeitslehre*, Leipzig 1883. Cf. also Couturat's book below, pp. 617f.; and A. Schönflies, 'Die Entwicklung der Lehre von den Punktmannigfaltigkeiten', *Jahrb. d. Deutschen Mathematiker-Vereinigung* **8**, 2, 1900.

[16] Couturat, *De l'infini mathématique*, Paris 1896. A nice and compact survey of the development of number concepts in O. Stolz, *Grössen und Zahlen*, Leipzig 1891.

[17] F. Klein, *Anwendungen der Differential- und Integralrechnung auf Geometrie, eine Revision der Prinzipien*, Leipzig 1902.

[18] Cf. Helmholtz, 'Zählen und Messen', *Philos. Aufsätze, E. Zeller gewidmet*, 1887, pp. 15f.

CHAPTER XX

PHYSIOLOGICAL SPACE IN CONTRAST WITH METRICAL SPACE

1. Physiological space, the space of our sense intuition found ready made when our consciousness is fully awake, is very different from metrical space which is conceptual. The concepts of geometry are mostly acquired through deliberate experience. The space of Euclidean geometry has the same properties everywhere and in all directions, and is unbounded and infinite. If with this we compare visual space, the 'seeing space' of Joh, Müller and Hering, which is familiar above all to the seeing observer, we find that it is neither homogeneous everywhere nor in all directions, nor infinite, nor unbounded[1]. The facts, discussed by me elsewhere[2], tell us that 'up' and 'down', 'near' and 'far' correspond to quite different sensations. Likewise for 'right' and 'left', though here the sensations are more similar, as a consequence of the facts of physiological symmetry.[3] The anisotropy shows itself in phenomena of physiological similarity[4]. The apparent swelling of stones at the entrance to a tunnel when a train moves in, their shrinking when the train moves out, quite noticeably reminds us of the daily experience that visual objects in visual space cannot be displaced without shrinkage or expansion as are the corresponding immutable geometrical objects. This is clear even from familiar objects at rest. A wide and deep cylindrical glass put over one's face or a cylindrical stick held horizontally against the arches of the eyebrows will in this unusual position appear noticeably conical and opened up towards the face like a trumpet, or thickened, in the case of the stick[5]. Visual space resembles the space of metageometry rather than that of Euclid. It is not only bounded, but rather narrowly at that. An experiment by Plateau shows that an after image projected on a surface does not shrink noticeably further if the latter recedes beyond 30 metres from the eye. Any naïve person who has to rely on immediate impressions, including the astronomers of antiquity, sees the sky roughly as a sphere of finite radius. Indeed the flattening of the celestial vault, which Ptolemy knew already and Euler discussed in modern times, tells us of an unequal extension of visual space in different directions.

O. Zoth[6] has started to give a physiological account of this fact by showing that the phenomenon depends on the angle of elevation of sight relative to the head. The narrow boundaries of visual space follow from the very possibility of panoramic painting. Finally we note that at the outset visual space is not at all metrical: its positions, distances and so on are qualitatively distinguished, not quantitatively. What we call visual measure develops only gradually on the basis of primitive physical and metrical experience.

2. The skin, too, a closed surface of complicated geometrical shape, conveys spatial sensations. We distinguish not only the quality of stimuli but also the spot that is stimulated by an additional sensation. If this last differs only from place to place and the more so the further the separation, then our biological needs are already met. The great anomalies of the skin's spatial sense as against metrical space were well expounded by E. H. Weber.[7] The distance between the two points of a compass at which contact in two adjacent places can just be distinguished is 50–60 times smaller on the tip of the tongue than in the middle of the back. The parts

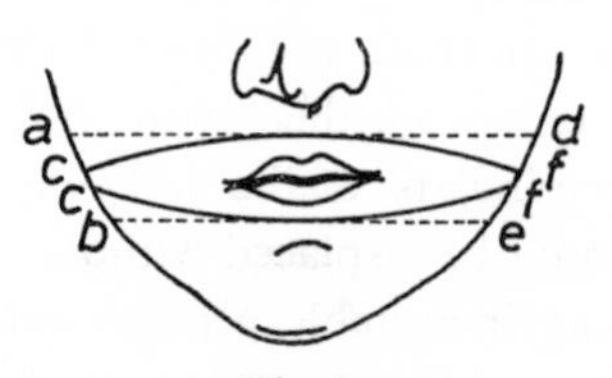

Fig. 9.

of the skin show large gradations of spatial sensitivity. A compass whose points grip the upper and lower lip between them seems noticeably to close if one laterally strokes the face with it (Figure 9). If the compass points are set at the distance of two adjacent finger tips and then slid along the inner surface of the hand on to the forearm, they seem to fall together completely there (Figure 10). In the figure the actual paths are dotted and the apparent ones drawn in full. The shapes of bodies that the skin touches are distinguished[8], but the skin's spatial sense is much inferior to the eye's[9]. The tip of the tongue can just discern the circular cross section of a pipe of 2 mm diameter. To the skin's space there corresponds a two-dimensional, finite unbounded (closed) Riemann space. By the

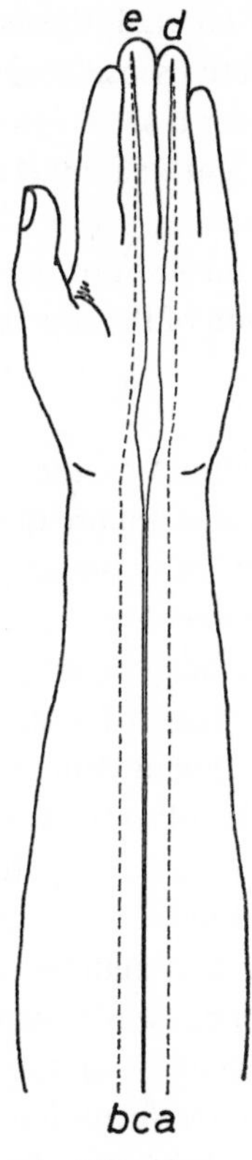

Fig. 10.

sensation of the motion of limbs, especially arms, hands and fingers something that corresponds to a third dimension is added. We gradually learn to interpret this system of sensations by the simpler and more intuitive framework of physics. Thus in the dark we can estimate the thickness of a table top tolerably well between thumb and index, or even if we use one finger of each hand. Tactile space has as little to do with metric space as does visual space, being anisotropic and inhomogeneous like the latter: the principal directions of organization: front–back, up–down, right–left, agree in being unequal in both physiological spaces.

3. That we find a sense of space undeveloped where it has no biological function can hardly surprise us. What would be the point of being in-

formed about the position of internal organs, since we have no influence on their function? Thus a spatial sense does not reach far into the nose. One cannot distinguish whether a smell introduced by one of two small tubes is felt on the left or the right[10]; but tactile sensitivity, according to E. Weber reaches as far as the eardrum[11] which decides whether the stronger feeling of sound comes from the right or the left. This helps to ascertain, albeit very roughly, the position of the source of sound, but is inadequate for more precise orientation.

4. Even though in some sensations the characteristics of space and location make themselves felt much more distinctly than in others, James is surely right in holding that every sensation has a certain spatiality.[12] To every sensation there corresponds a location because of the stimulated element, and since there are usually several or many elements, we can in a certain sense speak of the volume of a sensation. In his account James refers several times to Hering, who denotes the impression of glowing surfaces or illuminated spaces and so on as space-like. Musical notes are usually given as examples of quite unspatial sensations, but Hering's incidental remark[13] that deeper notes have a greater volume than the higher seems to me apposite. The highest audible tones of König's rods almost give the impression of a pin-prick, while deep notes seem to fill the whole head, or better, the whole of acoustic space. The possibility of localizing the source of sound even if imperfectly likewise suggests a relation between the sensation of notes and space. Steinhauser's parallel between bifocal vision and binaural hearing may not go very far, but there is a certain analogy, and localization is best achieved by means of the high notes of small volume and more sharply defined position.[14]

5. The physiological spaces that different senses embrace are only in part a common physical field. To the sense of touch, the whole skin is accessible, while only part of it is visible. On the other hand, the sense of sight as a sense of distance reaches physically much further as such. Spatial orientation is much more indeterminate and confined to a more restricted space by ear than by eye. However loose the original links between the different spatial sensations, they do enter into connection through association, and whatever system is of more practical importance at the moment is ready to complement and represent the others. The spa-

tial sensations of different senses may be quite related, but they will hardly be identical. There seems no need to strengthen and complete the evident and adequate associative bond by means of a general spatial sense.[15]

6. All spatial sensations have the function of correctly guiding self-preserving movements. This common function constitutes the associative bond between them. A sighted observer is mainly led by sensations and ideas of visual space, for these are most familiar and beneficial to him. A figure slowly drawn on his skin, in the dark or while he shuts his eyes, is translated, by means of the movements sensed, into a visual image by thinking of himself carrying out these movements. For example, if a figure that somebody draws on my forehead is to appear to me as R, then if he stands in front of me he must write Я. On the back of my head he would have to write R, and on the abdomen ꓤ, if I am to recognize these signs as R written by myself.[16] It is as though in the first two cases I imagine my head as transparent and myself as standing behind it in the same orientation and carrying out the writing movements. In the last case I imagine myself as writing on and reading off my own abdomen. It is quite difficult for a sighted person to find his way into the spatial ideas of a blind one. The fact that these too can reach a high degree of clarity is shown by the achievements of the blind geometer Saunderson. Still, for him it must have remained somewhat difficult to orientate himself, as is shown by the very simple way in which he divided his blackboard into squares. Into the corners and centres of the field he used firmly to embed pins, whose heads he connected by strings. However, just because of this simplicity, his original expositions must have been especially easy to understand for beginners. Thus he proved the proposition that the volume of a pyramid is a third of that of a prism of equal base and height by dividing a cube into six congruent pyramids each with a face of the cube as base and with the vertex at the centre of the cube.[17]

7. We may assume that the system of spatial sensations for all animals whose bodies have three principal directions, as man's does, is very similar to his system, although less developed. These animals differ with regard to up and down, and front and back. As regards left and right they seem to be the same, but the symmetry of geometry and mass, which ministers to speed of locomotion, must not deceive us about the anatomi-

cal and physiological asymmetry: however slight it may seem, it emerges clearly in the fact that animals closely related to symmetrical ones often assume very unsymmetrical shapes; witness the unsymmetrical plaice, or the symmetry of shell-less snails and the asymmetry of their shelled relatives.

8. If now we ask what is common to physiological and geometrical space, we find very few agreements. Both spaces are three-dimensional manifolds. To every point $A, B, C, D\ldots$ of geometrical space there corresponds a point $A', B', C', D'\ldots$ of physiological space. If C lies between B and D, then C' lies between B' and D'. We might say: to a continuous motion in geometrical space there corresponds a continuous motion of the coordinated point in physiological space. That the continuity we assume for the sake of convenience need not be real for either space has been pointed out elsewhere [18]. If further we may assume without scruple that physiological space is innate, it shows too slight an agreement with geometrical space to be considered as an adequate basis for an a priori development of geometry in Kant's sense. At best one might use it as a basis for a topology [19]. Why, then, is physiological space so different from geometrical space? How do we nevertheless manage to travel from the ideas of the first to those of the second? These are the questions we shall now try to answer as far as possible.

9. Let us start with a simple and general teleological reflection. Suppose a frog's skin is stimulated in various places by drops of acid. The animal will respond to every stimulus with a specific defensive movement corresponding to the stimulated spot. Qualitatively equal stimuli meeting different elementary organs and entering the animal body along different paths, set off reactions that through different paths reverberate back into the environment [20]. What holds for the skin holds also for sight and every other sense. Not only the movements of defence or flight but also those of attack are specialized for the stimulated spot and for the individuality of the organ affected. We may recall the frog snatching after flies, or the newly hatched chicks picking at grains. What has been said so far applies also to mere reflex reactions, for plants as well as for lower animals. However, if reflex reactions are to be influenced and modified for a purpose, and voluntary action is to replace them, then the stimuli must

become conscious as sensations leaving memory traces. Indeed, as self-observation shows, we recognize not only the quality of a stimulus, for example of burning, whatever sensitive spot may be affected by it, but also the different spot stimulated. Both factors determine our reactive movement. We may assume that in these cases a variable element attaches to sensations that are qualitatively alike, depending on the specific nature of the elementary organ, on the particular spot stimulated, or to use Hering's phrase, on the location of the attention. In spatial perception there is an especially marked expression of a most perfect mutual biological adaptation of a multiplicity of elementary organs.

10. We may imagine spatial perception physiologically founded as follows. The sensation supplied by an elementary organ partly depends on the kind (quality) of stimulus; let us call this part sensation proper. Moreover, suppose a part of the activity of elementary organs is determined only by its own individuality, so that it is the same whatever the stimulus, though it will vary from organ to organ; let us call this part organ sensation and regard it as identical with spatial sensation. We assume organ sensations to be the more varied the further the ontogenetic relation of elementary organs of common descent. Organ sensation or spatial sensation can arise only if there is stimulation of elementary organs, and it remains the same whenever the same organ or complex of organs is called into play. We may say that physiological space is a system of graduated organ sensations which would of course not exist without sensations proper; but if this system is aroused by varying sensations it forms a permanent register in which they are ranged. The only assumption we make about elementary organs are very similar to what we would very naturally find confirmed by experience as regards separate individuals of common descent but different degrees of affinity. Still, what we are trying to work out here is not a genuine theory of spatial perception, but merely a physiological paraphrase of observed psychological features. This paraphrase nevertheless seems to contain something that can be reconciled with innate views of physiological space, with the observations of E. H. Weber[21], and with his theory of spheres of sensation, with Lotze's[22] doctrine of local signs insofar as it is physiological, with the views of Hering and with the critical reflections of Stumpf[23]. This opens the prospect to a phylogenetic and ontogenetic understanding of spatial perceptions, and,

if one day these matters have been clarified, to a physical and physiological understanding of it in principle.

11. If the system of spatial sensations is to respond to immediate biological needs and to guide the self-preserving movements of the body, it can hardly be imagined other than we find it. Every system of sensations, including that of spatial sensations, is finite; an inexhaustible series of sensible qualities or intensities is physiologically unthinkable. Different bodily organs require different spatial sensitivities to guide their functions. Hence the macula lutea of the retina, the tip of the tongue and the fingers are well appointed with spatially sensitive organs, as against the lateral parts of the retina and the skin of the upper arms or back. Spatial sensations must relate to the limbs of the body and be orientated towards them, if they are to satisfy biological needs. It is important for us to distinguish up and down, front and back, left and right; that is, relations with regard to the body. A mere relation of locations to each other as in geometry would not help us. Moreover, it is appropriate that for nearer objects of sight that are biologically more important, there is a richer gradation of stereoscopic indications of depth, than for objects that lie further off and are less important, where there is greater economy of the limited store of indications. If, using criteria of appropriateness, we were to construct physiological space starting from geometrical space, it cannot turn out much different from what we find it to be.

12. The incongruity between physiological and geometrical space is never noticed by people who do not specially examine the matter, provided that geometrical space does not seem to them a monstrous falsification of innate space; a closer consideration of the conditions and development of human life will explain this. Spatial sensations guide our movements, but we rarely have reason to notice or analyse them accurately as such. We are much more interested in the goal of the movement. After acquiring the first primitive experiences about physical bodies, distances and so on, we become almost fully absorbed in attending to these. If man was like an immobile marine animal, unable to leave his location or greatly change his orientation, he would hardly ever have gained the idea of Euclidean space. His space would then be related to Euclidean space roughly as a triclinic to a tesseral medium, so that it would always remain anisotropic

and bounded. Arbitrary locomotion and orientation of the body as a whole promote the insight that we can move equally everywhere and in all directions, that space can be represented as uniform and isotropic, unbounded and infinite. The geometer says that from any point and in any direction the same constructions can be carried out. In uniform locomotion the same spatial changes constantly recur. Similarly for uniform change of orientation, for example rotation about a vertical axis. This reveals not only the uniformity of certain spatial experiences, but also the fact that they are inexhaustible, repeatable and can be continued. In place of the fixed spatial values of objects, which is all that a man finds who can move only his limbs, locomotion introduces flowing spatial values. Thus our spatial experiences gradually come nearer to Euclidean space, without reaching it completely along this path.

13. Just as spatial sensations determine the movements of individual limbs, they may at times lead to locomotion in general. A chick can look at an object and peck at it, or even be determined by this stimulus to turn and run towards it. A child behaves similarly when it looks and grabs at a target, crawling there if it is out of reach and finally one day getting up and walking there in a few steps. All such cases that continuously run into one another must be considered in the same way. Inducement to intense locomotion and change in orientation proceeds not only from optical stimuli, but can be imitated also by chemical, thermal, acoustic, galvanic[24] and other stimuli, even in blind animals. Indeed we have observed extensive locomotion and reorientation in naturally blind animals such as worms and in animals that become blind by adaptation (moles, cave animals), except that with blindness the perception of distance that determines movements is confined to a smaller region.

14. The main difficulty in analysing physiological space is that when we start thinking about these matters, we, having been educated, are already too familiar with scientific geometrical ideas that we import everywhere as self-evident. The best example for this is the well-known theory of optical direction lines which was able to maintain itself from Ptolemy to Kepler and Descartes, and was not definitively eliminated until Hering. The enquirer in this field must adopt an artificially naïve stance and try to forget much that he has learnt, if he is to gain an innocent eye.

Without entering into physiological details, let us add one simple general reflection.[25]

15. Certain stimuli are followed by certain movements determined by reflexes. These movements in turn excite peripheral stimuli that leave traces in the cerebral cortex as images of these movements. If for some reason, for example association, these images come alive again, they are apt to provoke the same movements again. The points of space are known to us physiologically as goals of various movements of grabbing, looking and locomotion. The images are no doubt tied to more or less sharply defined parts of the brain, that is they are somehow localized. It is unlikely that the whole brain is equally involved in all of them, as follows from the conditions of centrifugal outflow and centripetal inflow of stimuli. Thus we may perhaps regard the various goals as coordinated with the centres of the image groups in the cortex. As far as space is physiologically viewed, its points would thus be represented by places in the brain. Spatial sensations would correspond to the organ sensations of these places. One will of course assume that the view of space is mainly pre-formed by innate organization, but there remains nevertheless wide scope for individual development, which will be very different for blind and sighted, for a sculptor, painter, hunter, or musician.[26]

16. Kant asserts that we can never form the idea that there is no space although one can well imagine that there are no objects to be found in it. Today hardly anybody doubts that sensations proper and spatial sensations enter consciousness only together, and can vanish again. The same presumably holds of the corresponding ideas. If for Kant space is not a 'concept' but 'pure (mere) intuition a priori', enquirers today incline to regard geometrical space as a concept acquired through experience. The mere system of spatial sensations cannot be intuited, but we can neglect sensations proper as subsidiary, and if we do not sufficiently notice this process, which occurs readily and unobserved, the idea easily arises that a pure intuition has occurred. If spatial sensations are independent of the quality of the stimuli that help to excite them, then within the limits mentioned earlier we may make statements about these sensations independently of physical experience, as is incidentally likewise the case for any system of sensations, for example those of colour or sound. This

much of Kant's view remains correct, but this is not enough to form a basis for developing geometry, for there we decidedly require further concepts, and ones that rest on experience.[27]

17. Geometrical space is conceptually clearer, but physiological space is nearer to sensation. That is why when we are occupied with geometry, the properties of physiological space nevertheless often make themselves felt. In our figures we distinguish according to physiological factors points that are nearer from those that are further off, those that lie on the right from those on the left, those above from those below, although geometrical space has no relations to our bodies but only of points to each other. Amongst geometrical structures the straight line and the plane are marked by their physiological properties and are indeed the first objects of investigation. Symmetry is noticed above all through its physiological advantages and thus attracts the geometer's attention. Moreover there is no doubt that symmetry is involved in the choice of spatial division into right angles. That similarity was investigated before other geometrical affinities, is due likewise to physiological circumstances. Descartes' coordinate geometry is a liberation from physiological influences, but vestiges remain in the distinction between positive and negative coordinates as being counted respectively right and left, or up and down and so on. This is convenient and intuitive but not necessary. A fourth coordinate plane or the determination of a point by its distance from four non-coplanar base points frees space from the constant recurrence of physiological factors. The need to indicate 'right way round', 'left way round' and the distinction between properly congruent and symmetrically congruent figures is thus eliminated. Of course, we cannot eliminate the historical influences that the physiological view exerts on the development of geometry.

18. Even in its closest approximation to Euclidean space, physiological space remains markedly different. This manifests itself in physics too. A naïve person readily learns to overcome the difference between left and right or front and back, but not so for up and down, because his geotropy puts obstacles in the way of permanently interchanging these last two directions. To denote that something is impossible, Herodotus attributes to Sosides of Corinth the sentiment that sooner heaven will be below and

earth above (V, 92). What Lactantius urged against the theory of the antipodes, where people's heads and tree-tops point downwards, which went against the grain with Augustine and continued for centuries to seem inconceivable to the naïve, becomes quite understandable in terms of the properties of physiological space. We have less reason to marvel at the narrow minds of those who opposed the antipodean theory, than to admire the power of abstraction of Archytas of Tarentum, Aristarchus of Samos, and other thinkers of antiquity.

NOTES

[1] The expressions are here to be taken in Riemann's sense.

[2] *A* 4, pp. 86f.

[3] *Ibid.*, p. 88.

[4] *Ibid.*, p. 89.

[5] Since then a detailed and thorough account of the question here touched on has been published: F. Hillebrand, 'Theorie der scheinbaren Grösse bei binokularem Sehen' (*Denkschr. d. Wiener Akademie, math.-naturw. Cl.*, Vol. 72, 1902). The expression 'apparent size' is taken in the sense of Hering's 'seen size'. The phenomenon mentioned in the text stands out very clearly and measurably with the author's ingenious method of observation. R. v. Sterneck, 'Versuch einer Theorie der scheinbaren Entfernungen', *Ber. d. Wiener Akademie, math.-naturw. Cl.*, Vol. 114, *A*. IIa, p. 1685 (1905).

[6] O. Zoth, 'Über den Einfluss der Blickrichtung auf die scheinbare Grösse der Gestirne und die scheinbare Form des Himmelgewölbes', *Pflügers Archiv.* **78**, 1899. An extension of Hillebrand's experiments with regard to the direction of vision would now be desirable.

[7] E. H. Weber, 'Über den Raumsinn und die Empfindungskreise in der Haut und im Auge', *Ber. d. kgl. sächs. Gesellsch. d. Wissenschaften, math.-naturw. Cl.* 1852, pp. 85f.

[8] One must of course ensure that skin and imposed body are in close contact. When I had various objects placed in my apoplectically paralysed hand I failed to recognize some of them, from which it was concluded that my sense of touch was partially disturbed. However the inference was wrong. For immediately after this examination I asked somebody to close my paralysed hand and then at once recognized all objects placed in it.

[9] E. H. Weber, *l.c.* p. 125.

[10] *Ibid.*, p. 226.

[11] *Ibid.*, p. 127.

[12] James, *The Principles of Psychology* **II**, especially pp. 136f.

[13] My memory probably rests on an oral statement, since I can find no reference to this in Hering's writings.

[14] *A*, p. 206.

[15] Cf. however E. H. Weber, *l.c.*, p. 85.

[16] *Ibid.*, p. 99.

[17] Diderot, *Lettre sur les aveugles.*

[18] *W*, p. 76.

[19] Cf. Listing, *Vorstudien zur Topologie*, Göttingen 1847.

[20] I here follow a view put forward by R. Wlassak, though my version is somewhat modified and amplified. Cf. his splendid report Wlassak 'Die statischen Funktionen des Ohrlabyrinths', *Vierteljahrschr. f. wiss. Philosophie* **XVII**, I p. 29.
[21] E. H. Weber, *l.c.*
[22] Lotze has expounded his doctrine in several writings: *Medizinische Psychologie*, 1852; *Mikrokosmos*, 1856; in Wagner's *Handwörterbuch der Psychologie*; in the appendix to the book by Stumpf mentioned below.
[23] Stumpf, *Über den physiologischen Ursprung der Raumvorstellungen*, 1873.
[24] Loeb, *Vergleichende Gehirnphysiologie*, Leipzig 1899, pp. 118f.
[25] For details I must refer to physiological literature in general. Cf. also *A* 4, pp. 137–146, and the article in *The Monist* **XI**, April 1901, pp. 321–338.
[26] In the course of an individual person's development the sense of space probably undergoes important changes. When travelling by rail as a child I almost always experienced the phenomenon of micropsy. I saw far-off hills, mountains, and buildings and people on them as quite small and nearby models, as attractive Lilliputian landscapes, although I knew that this was not really so. Later I could not recapture this impression. Cf. *A*, p. 1944, an analogous observation about the sense of time. However, our spatial sense may also suffer very rapid temporary changes. After a serious illness during childhood I would, when tired by school hours, see others as very small and far-off. Some drugs, like hashish, are well known to produce strong temporary changes in spatial sense. Such events can hardly be reconciled with the view that the perception of space rests merely on an ordering of the elements of the visual organs and the brain, as though it consisted only in the order and environment of the organs. We should rather think in terms of qualities of sensation, corresponding to graduated chemical processes and possibly subject to chemical influences. Cf. Veraguth, 'Über Mikropsie und Makropsie', *Deutsche Zeitschr. f. Nervenheilkunde* (Strümpeli) **24**, (1903), 453; Koster, 'Zur Kenntnis der Mikropie und Makropie', *Graefes Archiv für Ophthalmologie* **42**, (1896), 134.
[27] On the various views as to Kant's position, cf. C. Siegel, *Über Raumvorstellung und Raumbegriff*, Leipzig J. A. Barth, 1905.

CHAPTER XXI

ON THE PSYCHOLOGY AND NATURAL DEVELOPMENT OF GEOMETRY*

1. For the animal organism, the relations of the different parts of its own body to one another and of physical objects to these different parts are primarily of the greatest importance. Upon these relations is based its system of physiological sensations of space. More complicated conditions of life, in which the simple and direct satisfaction of needs is impossible, result in an augmentation of intelligence. The physical, and particularly the spatial, behaviour of bodies toward one another may then acquire a mediate and indirect interest for transcending the interest of the momentary sensations. In this way, a spatial image of the world is created, at first instinctively, then in the practical arts, and finally scientifically, in the form of geometry. The mutual relations of bodies are geometrical in so far as they are determined by sensations of space, or find their expression in such sensations. Just as without sensations of heat there would have been no theory of heat, so also there would be no geometry without sensations of space; but both the theory of heat and geometry stand in additional need of experiences concerning bodies; that is to say, they must both go out beyond the narrow boundaries of the domain of sense that constitutes their peculiar foundation.

2. Isolated sensations have independent significance only in the lowest stages of animal life; as, for example, in reflex motions, in the removal of some disagreeable irritation of the skin, in the snapping reflex of the frog, etc. In the higher stages, attention is directed, not to space-sensation alone, but to those intricate and intimate complexes of other sensations with space-sensations which we call bodies. Bodies arouse our interest; they are the objects of our activities. But the character of our activities is coincidently determined by the place of the body, whether near or far, whether above or below, etc., – in other words, by the space-sensations characterizing it. The mode of reaction is thus determined by which the body can be reached, whether by extending the arms, by a few or many steps, by hurling missiles, or what not. The quantity (number) of sensitive

elements which a body excites, the number of places which it covers, that is to say, the volume of the body, is, all other things being the same, proportional to its capacity for satisfying our needs, and possesses consequently a biological import. Although our sensations of sight and touch are primarily produced only by the surfaces of bodies, nevertheless powerful associations impel primitive man especially to imagine more, or, as he thinks, to perceive more, than he actually observes. He imagines to be filled with matter the places enclosed by the surface, which alone he perceives; and this is especially the case when he sees or seizes bodies with which he is in some measure familiar. It requires considerable power of abstraction to bring to consciousness the fact that we perceive only the surface of bodies – a power which cannot be ascribed to primitive man.

3. Of importance in this regard are also the peculiar distinctive shapes of objects of prey and utility. Certain definite forms, that is, certain specific combinations of space-sensations, which man learns to know through intercourse with his environment, are unequivocally characterized even by purely physiological features. The straight line and the plane are distinguished above other forms by their physiological simplicity, as are likewise the circle and the sphere. The affinity of symmetric and geometrically similar forms is revealed by purely physiological properties. The variety of shapes with which we are acquainted from our physiological experience is far from being inconsiderable. Finally, through employment with bodily objects, physical experience also contributes its quota of wealth to the general store.

4. Crude physical experience impels us to attribute to bodies a certain constancy. Unless there are special reasons for not doing so, the same constancy is also ascribed to the individual attributes of the complex 'body'. We also regard the colour, hardness, shape, etc., of the body as constant; particularly we look upon the body as constant with respect to space, as indestructible. This assumption of spatial constancy, of spatial substantiality, finds its direct expression in geometry. Our physiological and psychological organization is independently predisposed to emphasize constancy; inasmuch as general physical constancies must necessarily have found lodgement in our organization, which is itself physical, while in the adaptation of the species very definite physical constancies were at

work. Since memory revives the images of bodies perceived before in their original forms and dimensions, it supplies the condition for the recognition of the same bodies, thus furnishing the first foundation for the impression of constancy. But geometry is still in need of certain individual experiences.

5. Let a body *K* move away from an observer *A* by being suddenly transported from the environment *FGH* to the environment *MNO*. To the optical observer *A* the body *K* decreases in size and assumes generally a different form. But to an optical observer *B* who moves along with *K* and retains the same position with respect to *K*, *K* remains unaltered. An analogous sensation is experienced by the tactual observer, although the perspective diminution is here wanting for the reason that the sense of touch is not a telepathic sense. The experiences of *A* and *B* must now be harmonized and their contradictions eliminated – a requirement which becomes especially imperative when the same observer plays alternately the part of *A* and of *B*. And the only method by which they can be harmonized is to attribute to *K* certain constant spatial properties independently of its position with respect to other bodies. The space-sensations determined by *K* in the observer *A* are recognized as dependent on other space-sensations (the position of *K* with respect to the body of the observer *A*). But these same space-sensations determined by *K* in *A* are independent of other space-sensations, characterizing the position of *K* with respect to *B*, or with respect to *FGH*... *MNO*. In this independence lies the constancy with which we are here concerned. The fundamental assumption of geometry thus reposes on an experience, although of the idealized kind.

6. For the experiences in question to assume conspicuous and perfectly determinate form the body *K* must be a so-called rigid body. If the space-sensations associated with three distinct acts of sense-perception remain unaltered, then the condition is given for the invariability of the complex of space-sensations determined by a rigid body. This determination of the space sensations produced by a body by means of three space-sensational elements accordingly characterizes the rigid body from the point of view of the physiology of the senses. This holds good for both the visual and the tactual sense. In employing

this designation we are not thinking of the physical conditions of rigidity, in defining which we should be compelled to enter different sensory domains, but of the fact given merely to our spatial sense. Indeed, we are now regarding every body as rigid which possesses the property assigned, even liquids, as long as their parts are not in motion with respect to one another.

7. Correct as the oft-repeated contention is that geometry is concerned, not with physical, but with ideal objects, it nevertheless cannot be doubted that geometry has sprung from the interest centring in the spatial relations of physical bodies. It bears the distinctest marks of this orgin, and the course of its development is fully intelligible only on a consideration of these traces. Our knowledge of the spatial behaviour of bodies is based upon the comparison of the space-sensations produced by them. Even without the least artificial or scientific assistance we acquire abundant experience of space. We can judge approximately whether rigid bodies which we perceive alongside one another in different positions at different distances, will, when brought successively into the same position, produce approximately the same or dissimilar space-sensations. We know pretty well whether one body will coincide with another, whether a pole lying flat on the ground will reach to a certain height. Our sensations of space are, however, subject to physiological circumstances which can never be absolutely identical for the members compared. In every case, rigorously viewed, a memory-trace of a sensation is necessarily compared with a real sensation. If, therefore, it is a question of the exact spatial relationship of bodies to one another, we must provide characteristics that depend as little as possible on the physiological conditions, so difficult to control. This is accomplished by comparing bodies with bodies. Whether a body *A* coincides with another body *B*, whether it can be made to occupy exactly the space filled by the other, that is, whether under like circumstances both bodies produce the same space-sensations, can be estimated with great precision. We regard such bodies as spatially or geometrically equal in every respect – as congruent. The character of the sensations is here no longer authoritative; it is now solely a question of their equality or inequality. If both bodies are rigid bodies, we can apply to the second body *B* all the experiences which we have gathered in connection with the first, more convenient, and more easily transportable, standard body *A*. We shall revert later to the circum-

stance that it is neither necessary nor possible to employ a special body of comparison, or standard, for every body. The most convenient bodies of comparison, though applicable only after a crude fashion – bodies whose invariance during transportation we always have before our eyes – are our hands and feet, our arms and legs. The names of the oldest measures show distinctly that originally we made our measurements with hand-breadths, forearms ells, feet, paces, etc. Nothing but a period of greater exactitude in measurement began with the introduction of conventional and carefully preserved physical standards; the principle remains the same. The measure enables us to compare bodies which are difficult to move or practically immovable.

8. As has been remarked, it is not the spatial, but predominantly the material, properties of bodies that possess the strongest interest. This fact certainly finds expression even in the beginnings of geometry. The volume of a body is instinctively taken into account as representing the quantity of its material properties, and so comes to form an object of contention long before its geometric properties receive anything approaching to profound consideration. It is here, however, that the comparison, the measurement of volumes acquires its initial import, and thus takes its place among the first and most important problems of primitive geometry. The first measurements of volume were doubtless of liquids and fruit, and were made with hollow measures. The object was to ascertain conveniently the quantity of like matter, or the quantity (number) of homogeneous, similarly-shaped (identical) bodies. Thus, conversely, the capacity of a store-room (granary) was in all likelihood originally estimated by the quantity or number of homogeneous bodies which it was capable of containing. The measurement of volume by a unit of volume is in all probability a much later conception, and can only have developed on a higher stage of abstraction.

9. Estimates of areas were also doubtless made from the number of fruit-bearing or useful plants which a field would accommodate, or from the quantity of seed that could be sown on it; or possibly also from the labour which such work required. The measurement of a surface by a surface was readily and obviously suggested in this connection when fields of the same size and shape lay near one another. There one

could scarcely doubt that the field made up of *n* fields of the same size and form possessed also *n*-fold argicultural value. We shall not be inclined to underrate the significance of this intellectual step when we consider the errors in the measurement of areas which the Egyptians[1] and even the Roman *agrimensores*[2] commonly committed. Even with a people so splendidly endowed with geometrical talent as the Greeks, and in a late period, we meet with the sporadic expression of the idea that surfaces having equal perimeters were equal in area. When the Persian 'Overman' Xerxes[3] wished to count the army which he had to 'feed', and which he drove under the lash across the Hellespont against the Greeks, he adopted the following procedure: 10000 men were drawn up closely packed together. The area which they covered was surrounded with an enclosure, and each successive division of the army, or rather, herd of slaves, that was driven into and filled the pen, counted for another 10000. We meet here with the converse application of the idea by which a surface is measured by the quantity (number) of equal, identical, immediately adjacent bodies which cover it. In abstracting, first instinctively and then consciously, from the height of these bodies, the transition is made to measuring surfaces by means of a unit of surface. The analogous step to measuring volumes by volume demands a far more practised, geometrically schooled intuition. It is effected later and is even at this day less easy to the masses.

10. The oldest estimates of long distances, which were computed by day's journeys, hours of travel, etc, were doubtless based upon the effort, labour, and expenditure of time necessary for covering these distances. But when lengths are measured by the repeated application of the hand, the foot, the arm, the rod, or the chain, then, accurately viewed, the measurement is made by the enumeration of like bodies, and we have again really a measurement of volume. The singularity of this conception will disappear in the course of this exposition. If, now, we abstract, first instinctively and then consciously, from the two transverse dimensions of the bodies employed in the enumeration, we reach the measuring of a line by a line.

11. A surface is commonly defined as the boundary of a space. Thus, the surface of a metal sphere is the boundary between the metal and the

air; it is not part either of the metal or of the air; two dimensions only are ascribed to it. Analogously, the one-dimensional line is the boundary of a surface; for example, the equator is the boundary of the surface of a hemisphere. The dimensionless point is the boundary of a line; for example, of the arc of a circle. A point, by its motion, generates a one-dimensional line, a line a two-dimensional surface, and a surface a three-dimensional solid space. No difficulties are presented by this concept to minds at all skilled in abstraction. It suffers, however, from the drawback that it does not exhibit, but on the contrary artificially conceals, the natural and actual way in which the abstractions have been reached. A certain discomfort is therefore felt when the attempt is made from this point of view to define the measure of surface or unit of area after the measurement of lengths has been discussed.[4]

12. A more homogeneous conception is reached if every measurement be regarded as a counting of space by means of immediately adjacent, spatially identical, or at least hypothetically identical, bodies, whether we be concerned with volumes, with surfaces, or with lines. Surfaces may be regarded as corporeal sheets, having everywhere the same constant thickness which we may make small at will, vanishingly small; lines, as strings or threads of constant, vanishingly small thickness. A point then becomes a small corporeal space from the extension of which we purposely abstract, whether it be part of another space, of a surface, or of a line. The bodies employed in the enumeration may be of any smallness or any form which conforms to our needs. Nothing prevents our idealizing in the usual manner these images, reached in the natural way indicated, by simply leaving out of account the thickness of the sheets and the threads. The usual and somewhat timid mode of presenting the fundamental notions of geometry is doubtless due to the fact that the infinitesimal method which freed mathematics from the historical and accidental shackles of its early elementary form, did not begin to influence geometry until a later period of development, and that the frank and natural alliance of geometry with the physical sciences was not restored until still later, through Gauss. But why the elements shall not now partake of the advantages of our better insight, is not to be clearly seen. Even Leibniz adverted to the fact that it would be more rational to begin with the solid in our geometrical definitions.[5]

13. The measurement of spaces, surfaces, and lines by means of solids is a conception from which our refined geometrical methods have become entirely estranged. Yet this idea is not merely the forerunner of the present idealized methods, but it plays an important part in the psychology of geometry, and we find it still powerfully active at a late period of development in the workshop of the investigator and inventor in this domain. Cavalieri's Method of Indivisibles appears best comprehensible through this idea. Taking his own illustration, let us consider the surfaces to be compared (the quadratures) as covered with equidistant parallel threads of any number we will, after the manner of the warp of woven fabrics, and the spaces to be compared (the cubatures) as filled with parallel sheets of paper. The total length of the threads may then serve as measure of the surfaces, and the total area of the sheets as measure of the volumes, and the accuracy of the measurement may be carried to any point we wish. The number of like equidistant bodies, if close enough together and of the right form, can just as well furnish the numerical measures of surfaces and solid spaces as the number of identical bodies absolutely covering the surfaces or absolutely filling the spaces. If we cause these bodies to shrink until they become lines (straight lines) or until they become surfaces (planes), we shall obtain the division of surfaces into surface elements and of spaces into space-elements, and coincidently the customary measurement of surfaces by surfaces and of spaces by spaces. Cavalieri's defective exposition, which was not adapted to the state of the geometry of his time, has evoked from the historians of geometry some very harsh criticisms of his beautiful and prolific procedure.[6] The fact that a Helmholtz, his critical judgement yielding in an unguarded moment to his fancy, could, in his great youthful work[7] regard a surface as the sum of the lines (ordinates) contained in it, is merely proof of the great depth to which this original natural conception reaches, and of the facility with which it reasserts itself.[8]

14. We have then, first, the general experience that movable bodies exist, to which, in spite of their mobility, a certain spatial constancy in the sense above described, a permanently identical property, must be attributed – a property which constitutes the foundation of all notions of measurement. But in addition to this there has been gathered instinctively, in the pursuit of the trades and the arts, a considerable variety of special

experiences, which have contributed their share to the development of geometry. Appearing in part in unexpected form, in part harmonizing with one another, and sometimes, when incautiously applied, even becoming involved in what appears to be paradoxical contradictions, these experiences disturb the course of thought and incite it to the pursuit of the orderly logical connection of these experiences. We shall now devote our attention to some of these processes.

15. Even though the well-known statement of Herodotus [9] were wanting in which he ascribes the origin of geometry to land-surveying among the Egyptians; and even though the account were totally lost [10] which Eudemus has left regarding the early history of geometry, and which is known to us from an extract in Proclus, it would be impossible for us to doubt that a pre-scientific period of geometry existed. The first geometrical knowledge was acquired accidentally and without design in the way of pratical experience, in connection with the most varied employments. It was gained at a time when the scientific spirit, or interest in the interconnection of the experiences in question, was but little developed. This is plain even in our meagre history of the beginnings of geometry, but still more so in the history of primitive civilization at large, where technical geometrical appliances are known to have existed at so early and barbaric a day as to exclude absolutely the assumption of scientific effort.

16. All savage tribes practise the art of weaving, and here, as in their drawing, painting, and wood-cutting, occur preferably ornamental themes consisting of the simplest geometrical forms. For such forms, like the drawings of our children, answer to the simplified, typical, schematic conception of the objects which they are desirous of imitating, while it is also these forms which are most easily produced, with their primitive implements and manual dexterity. Such an ornament consisting of a series of similarly-shaped triangles alternately inverted, or of a series of parallelograms (Figure 11), clearly suggests the idea, that the sum of the three angles of a triangle, when their vertices are placed together, makes up two right angles. Also this fact could not possibly have escaped the clay and stone workers of Assyria, Egypt, Greece, etc., in constructing the customary mosaics and pavements from differently coloured stones

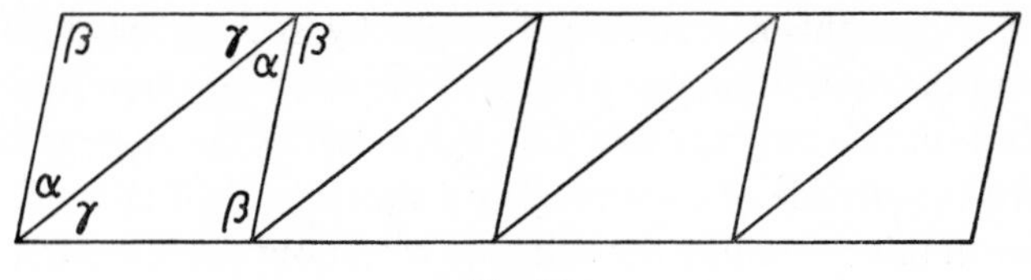

Fig. 11.

of the same shape. The theorem of the Pythagoreans that the plane space about a point can be completely filled by only three regular polygons, viz., by six equilateral triangles, by four squares, and by three regular hexagons, points to the same source.[11] A like origin is revealed also in the early Greek method of demonstrating the theorem regarding the angle-sum of any triangle by dividing it (by drawing the altitude) into two right-angled triangles and completing the rectangles corresponding to the parts so obtained.[12] The same experiences happen on many other occasions. If a surveyor walk round a polygonal piece of land, he will observe, on arriving at the starting-point, that he has performed a complete revolution, consisting of four right angles. In the case of a triangle, accordingly, of the six right angles constituting the interior and exterior angles (Figure 12) there will remain, after subtracting the three exterior angles of revolution, *a*, *b*, *c*, two right angles as the sum of the interior angles. This deduction of the theorem was employed by Thibaut[13], a contemporary of

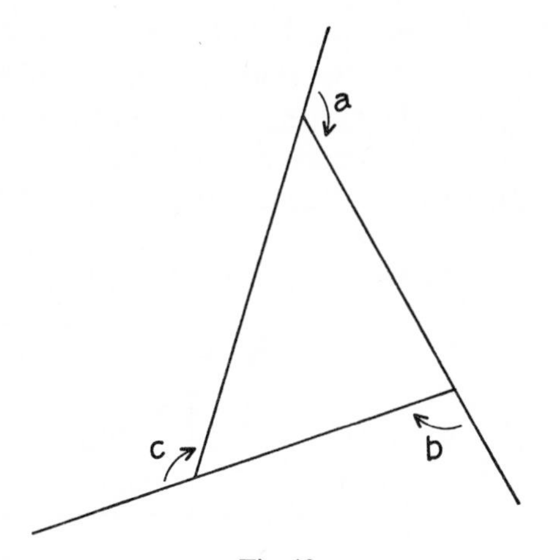

Fig. 12.

Gauss. If a draughtsman draw a triangle by successively turning his ruler round the interior angles, always in the same direction (Figure 13), he will find on reaching the first side again that if the edge of his ruler lay toward the outside of the triangle on starting, it will now lie toward the inside. In this procedure the ruler has swept out the interior angles of

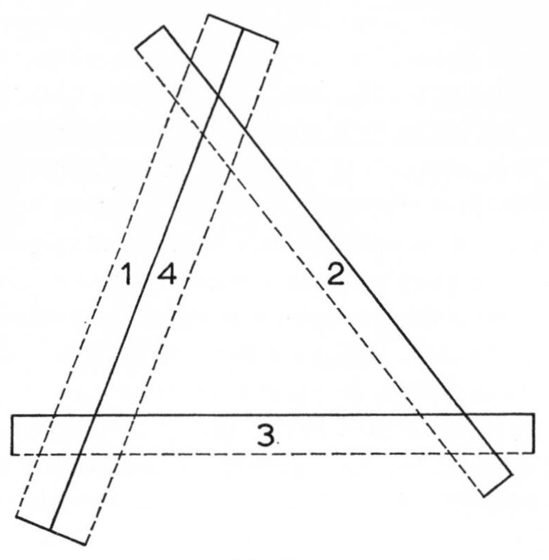

Fig. 13.

the triangle in the same direction, and in doing so has performed half a revolution.[14] Tylor[15] remarks that cloth or paperfolding may have led to the same results. If we fold a triangular piece of paper in the manner shown in Figure 14, we shall obtain a double rectangle, equal in area to one half the triangle, where it will be seen that the sum of the angles of the

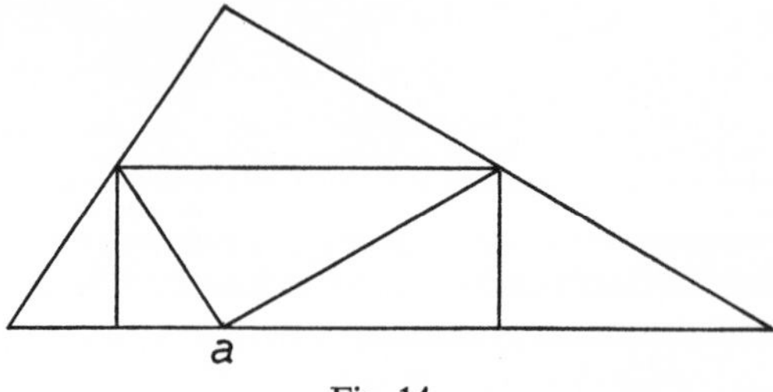

Fig. 14.

triangle coinciding at *a* is two right angles. Although some very astonishing results may be obtained by paper-folding, it can scarcely be assumed that these processes were historically very productive for geometry. The material is of too limited application, and artisans employed with it have too little incentive to exact observation.[16]

17. The knowledge that the angle-sum of the plane triangle is equal to a determinate quantity, namely, to two right angles, has thus been reached by experience, not otherwise than the law of the lever and Boyle and Mariotte's law of gases. It is true that neither the unaided eye nor measurements with the most delicate instruments can demonstrate absolutely that the sum of the angles of a plane triangle is exactly equal to two right angles. But the case is precisely the same with the law of the lever and with Boyle's law. All these theorems are therefore idealized and schematized experiences; for real measurements will always show slight deviations from them. But whereas the law of gases has been proved by further experimentation to be approximate only and to stand in need of modification when the facts are to be represented with great exactness, the law of the lever and the theorem regarding the angle-sum of a triangle have remained in as exact accord with the facts as the inevitable errors of experimenting would lead us to expect; and the same statement may be made of all the consequences that have been based on these two laws as preliminary assumptions.

18. Equal and similar triangles placed in paving alongside one another with their bases in one and the same straight line must also have led to a very important piece of geometrical knowledge. (Figure 15.) If a triangle be displaced in a plane along a straight line (without rotation), all its points, including those of its bounding lines, will describe equal paths. The same bounding line will furnish, therefore, in any two different positions, a system of two straight lines equally distant from one another at all points, and the operation coincidently vouches for the equality of the angles made by the line of displacement on corresponding sides of the two straight lines. The sum of the interior angles on the same side of the line of displacement was consequently determined to be two right angles, and thus Euclid's theorem of parallels was reached. We may add that the possibility of extending a pavement of this kind indefinitely,

necessarily lent increased obviousness to this discovery. The sliding of a triangle along a ruler has remained to this day the simplest and most natural method of drawing parallel lines. It is scarcely necessary to remark that the theorem of parallels and the theorem of the angle-sum of a triangle are inseparably connected and represent merely different aspects of the same experience.

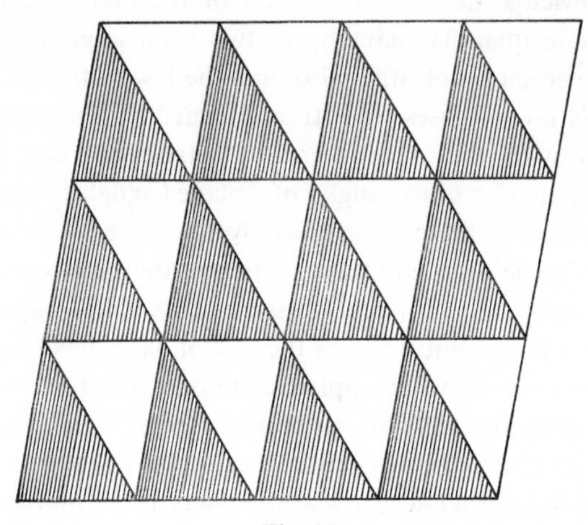

Fig. 15.

19. The stone masons above referred to must have readily made the discovery that a regular hexagon can be composed of equilateral triangles. Thus resulted immediately the simplest instances of the division of a circle into parts – namely its division into six parts by the radius, its division into three parts, etc. Every carpenter knows instinctively and almost without reflection that a beam of rectangular symmetric cross-section may, owing to the perfect symmetry of the circle, be cut out from a cylindrical tree-trunk in an infinite number of different ways. The edges of the beam will all lie in the cylindrical surface, and the diagonals of a section will pass through the centre. It was in this manner, according to Hankel[17] and Tylor[18], that the discovery was probably made that all angles inscribed in a semicircle are right angles.

20. A stretched thread furnishes the distinguishing visualization of the straight line. The straight line is characterized by its physiological sim-

plicity. All its parts induce the *same* sensation of direction; every point evokes the mean of the space-sensations of the neighbouring points; every part, however small, is similar to every other part, however great. But, though it has influenced the definitions of many writers[19], the geometer can accomplish little with this physiological characterization. The visual image must be enriched by physical experience concerning corporeal objects to be geometrically available. Let a string be fastened by one extremity at *A*, and let its other extremity be passed through a ring fastened at *B*. If we pull on the extremity at *B*, we shall see parts of the string which before lay between *A* and *B* pass out at *B*, while at the same time the string will approach the form of a straight line. A smaller number of like parts of the string, identical bodies, suffices to compose the straight line joining *A* and *B* than to compose a curved line. It is erroneous to assert that the straight line is recognized as the shortest line by mere visualization. It is quite true we can, so far as quality is concerned, reproduce in imagination with perfect accuracy and reliability, the simultaneous change of form and length which the string undergoes. But this is nothing more than a reviving of a prior experience with bodies – an experiment in thought. The mere passive contemplation of space would never lead to such a result. Measurement is experience involving a physical reaction, a superposition-experiment. Visualized or imagined lines having different directions and lengths cannot be applied to one another forthwith. The possibility of such a procedure must be actually experienced with material objects accounted as unalterable. It is erroneous to attribute to animals an instinctive knowledge of the straight line as the shortest distance between two points. If a stimulus attract an animal's attention, and if the animal has so turned that its plane of symmetry passes through the stimulating object, then the straight line is the path of motion uniquely determined by the stimulus. This is distinctly shown in Loeb's investigations on the tropisms of animals.

21. Further, visualization alone does not prove that any two sides of a triangle are together greater than the third side. It is true that if two sides be laid upon the base by rotation round the vertices of the basal angles, it will be seen by an act of imagination alone that the two sides with their free ends moving in arcs of circles will ultimately overlap, thus more than filling up the base. But we should not have attained to this re-

presentation had not the procedure been actually witnessed in connection with corporeal objects. Euclid[20] deduces this truth circuitously and artificially from the fact that the greater side of every triangle is opposite to the greater angle. But the source of our knowledge here also is experience – experience of the motion of the side of a physical triangle; this source has, however, been laboriously concealed by the form of the deduction – and this not to the enhancement of perspicuity and brevity.

22. But the properties of the straight line are not exhausted with the preceding empirical truths. If a wire of any arbitrary shape be laid on a board in contact with two upright nails, and slid along so as to be always in contact with the nails, the form and position of the parts of the wire between the nails will be constantly changing. The straighter the wire is, the slighter the alteration will be. A straight wire submitted to the same operation slides in itself. Rotated round two of its own fixed points, a crooked wire will keep constantly changing its position, but a straight wire will maintain its position, it will rotate within itself.[21] When we define, now, a straight line as the line which is completely determined by two of its points, there is nothing in this concept except the idealization of the empirical notion derived from the physical experience mentioned – a notion by no means directly furnished by the physiological act of visualization.

23. The plane, like the straight line, is physiologically characterized by its simplicity. It appears the same at all parts[22]. Every point evokes the mean of the space-sensations of the neighbouring points. Every part, however small, is like every other part, however great. But experiences gained in connection with physical objects are also required, if these properties are to be put to geometrical account. The plane, like the straight line, is physiologically symmetrical with respect to itself, if it coincides with the median plane of the body or stands at right angles to the same. But to discover that symmetry is a permanent geometrical property of the plane and the straight line, both constructs must be given as movable, unalterable physical objects. The connection of physiological symmetry with metrical properties is in need also of special metrical demonstration.

24. Physically a plane is constructed by rubbing three bodies together

until three surfaces, *A*, *B*, *C*, are obtained, each of which exactly fits the others – a result which can be accomplished, as Figure 16 shows, with neither convex nor concave surfaces, but with plane surfaces only. The convexities and concavities are, in fact, removed by the rubbing. Similarly, a truer straight line can be obtained with the aid of an imperfect ruler, by first placing it with its ends against the points *A*, *B*, then

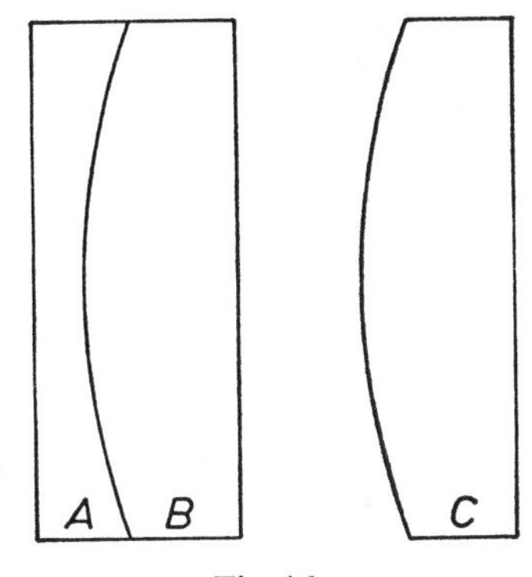

Fig. 16.

turning it through an angle of 180° out of its plane and again placing it against *A*, *B*, afterwards taking the mean between the two lines so obtained as a more perfect straight line, and repeating the operation with the line last obtained. Having produced by rubbing, a plane, that is to say, a surface having the same form at all points and on both sides, experience furnishes additional results. Placing two such planes one on the other, it will be learned that the plane is displaceable onto itself, and rotatable within itself, just as a straight line is. A thread stretched between any two points in the plane falls entirely within the plane. A piece of cloth drawn tight across any bounded portion of a plane coincides with it. Hence the plane represents the minimum of surface within its boundaries. If the plane be laid on two sharp points, it can still be rotated around the straight line joining the points, but any third point outside of this straight line fixes the plane, that is, determines it completely.

In the letter to Vitale Giordano, above referred to, Leibniz makes the frankest use of this experience with corporeal objects, when he defines a plane as a surface which divides an unbounded solid into two congruent parts, and a straight line as a line which divides an unbounded plane into two congruent parts.

25. If attention be directed to the symmetry of the plane with respect to itself, and two points be assumed, one on each side of it, each symmetrical to the other, it will be found that every point in the plane is equidistant from these two points, and Leibniz's definition of the plane is reached.[23] The uniformity and symmetry of the straight line and the plane are consequences of their being absolute minima of length and area respectively. For the boundaries given the minimum must exist, no other collateral condition being involved. The minimum is unique, single in its kind; hence the symmetry with respect to the bounding points. Owing to the absoluteness of the minimum, every portion, however small, again exhibits the same minimal property; hence the uniformity.

26. Empirical truths organically connected may make their appearance independently of one another, and doubtless were so discovered long before the fact of their connection was known. But this does not preclude their being afterwards recognized as involved in, and determined by, another, as being deducible from one another. For example, supposing we are acquainted with the symmetry and uniformity of the straight line and the plane, we easily deduce that the intersection of two planes is a straight line, that any two points of the plane can be joined by a straight line lying wholly within the plane, etc. The fact that only a minimum of inconspicuous and unobtrusive experiences is requisite for such deductions should not lure into the error of regarding this minimum as wholly superfluous, and of believing that visualization and reasoning are alone sufficient for the construction of geometry.

27. Like the concrete visual images of the straight line and the plane, so also our visualizations of the cricle, the sphere, the cylinder, etc., are enriched by metrical experiences, and in this manner first rendered amenable to fruitful geometrical treatment. The same economic impulse that prompts our children to retain only the typical features in their concepts and drawings, leads us also to the schematization and conceptual idealization of the images derived from our experience. Although we never come across in nature a perfect straight line or an exact circle, in

our thinking we nevertheless designedly abstract from the deviations which thus occur. Geometry, therefore, is concerned with ideal objects produced by schematization of experiential objects.

28. I have remarked elsewhere that it is wrong in elementary geometrical instruction to cultivate predominantly the logical side of the subject, and to neglect to throw open to young students the wells of knowledge contained in experience. It is gratifying to note that the Americans, who are less dominated than we by tradition, have recently broken with this system and are introducing a sort of experimental geometry as introductory to systematic geometric instruction.[24]

29. No sharp line can be drawn between the instinctive, the technical, and the scientific acquisition of geometric notions. Generally speaking, we may perhaps say that with the division of labour in the industrial and economic fields, with increasing employment with specific objects, the instinctive acquisition of knowledge falls into the background, and the technical begins. Finally, when measurement becomes an aim and profession in itself, the connection obtaining between the various operations of measuring acquires a powerful economic interest, and we reach the period of the scientific development of geometry, to which we now proceed.

30. The knowledge that the measures of geometry depend on one another, was reached in divers ways. After surfaces came to be measured by surfaces, certain other progress was almost inevitable. In a parallelogrammatic field permitting of division into equal partial parallelogrammatic fields so that n rows of partial fields each containing m fields lay alongside one another, the counting of these fields was unnecessary. By multiplying together the numbers measuring the sides, the area of the field was found to be equal to mn such fields, and the area of each of the two triangles formed by drawing the diagonal was readily discovered to be equal to $mn/2$ such fields. This was the first and simplest application of arithmetic to geometry. Coincidently, the dependence of measures of area on other measures, linear and angular, was discovered. The area of a rectangle was found to be larger than that of an oblique parallelogram having sides of the same length; the area, consequently, depended not

only on the length of the sides, but also on the angles. On the other hand, a rectangle constructed of strips of wood running parallel to the base, can, as is easily seen, be converted by displacement into any parallelogram of the same height and base without altering its area. Quadrilaterals having their sides given are still undetermined in their angles, as every carpenter knows. He adds diagonals, and converts his quadrilateral into triangles, which, the sides being given, are rigid, that is to say, are unalterable as to their angles also. With the perception that measures were dependent on one another, the real problem of geometry was introduced. Steiner has aptly and justly entitled his principal work "Systematic Development of the Dependence of Geometrical Figures on One Another".[25] In Snell's original unappreciated treatise on Elementary Geometry, the problem in question is made obvious even to the beginner.[26]

31. A plane physical triangle is constructed of wires. If one of the sides be rotated around a vertex, so as to increase the interior angle at that point, the side moved will be seen to change its position and the side opposite to grow larger with the angle. New pieces of wire besides those before present will be required to complete the last-mentioned side. This and other similar experiments can be repeated in thought, but the mental experiment is never anything more than a copy of the physical experiment. The mental experiment would be impossible if physical experience had not antecedently led us to a knowledge of spatially unalterable physical bodies[27] – to the concept of measure. By experiences of this character, we are conducted to the truth that of the six metrical magnitudes discoverable in a triangle (three sides and three angles) three, including at least one side, suffice to determine the triangle. If one angle only be given among the parts determining the triangle, the angle in question must be either the angle included by the given sides, or that which is opposite to the greater side – at least if the determination is to be unique. Having reached the perception that a triangle is determined by three sides and that its form is independent of its position, it follows that in an equilateral triangle all three angles and in an isoceles triangle the two angles opposite the equal sides, must be equal, in whatever manner the angles and sides may depend on one another. This is logically certain. But the empirical foundation on which it rests is for that reason not a whit more superfluous than it is in the analogous cases of physics.

32. The mode in which the sides and angles depend on one another is first recognized, naturally, in special instances. In computing the areas of rectangles and of the triangles formed by their diagonals, the fact must have been noticed that a rectangle having sides 3 and 4 units in length gives a right-angled triangle having sides, 3, 4, and 5 units in length. Rectangularity was thus shown to be connected with a definite, rational ratio between the sides. The knowledge of this truth was employed to stake off right angles, by means of three connected ropes respectively 3, 4, and 5 units in length[28]. The equation $3^2+4^2=5^2$, the analogue of which was proved to be valid for all right-angled triangles having sides of length a, b, c (the general formula being $a^2+b^2=c^2$), now riveted the attention. It is well known how profoundly this relation enters into metrical geometry, and how all indirect measurements of distance may be traced back to it. We shall endeavour to disclose the foundation of this relation.

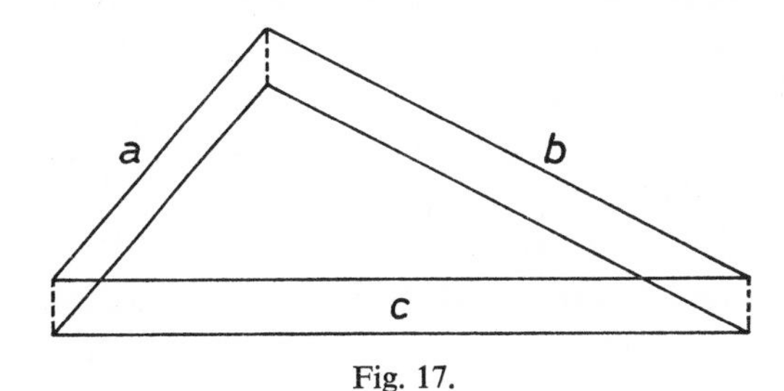

Fig. 17.

33. It is to be remarked first that neither the Greek geometrical nor the Hindu arithmetical deductions of the so-called Pythagorean Theorem could avoid the consideration of areas. One essential point on which all the deductions rest and which appears more or less distinctly in different forms in all of them, is the following: If a triangle, a, b, c (Figure 17) be slid along a short distance in its own plane, it is assumed that the space which it leaves behind is made up for, or compensated for, by the new space on which it enters. That is to say, the area swept out by two of the sides during the displacement is equal to the area swept out by the third side. The basis of this conception is the assumption of the conservation of the area of the triangle. If we consider a surface as a body of very minute but unvarying thickness of third dimension (which for that reason is uninfluential in the present connection), we shall again have the

conservation of the volume of bodies as our fundamental assumption. The same conception may be applied to the translation of a tetrahedron, but it does not lead in this instance to new points of view. Conservation of volume is a property which rigid and liquid bodies possess in common and was idealized by the old physics as impenetrability. In the case of rigid bodies, we have the additional attribute that the distances between all the parts are preserved, while in the case of liquids, the properties of rigid bodies exist only for the smallest time and space elements.

34. If an oblique-angled triangle having the sides a, b, and c be displaced in the direction of the side b, only a and c will, by the principle above stated, describe equivalent parallelograms, which are alike in an equal pair of parallel sides on the same parallels. If a make with b a right angle, and the triangle be displaced at right angles to c, the distance c, the side c will describe the square c^2, while the two other sides will describe parallelograms the combined areas of which are equal to the area of the square. The two parallelograms are, by the observation which just precedes, equivalent respectively to a^2 and b^2 – and with this the Pythagorean theorem is reached. The same result may also be attained (Figure 18) by first sliding the triangle a distance a at right angles to a, and then a distance b at right angles to b, where a^2+b^2 will be equal to

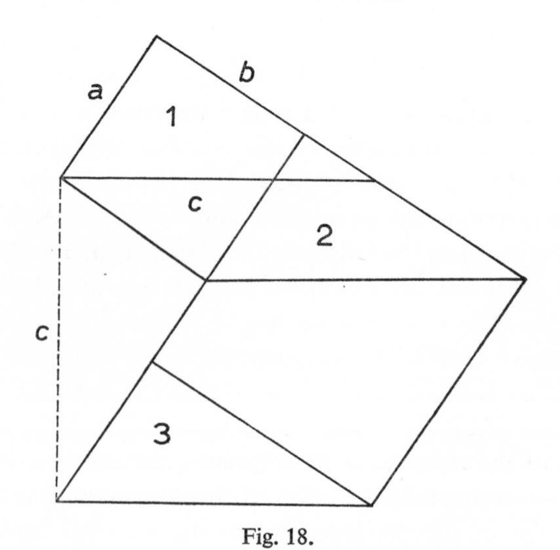

Fig. 18.

the sum of the surfaces swept out by c, which is obviously c^2. Taking an oblique-angled triangle, the same procedure just as easily and obviously gives the more general proposition, $c^2 = a^2 + b^2 - 2ab \cos \gamma$.

35. The dependence of the third side of the triangle on the two other sides is accordingly determined by the area of the enclosed triangle; or, in our conception, by a condition involving volume. It will also be directly seen that the equations in question express relations of area. It is true that the angle included between two of the sides may also be regarded as determinative of the third side, in which case the equations will apparently assume an entirely different form. Let us look a little more closely at these different measures. If the extremities of two straight lines of lengths a and b meet in a point, the length of the line c joining their free extremities will be included between definite limits. We shall have $c \leqslant a+b$, and $c \geqslant a-b$. Visualization alone cannot inform us of this fact; we can learn it only from experimenting in thought – a procedure which reposes on physical experience and reproduces it. This will be seen by holding a fast, for exampling, and turning b, first, until it forms the prolongation of a, and, secondly, until it coincides with a. A straight line is primarily a unique concrete image characterized by physiological properties – an image which we have obtained from a physical body of a definite specific character, which in the form of a string or wire of indefinitely small but constant thickness interposes a minimum of volume between the positions of its extremities – which can be accomplished only in one uniquely-determined manner. If several straight lines pass through a point, we distinguish between them physiologically by their directions. But in abstract space obtained by metrical experiences with physical objects, differences of direction do not exist. A straight line passing through a point can be completely determined in abstract space only by assigning a second physical point on it. To define a straight line which is constant in direction, or an angle as a difference between directions, or parallel straight lines as straight lines having the same direction, is to define these concepts physiologically.

36. Different methods are at our disposal when we come to characterize or determine geometrically angles which are visually given. An angle is determined when the distance is assigned between any two fixed points lying each on a separate side of the angle outside the point of intersection.

To render the definition uniform, points situated at the same fixed and invariable distance from the vertex might be chosen. The inconvenience that then equimultiples of a given angle placed alongside one another in the same plane with their vertices coincident, would not be measured by the same equimultiples of the distance between those points, is the reason that this method of determining angles was not introduced into elementary geometry.[29] A simpler measure, a simpler characterization of an angle, is obtained by taking the aliquot part of the circumference or the area of a circle which the angle intercepts when laid in the plane of the circle with its vertex at the centre. The convention here involved is more convenient.[30]

In employing an arc of a circle to determine an angle, we are again merely measuring a volume – viz., the volume occupied by a body of simple definite form introduced between two points on the arms of the angle equidistant from the vertex. But a circle can be characterized by simple rectilinear distances. It is a matter of perspicuity, immediacy, and of the facility and convenience resulting therefrom, that two measures, viz., the rectilinear measure of length and the angular measure, are principally employed as fundamental measures, and the others derived from them. It is in no sense necessary. For example (Figure 19), it is possible without a special angular measure to determine the straight line that cuts another straight line at right angles by making all its points

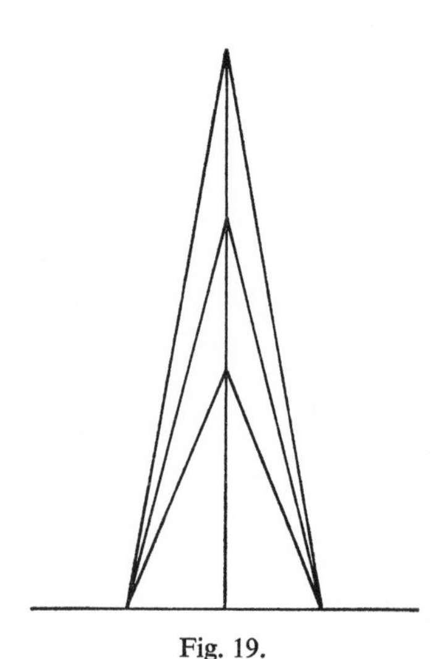

Fig. 19.

equidistant from two points in the first straight line lying at equal distances from the point of intersection. The bisector of an angle can be determined in a quite similar manner, and by continued bisection an angular unit can be derived of any smallness we wish. A straight line parallel to another straight line can be defined as one, all of whose points can be translated by congruent curved or straight paths into points of the first straight line.[31] It is quite possible to start with the straight length alone as our fundamental measure. Let a fixed physical point *a* be given. Another point, *m*, has the distance r_a from the first point. Then this last point can still lie in any part of the spherical surface described about *a* with radius r_a. If we know yet a further fixed point *b*, from which *m* is removed by the distance r_b, the triangle *abm* will be rigid, determined; but *m* can still revolve round in the circle described by the rotation of the triangle around the axis *ab*. If now the point *m* be held fast in any position, then also the whole rigid body to which the three points in question, *a*, *b*, *m*, belong will be fixed.

37. A point *m* is spatially determined, accordingly, by the distances r_a, r_b, r_c from at least three fixed points in space, *a*, *b*, *c*. But this determination is still not unique, for the pyramid with the edges r_a, r_b, r_c, in the vertex of which *m* lies, can be constructed as well on the one as on the other side of the plane *a*, *b*, *c*. If we were to fix the side, say by a special sign, we should be resorting to a physiological determination, for geometrically the two sides of the plane are not different. If the point *m* is to be uniquely determined, its distance, r_d, from a fourth point, *d*, lying outside the plane *abc*, must be given in addition. Another point, *m'*, is determined with like completeness by four distances, r'_a, r'_b, r'_c, r'_d. Hence, the distance of *m* from *m'* is also given by this determination. And the same holds true of any number of other points as severally determined by four distances. Between four points $\frac{4(4-1)}{1\cdot 2}=6$ distances are conceivable, and precisely this number must be given to determine the form of the point complex. For $4+z=n$ points, $6+4z$ or $4n-10$ distances are needed for the determination, while a still larger number, viz., $\frac{n(-1)}{1\cdot 2}$ distances exist, so that the excess of the distances is also coincidently determined.[32]

38. If we start from three points and prescribe that the distances of all points to be further determined shall hold for one side only of the plane determined by the three points, then $3n-6$ distances will suffice to determine the form, magnitude, and position of a system of n points with respect to the three initial points. But if there be no condition as to the side of the plane to be taken – a condition which involves sensuous and physiological, but not abstract metrical characteristics – the system of points, instead of the intended form and position, may assume that symmetrical to the first, or be combined of the points of both. Symmetric geometrical figures are, owing to our symmetric physiological organization, very easily taken for the same, whereas metrically and physically they are entirely different. A screw with its spiral winding to the right and one with its spiral winding to the left, two bodies rotating in contrary directions, etc., appear very much alike to the eye. But we are for this reason not permitted to regard them as geometrically or physically equivalent. Attention to this fact would avert many paradoxical questions. Think only of the trouble that such problems gave Kant! Sensory physiological attributes are determined by relationship to our body, to a corporeal system of specific constitution; while metrical attributes are determined by relations to the world of physical bodies at large. The latter can be ascertained only by experiments of coincidence – by measurements.

39. As we see, every geometrical measurement is at bottom reducible to measurements of volumes, to the numeration of bodies. Measurements of lengths, like measurements of areas, repose on the comparison of the volumes of very thin strings, sticks, and leaves of constant thickness. This is not at variance with the fact that measures of area may be arithmetically derived from measures of length, or solid measures from measures of length alone, or from these in combination with measures of area. This is merely proof that different measures of volume are dependent on one another. To ascertain the forms of this interdependence is the fundamental object of geometry, as it is the province of arithmetic to ascertain the manner in which the various numerical operations, or ordinating activities of the mind, are connected together.

40. It is extremely probable that the experiences of the visual sense were the reason for the rapidity with which geometry developed. But

our great familiarity with the properties of rays of light gained from the present advanced state of optical technique, should not mislead us into regarding our experiential knowledge of rays of light as the principal foundation of geometry. Rays of light in dust or smoke-laden air furnish admirable visualizations of straight lines. But we can derive the metrical properties of straight lines from rays of light just as little as we can derive them from imagined straight lines. For this purpose experiences with physical objects are absolutely necessary. The rope-stretching of the practical geometers is certainly older than the use of the theodolite. But once the physical straight line is known, the ray of light furnishes a very distinct and handy means of reaching new points of view. A blind man could scarcely have invented modern synthetic geometry. But the oldest and the most powerful of the experiences lying at the basis of geometry are just as accessible to the blind man, through his sense of touch, as they are to the person who can see. Both are acquainted with the spatial permanency of bodies despite their mobility; both acquire a conception of volume by taking hold of objects. The creator of primitive geometry disregards, first instinctively and then intentionally and consciously, those physical properties that are unessential for his operations and that for the moment do not concern him. In this manner, and by gradual growth, the idealized concepts of geometry arise on the basis of experience.

41. Our geometrical knowledge is thus derived from various sources. We are physiologically acquainted, from direct visual and tactual contact, with many and various spatial forms. With these are associated physical (metrical) experiences (involving comparison of the space-sensations evoked by different bodies under the same circumstances), which experiences are in their turn also but the expressions of other relations obtaining between sensations. These diverse orders of experience are so intimately interwoven with one another that they can be separated only by the most thoroughgoing scrutiny and analysis. Hence originate the widely divergent views concerning geometry. Here it is based on pure visualization (*Anschauung*), there on physical experience, according as the one or the other factor is overrated or disregarded. But both factors entered into the development of geometry and are still active in it to-day; for, as we have seen, geometry by no means exclusively employs purely metrical concepts.

42. If we were to ask an unbiassed, candid person under what form he pictured space, referred, for example, to the Cartesian system of co-ordinates, he would doubtless say: I have the image of a system of rigid (fixed form), transparent, penetrable, contiguous cubes, having their bounding surfaces marked only by nebulous visual and tactual percepts – a kind of phantom cubes. Over and through these phantom constructs the real bodies or their phantom counterparts move, conserving their spatial permanency (as above defined), whether we are pursuing practical or theoretical geometry, or phoronomy. Gauss's famous investigation of curved surfaces, for instance, is really concerned with the application of infinitely thin laminate and hence flexible bodies to one another. That diverse orders of experience have co-operated in the formation of the fundamental conceptions under consideration, cannot be gainsaid.

43. Yet, varied as the special experiences are from which geometry has sprung, they may be reduced to a minimum of facts: Movable bodies exist having definite spatial permanency – viz., rigid bodies exist. But the movability is characterized as follows: we draw from a point three lines not all in the same plane but otherwise undetermined. By three movements parallel to these straight lines any point can be reached from any other. Hence, three measurements or dimensions, physiologically and metrically characterized as the simplest, are sufficient for all spatial determinations. These are the fundamental facts.

44. Physical metrical experiences, like all experiences forming the basis of experimental sciences, are conceptualized – idealized. The need of representing the facts by simple perspicuous concepts under easy logical control, is the reason for this. Absolutely rigid, spatially invariable bodies, perfect straight lines and planes, no more exist than a perfect gas or a perfect liquid. Nevertheless, we work preferably and also more readily with these concepts than with others that conform more closely to the properties of the objects, deferring the consideration of the deviations. Theoretical geometry does not even need to consider these deviations, inasmuch as it assumes objects that fulfil the requirements of the theory absolutely, just as theoretical physics does. But in the case of practical geometry, where we are concerned with actual objects, we are obliged, as

in practical physics, to consider the deviation from the theoretical assumptions. But geometry has the additional advantage that every deviation of its object from the assumptions of the theory which may still be detected can be removed; whereas physics for obvious reasons cannot construct more perfect gases than actually exist in nature. For, in the latter case, we are concerned not with a single arbitrarily constructible spatial property alone, but with a relation, occurring in nature and independent of our will, between pressure, volume, and temperature.

45. The choice of the concepts is suggested by the facts; yet, seeing that this choice is the outcome of our voluntary reproduction of the facts in thought, some free scope is left in the matter. The importance of the concepts is estimated by their range of application. This is why the concepts of the straight line and the plane are placed in the foreground, for every geometrical object can be split up with sufficient approximateness into elements bounded by planes and straight lines. The particular properties of the straight line, plane, etc., which we decide to emphasize, are matters of our own free choice, and this truth has found expression in the various definitions that have been given of the same concept.[33]

46. The fundamental truths of geometry have thus, unquestionably, been derived from physical experience, if only for the reason that our visualizations and sensations of space are absolutely inaccesssible to measurement and cannot possibly be made the subject of metrical experience. But it is no less indubitable that when the relations connecting our visualizations of space with the simplest metrical experiences have been made familiar, then geometrical facts can be reproduced with great facility and certainty in the imagination alone – that is by purely mental experiment. The very fact that a continuous change in our space-sensation corresponds to a continuous metrical change in physical bodies, enables us to ascertain by imagination alone the particular metrical elements that depend on one another. Now, if such metrical elements are observed to enter different constructions having different positions in precisely the same manner, then the metrical results will be regarded as equal. The case of the isosceles and equilateral triangles, above mentioned, may serve as an example. The geometric mental experiment has advantage over the physical, only in the respect that it can be performed with far simpler

experiences and with such as have been more easily and almost unconsciously acquired.

47. Our sensory imaginings and visualizations of space are qualitative, not quantitative nor metrical. We derive from them coincidences and differences of extension, but never real magnitudes. Conceive, for example, Figure 20, a coin rolling clockwise down and around the rim of another

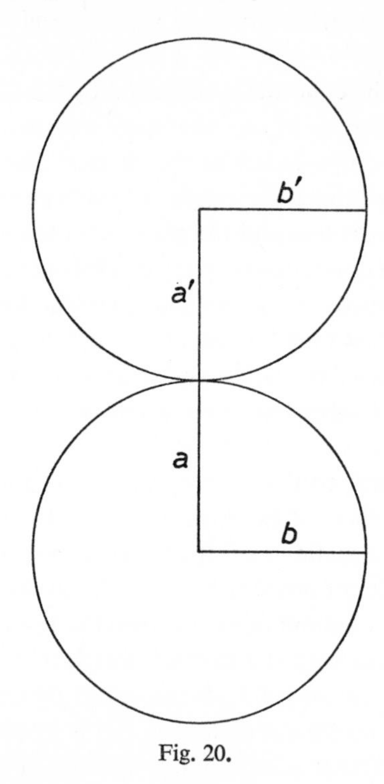

Fig. 20.

fixed coin of the same size, without sliding. Be our imagination as vivid as it will, it is impossible by a pure feat of reproductive imagery alone, to determine here the angle described in a full revolution. But if it be considered that at the beginning of the motion the radii a, a' lie in one straight line, but that after a quarter revolution the radii b, b' lie in a straight line, it will be seen at once that the radius a' now points vertically upwards and has consequently performed half a revolution. The measure of the revolution is obtained from metrical concepts, which fixate idealized experiences on definite physical objects, but the direction of the revolution is retained in the sensory imagination. The metrical concepts simply determine that in equal circles equal angles are subtended by equal arcs, that the radii to the point of contact lie in a straight line, etc.

48. If I picture to myself a triangle with one of its angles increasing, I shall also see the side opposite the angle increasing. The impression thus arises that the interdependence in question follows a priori from a feat of imagination alone. But the imagination has here merely reproduced a fact of experience. Measure of angle and measure of side are two physical concepts applicable to the same fact, – concepts that have grown so familiar to us that they have come to be regarded as merely two different attributes of the same imagined group of facts, and hence appear as linked together of sheer necessity. Yet we should never have acquired these concepts without physical experience. Compare Section 21.

49. The combined action of the sensory imagination with idealized concepts derived from experience is apparent in every geometrical deduction. Let us consider, for example, the simple theorem that the perpendicular bisectors of the sides of a triangle ABC meet in a common point. Experiment and imagination both doubtless led to the theorem. But the more carefully the construction is executed, the more one becomes convinced that the third perpendicular does not pass exactly through the point of intersection of the first two, and that in any actual construction, therefore, three points of intersection will be found closely adjacent to one another. For in reality neither perfect straight lines nor perfect perpendiculars can be drawn; nor can the latter be erected exactly at the midpoints; and so on. Only on the assumption of such ideal conditions does the perpendicular bisector of AB contain all points equally distant from A and B, and the perpendicular bisector of BC all points equidistant from B and C. From which it follows that the point of intersection of the two is equidistant from A, B, and C, and by reason of its equidistance from A and C is also a point of the third perpendicular bisector, of AC. The theorem asserts therefore that the more accurately the assumptions are fulfilled the more nearly will the three points of intersection coincide.

50. The importance of the combined action of the sensuous imagination [viz., of *Anschauung* or intuition as we have called it] and of concepts, will doubtless have been rendered clear by these examples. Kant says: "Thoughts without contents are empty, intuitions without concepts are blind." (*KdrV* A51/B75) Possibly we might more appropriately say: "Concepts without intuitions are blind, intuitions without concepts are lame." For it would appear to be not so absolutely correct to call intuitions [viz., sensory images] blind and concepts empty. When Kant further says that "there is in every branch of natural knowledge only so much science as there is mathematics contained in it ," (*Metaphysische Anfangsgründe der Naturwissenschaft, Vorwort*) one might possibly also assert of all sciences including mathematics "that they are only in so far sciences as they operate with concepts". For our logical mastery extends only to those concepts of which we have ourselves determined the contents.

51. The two facts that bodies are rigid and movable would be sufficient for an understanding of any geometrical fact, no matter how complicated – sufficient, that is to say, to derive it from the two facts mentioned. But geometry is obliged, both in its own interests and in its role as an auxiliary science, or in the pursuit of practical ends, to answer questions that recur repeatedly in the same form. Now it would be uneconomical, in such a contingency, to begin each time with the most elementary facts and to go to the bottom of each new case that presented itself. It is preferable, rather, to select a certain few simple, familiar, and indubitable theorems, in our choice of which caprice is by no means excluded [34], and to formulate from these, once for all, for application to practical ends, general propositions answering the questions that most frequently recur. From this point of view we understand at once the form geometry has assumed – the emphasis, for example, that it lays upon its propositions concerning triangles. For the designated purpose, it is desirable to collect the most general possible propositions, having the widest range of application. From history we know that propositions of this character have been obtained by comprehending various special cases of knowledge under single general cases. We are forced even today to resort to this procedure when we treat the relationship of two geometrical figures, or when the different special cases of form and position compel us to modify our modes of deduction. We may cite as the most familiar instance of this in

elementary geometry, the mode of deducing the relation obtaining between angles at the centre and angles at the circumference.

Kroman[35] has put the question, Why do we regard a demonstration made with a special figure (a special triangle) as universally valid for all figures? and finds his answer in the supposition that we are able by rapid variations to impart all possible forms to the figure in thought and so convince ourselves of the admissibility of the same mode of inference in all special cases. History and introspection declare this idea to be in all essential respects correct. But we may not assume, with Kroman, that in each special case every individual student of geometry acquires this complete synoptic view "with the rapidity of lightning," and attains forthwith to the lucidity and intensity of geometric conviction in question. Frequently the required operation is absolutely impracticable, and errors prove that in other cases it was actually not performed but that the enquirer rested content with a conjecture based on analogy[36]. But that which the individual does not or cannot achieve in a jiffy, he may achieve in the course of his life. Whole generations labour on the verification of geometry. And the conviction of its certitude is unquestionably strengthened by their collective exertions[37]. I once knew an otherwise excellent teacher who compelled his students to perform all their demonstrations with incorrect figures, on the theory that it was the logical connection of the concepts, not the figure, that was essential. But the experiences imbedded in the concepts cleave to our sensory images. Only the actually visualized or imagined figure can tell us what particular concepts are to be employed in a given case. The method of this teacher is admirably adapted for rendering palpable the degree to which logical operations share in reaching truth. But to employ it habitually is to miss utterly the truth that concepts draw their ultimate power from sensory sources.

The view that a new insight can be captured once and for all by fortunate arrangements of syllogisms cannot be maintained if the facts are accurately observed: it holds neither for an individual learner or enquirer, nor for a people or humanity as a whole, neither for geometry nor for any other science. On the contrary, the history of science shows that a correct new insight correctly reduced to its foundations may become more or less confused in time, appear incompletely or in distorted form or even be altogether lost to some enquirers, only to reappear in full blaze later. A single discovery and utterance of an insight is not enough. Often it takes years and centuries to develop general thinking habits to the point where the insight in question can become common property and stay permanently alive. This is shown with especial elegance by Duhem[38] in his detailed investigations on the history of statics.

NOTES

* Translated from Professor Mach's manuscript by Thomas J. McCormack for *The Monist*, 1902.

[1] Eisenlohr, *Ein mathematisches Handbuch der alten Aegypter: Papyrus Rhind*, Leipzig 1877.

[2] M. Cantor, *Die römischen Agrimensoren*, Leipzig 1875.

[3] Herodotus, VII 22, 56, 103, 223.

[4] Hölder, *Anschauung und Denken in der Geometrie*, Leipzig 1900, p. 18; W. Killing, *Einführung in die Grundlagen der Geometrie*, Paderborn 1898, II pp. 22f.

[5] Letter to Vitale Giordano, *Leibnitzens mathematische Schriften*, ed. Gerhardt, Berlin 1849, Section I, Volume I, p. 199.

[6] Weissenborn, *Principien der höheren Analysis in ihrer Entwickelung*, Halle 1856; Gerhardt, *Entdeckung der Analysis*, Halle 1855, p. 18; Cantor, *Geschichte der Mathematik*, Leipzig 1892, Vol. II.

[7] Helmholtz, *Erhaltung der Kraft*, Berlin 1847, p. 14.

[8] The following simple illustration (Figure 21) of Cavalieri's method may be helpful to readers not thoroughly conversant with geometry: Imagine a right circular cylinder of horizontal base cut out of a stack of paper sheets resting on a table, and conceive

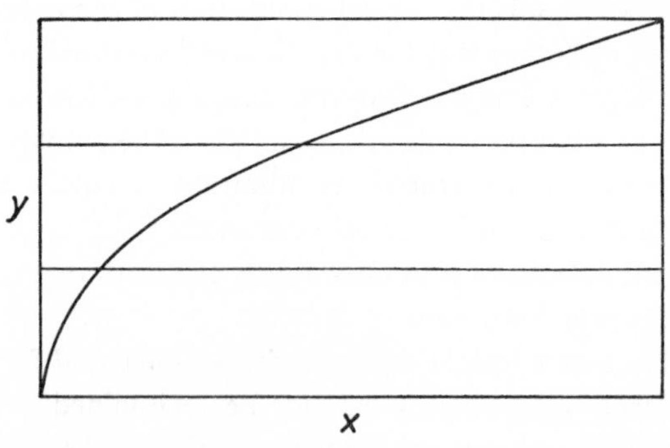

Fig. 21.

inscribed in the cylinder a cone of the same base and altitude. Whereas the sheets cut out by the cylinder are all equal, those forming the cone increase in size as the squares of their distances from the vertex. Now from elementary geometry we know that the volume of such a cone is one third that of the cylinder. This result may be applied at once to the quadrature of the parabola. Let a rectangle be described about a portion of a parabola, its sides coinciding with the axis and the tangent to the curve at the origin.

Conceiving the rectangle to be covered with a system of threads running parallel to x, every thread of the rectangle will be divided into two parts, of which that lying outside the parabola is proportional to y^2. Therefore, the area outside the parabola is to the total area of the rectangle precisely as is the volume of the cone to that of the cylinder, viz., as 1 is to 3.

It is significant of the naturalness of Cavalieri's view that the writer of these lines, hearing of the higher geometry when a student at the Gymnasium, but without any training in it, lighted on very similar conceptions – a performance not attended with any difficulty in the nineteenth century. By the aid of these he made a number of little discoveries, which were of course already long known, found Guldin's theorem, calculated some of Kepler's solids of rotation, etc.

[9] Herodotus, II 109.

[10] James Gow, *A Short History of Greek Mathematics*, Cambridge 1884, p. 134.

[11] This theorem is attributed to Pythagoras by Proclus. Cf. Gow, *A Short History of Greek Mathematics*, p. 143 footnote.

[12] Hankel, *Geschichte der Mathematik*, Leipzig 1874, p. 96.

[13] Thibaut, *Grundriss der reinen Mathematik*, Göttingen 1809, p. 177. The objections which may be raised to this and the following deductions will be considered later.

[14] Noticed by the writer too while drawing.

[15] Tylor, *Einleitung in das Studium der Anthropologie*, Brunswick 1883, p. 383 (German translation of *Anthropology, An Introduction to the Study of Man*).

[16] See for example Sundara Row's *Geometric Exercises in Paper-Folding*, Chicago 1901.

[17] Hankel, *Gesch. d. Mathem.* pp. 206–7.

[18] Tylor, *l.c.*

[19] Euclid, *Elements* I, Definition 3.

[20] Euclid, *Elements* I, Proposition 20.

[21] In a letter to Vitale Giordano (pp. 195–6 of the work cited in note 5 of this chapter) Leibniz makes use of the above-mentioned property of a straight line for its definition. The straight line shares the property of displaceability in itself with the circle and the circular cylindrical spiral. But the property of rotatability within itself and that of being determined by *two* points are exclusively its own.

[22] Euclid, *Elements* I, Definition 7.

[23] Leibniz's 'Geometrical Characterization' in his letter to Huygens of 8 September 1679 (Gerhardt, Sec. II, Vol. I, p. 23).

[24] W. T. Campbell, *Observational Geometry*, New York 1899; W. W. Spear, *Advanced Arithmetic*, Boston 1899.

[25] J. Steiner, *Systematische Entwicklung der Abhängigkeit der geometrischen Gestalten von einander.*

[26] Snell, *Lehrbuch der Geometrie*, Leipzig 1869.

[27] The entire construction of the Euclidean geometry shows traces of this foundation. It is still more apparent in Leibniz's 'geometrical characterization' mentioned earlier. We shall revert to this topic.

[28] Cantor, *Gesch. d. Math.*, Leipzig 1880, I, pp. 53, 56.

[29] A closely allied principle of measurement is however applied in trigonometry.

[30] So also the superficial portion of a sphere intercepted by the including planes is used as the measure of a solid angle.

[31] If this form had been adopted, the doubts as to the Euclidean theorem of parallels would probably have arisen much later.

[32] An interesting attempt at basing both Euclidean and non-Euclidean geometry on the mere concept of distance is due to Tilly, *Essai sur les principes fondamentaux de la géometrie et de la mécanique* (*Mémoires de la société des sciences physiques et naturelles de Bordeaux*, 1880).
[33] Compare, for example, the definitions of the straight line given by Euclid and by Archimedes.
[34] Zindler, 'Zur Theorie der mathematischen Erkenntniss'. *Wiener Berichte* (Phil.-hist. Abt.) 118 (1889).
[35] Kroman, *Unsere Naturkenntnis*, Copenhagen 1883, pp. 74ff.
[36] Hoelder, *Anschauung und Denken in der Geometrie*, p. 12.
[37] Gerken, who in his programmatic dissertation *Die philosophischen Grundlagen der Mathematik* (Perleberg 1887, p. 27) advances views similar to Kroman's, invokes Beneke in doing so. The latter in several places of his *Logik als Kunstlehre des Denkens* gives detailed discussion to mathematical knowledge, for example Vol. II pp. 51f. On pp. 52, 53 he argues that one can in fact compare infinitely many cases; his example is the angular sum of a triangle, where with fixed base the vertex is taken along the circumference of the circumscribed circle. However, Beneke does not speak of the dubious notion of 'lightning rapidity'. For explanations deviating from this point of view cf. C. Siegel, 'Versuch einer empiristischen Darstellung der räumlichen Grundgebilde etc....' (*Vierteljahrschr. f. wiss. Philosophie*, 1900, especially p. 203).
[38] Duhem, *Les origines de la statique*, Paris 1905, especially Vol. I, pp. 181f.

CHAPTER XXII

SPACE AND GEOMETRY FROM THE POINT OF VIEW OF PHYSICAL ENQUIRY*[1]

1. Our notions of space are rooted in our physiological constitution. Geometric concepts are the product of the idealization of physical experiences of space. Systems of geometry, finally, originate in the logical classification of the conceptual materials so gathered. All three factors have left their traces in unmistakeable modern geometry. Epistemological enquiries regarding space and geometry accordingly concern the physiologist, the psychologist, the physicist, the mathematician, the philosopher, and the logician alike, and they can be gradually carried to their definitive solution only by the consideration of the widely disparate points of view which are here offered.

Awakening in early youth to full consciousness, we find ourselves in possession of the notion of a space surrounding and encompassing our body, in which space move various bodies, partly altering and partly retaining their size and shape. It is impossible for us to ascertain how this notion has been begotten. Only the most thoroughgoing analysis of experiments purposefully and methodically planned has enabled us to conjecture that inborn idiosyncrasies of the body have co-operated to this end with simple and crude experiences of a purely physical character.

An object seen or touched is distinguished not only by a sensory quality (as 'red', 'rough', 'cold', etc.), but also by a locative quality (as 'to the left', 'above', 'before', etc.). The sensory quality may remain the same, while the locative quality continuously changes; that is, the same sensory object may move in space. Phenomena of this kind being again and again induced by physico-physiological circumstances, it is found that however varied the accidental sensory qualities may be, the same order of locative qualities invariably occurs, so that the latter appear perforce as a fixed and permanent system or register in which the sensory qualities are entered and classified. Now, although these qualities of sensation and locality can be excited only in conjunction with one another, and can make their appearance only concomitantly, the impression nevertheless easily arises that the more familiar system of locative qualities is given antecedently to the sensory qualities (Kant).

2. Extended objects of vision and of touch consist of more or less distinguishable sensory qualities, conjoined with adjacent distinguishable, continuously graduated locative qualities. If such objects move, particularly in the domain of our hands, we perceive them to shrink or swell (in whole or in part), or we perceive them to remain the same; in other words, the contrasts characterizing their bounding locative qualities change or remain constant. In the latter instance, we call the objects rigid. By the recognition of permanency as coincident with spatial displacement, the various constituents of our intuition of space are rendered comparable with one another – at first in the physiological sense. By the comparison of different bodies with one another, by the introduction of physical measures, this comparability is rendered quantitative and more exact, and so transcends the limitations of individuality. Thus, in the place of an individual and non-transmittable intuition of space are substituted the universal concepts of geometry, which hold good for all men. Each person has his own individual intuitive space; geometric space is common to all. We must sharply distinguish between the space of intuition and metric space, which contains physical experiences.

3. The need for a thorough-going epistemological elucidation of the foundations of geometry induced Riemann[2], about the middle of the last century, to propound the question of the nature of space; the attention of Gauss, Lobachevsky, and the two Bolyais having before been drawn to the empirico-hypothetical character of certain of the fundamental assumptions of geometry. In characterizing space as a special case of a multiply extended 'magnitude', Riemann had doubtless in mind certain geometric constructs which may similarly be imagined to fill all space – for example, the system of Cartesian co-ordinates. Riemann further asserts that "the propositions of geometry cannot be deduced from general conceptions of magnitude, but that the peculiar properties by which space is distinguished from other conceivable triply extended magnitudes can be derived from experience only.... These facts, like all facts, are in no wise necessary, but possess empirical certitude only – they are hypotheses." Like the fundamental assumptions of every natural science, so also, on Riemann's theory, the fundamental assumptions of geometry, to which experience has led us, are merely idealizations of experience.

In this physical conception of geometry, Riemann takes his stand on the same ground as his master Gauss, who once expressed the conviction that it was impossible to establish the foundations of geometry entirely a priori[3], and who further asserted that "we must in humility confess that if number is exclusively a product of the mind, space possesses in addition a reality outside of our mind, of which reality we cannot fully dictate a priori the laws."[4]

4. Every enquirer knows that the knowledge of an object he is investigating is materially augmented by comparing it with related objects. Quite naturally therefore Riemann looks about him for objects which offer some analogy to space. Geometric space is defined by him as a triply extended continuous manifold, the elements of which are the points determined by every possible three co-ordinate values. He finds that "the places of objects of sense and colours are probably the only concepts [*sic*] whose modes of determination form a multiply extended manifold." To this analogy others were added by Riemann's successors and elaborated by them, but not always, I think, felicitously.[5]

5. Comparing sensation of space with sensation of colour, we discover that to the continuous series 'above and below', 'right and left', 'near and far', correspond the three sensational series of mixed colours, black-white, red-green, blue-yellow. The system of sensed (seen) places is a triple continuous manifold like the system of colour-sensations. The objection which is raised against this analogy, viz., that in the first instance the three variations (dimensions) are homogeneous and interchangeable with one another, while in the second instance they are heterogeneous and not interchangeable, does not hold when space-sensation is compared with colour-sensation. For from the psycho-physiological point of view 'right and left' as little permit of being interchanged with 'above and below' as do red and green with black and white. It is only when we compare geometric space with the system of colours that the objection is apparently justified. But there is still a great deal lacking to the establishment of a complete analogy between the space of intuition and the system of colour-sensation. Whereas nearly equal distances in sensory space are immediately recognized as such, a like remark cannot be made of differences of colours, and in this latter province it is not possible to compare physiologically the different

portions with one another. And, furthermore, even if there be no difficulty, by resorting to physical experience, in characterizing every colour of a system by three numbers, just as the places of geometric space are characterized, and so in creating a metric system similar to the latter, it will nevertheless be difficult to find something which corresponds to distance or volume and which has an analogous physical significance for the system of colours.

6. There is always an arbitrary element in analogies, for they are concerned with the coincidences to which the attention is directed. But between space and time doubtless the analogy is fully conceded, whether we use the words in their physiological or in their physical senses. In both meanings of the terms, space is a triple, and time a simple, continuous manifold. A physical event, precisely determined by its conditions, of moderate, not too long or too short duration, seems to us physiologically now and at any other time as having the same duration. Physical events which at any time are temporarily coincident are likewise temporarily coincident at any other time. Temporal congruence exists, therefore, just as much as does spatial congruence. Unalterable physical temporal objects exist, therefore, as much as unalterable physical spatial objects (rigid bodies). There is not only spatial but there is also temporal substantiality. Galileo employed corporeal phenomena, like the beats of the pulse and breathing, for the determination of time, just as anciently the hands and the feet were employed for the estimation of space.

7. The simple manifold of tonal sensations is likewise analogous to the triple manifold of space-sensations[6]. The comparability of the different parts of the system of tonal sensations is given by the possibility of directly sensing the musical interval. A metric system corresponding to geometric space is most easily obtained by expressing tonal pitch in terms of the logarithm of the rate of vibration. For the constant musical interval we have here the expression,

$$\log \frac{n'}{n} = \log n' - \log n = \log \tau - \log \tau' = \text{const.},$$

where n', n denote the rates, and τ', τ the periods of vibration of the higher and the lower note respectively. The difference between the logarithms here represents the constancy of the lenght on displacement. The unalterable, substantial physical object which we sense as an interval is for the ear temporally determined, whereas the analogous object for the senses of sight and touch is spatially determined. Spatial measure seems to us simpler solely because we have chosen for the fundamental measure of geometry distance itself, which remains unalterable for sensation, whereas in the province of tones we have reached our measure only by a long and circuitous physical route.

8. Having dwelt on the coincidences of our analogized constructs, it now remains for us to emphasize their differences. Conceiving time and space as sensational manifolds, the objects whose motions are made perceptible by the alteration of temporal and spatial qualities are characterized by other sensational qualities, as colours, tactual sensations, tones, etc. If the system of tonal sensations is regarded as analogous to the optical space of sense, the curious fact results that in the first province the spatial qualities occur alone, unaccompanied by sensational qualities corresponding to the objects, just as if one could see a place or motion without seeing the object which occupied this place or executed this motion. Conceiving spatial qualities as organic sensations which can be excited only concomitantly with sensational qualities[7], the analogy in question does not appear particularly attractive. For the manifold-mathematician, essentially the same case is presented whether an object of definite colour moves continuously in optical space, or whether an object spatially fixed passes continuously through the manifold of colours. But for the physiologist and psychologist the two cases are widely different, not only because of what was above adduced, but also, and specifically, because of the fact that the system of spatial qualities is very familiar to us, whereas we can represent to ourselves a system of colour-sensations only laboriously and artificially, by means of scientific devices. Colour appears to us as an excerpted member of a manifold the arrangement of which is in no wise familiar to us.

9. The manifolds here analogized with space are, like the colour system, also threefold, or they represent a smaller number of variations. Space contains surfaces as two-fold and lines as one-fold manifolds, to which the mathematician, generalizing, might also add points as zero-

fold manifolds. There is also no difficulty in conceiving analytical mechanics, with Lagrange, as an analytical geometry of four dimensions, time being considered the fourth co-ordinate. In fact, the equations of analytical geometry, in their conformity to the co-ordinates, suggest very clearly to the mathematician the extension of these considerations to an unlimited larger number of dimensions. Similarly, physics would be justified in considering an extended material continuum, to each point of which a temperature, a magnetic, electric, and gravitational potential were ascribed, as a portion or section of a multiple manifold. Employment with such symbolic representations must, as the history of science shows us, by no means be regarded as entirely unfruitful. Symbols which initially appear to have no meaning whatever, acquire gradually, after subjection to what might be called intellectual experimentation, a lucid and precise significance. Think only of the negative, fractional, and variable exponents of algebra, or of the cases in which important and vital extensions of ideas have taken place which otherwise would have been totally lost or have made their appearance at a much later date. Think only of the so-called imaginary quantities with which mathematicians long operated, and from which they even obtained important results ere they were in a position to assign to them a perfectly determinate and even visualizable meaning. But symbolic representation has likewise the disadvantage that the object represented is very easily lost sight of, and that operations are continued with the symbols to which frequently no object whatever corresponds.[8]

10. It is easy to rise to Riemann's conception of an n-fold continuous manifold, and it is even possible to realize and visualize portions of such a manifold. Let $a_1, a_2, a_3, a_4 \ldots a_{n+1}$ be any elements whatsoever (sensational qualities, substances, etc.). If we conceive these elements intermingled in all their possible relations, then each single mixture will be represented by the expression

$$\alpha_1 a_1 + \alpha_2 a_2 + \alpha_3 a_3 + \cdots \alpha_{n+1} a_{n+1} = 1\,,$$

where the coefficients α satisfy the equation

$$\alpha_1 + \alpha_2 + \alpha_3 + \cdots \alpha_{n+1} = 1\,.$$

Inasmuch, therefore, as n of these coefficients α may be selected at pleasure, the totality of the mixtures of $n+1$ elements will represent an n-fold continuous manifold[9]. As co-ordinates of a point of this manifold, we may regard expressions of the form

$$\frac{\alpha_m}{\alpha_1}, \quad \text{or} \quad f\left(\frac{\alpha_m}{\alpha_1}\right), \quad \text{for example,} \quad \log\left(\frac{\alpha_m}{\alpha_1}\right).$$

But in choosing a definition of distance, or of any other notion analogous to geometrical concepts, we shall have to proceed very arbitrarily unless experiences of the manifold in question inform us that certain metric concepts have a real meaning, and are therefore to be preferred, as is the case for geometric space with the definition[10] derived from the voluminal constancy of bodies for the element of distance $ds^2 = dx^2 + dy^2 + dz^2$, and as is likewise the case for sensations of tone with the logarithmic expression mentioned above. In the majority of cases where such an artificial construction is involved, fixed points of this sort are wanting, and the entire consideration is therefore an ideal one. The analogy to space loses thereby in completeness, fruitfulness, and stimulating power.

11. In yet another direction Riemann elaborated ideas of Gauss; beginning with the latter's investigations concerning curved surfaces. Gauss's measure of the curvature[11] of a surface at any point is given by the expression $k = d\sigma/ds$, where ds is an element of the surface and $d\sigma$ is the superficial element of the unit-sphere, the limiting radii of which are parallel to the limiting normals of the element ds. This measure of curvature may also be expressed in the form $k = 1/\rho_1\rho_2$, where $\rho_1\rho_2$ are the principal radii of curvature of the surface at the point in question. Of special interest are the surfaces whose measure of curvature for all points has the same value – the surfaces of constant curvature. Conceiving the surfaces as infinitely thin, non-distensible, but flexible bodies, it will be found that surfaces of like curvature may be made to coincide by bending – as for example a plane sheet of paper wrapped round a cylinder or cone – but cannot be made to coincide with the surface of a sphere. During such deformation, nay, even on crumpling, the proportional parts of figures drawn in the surface remain invariable as to lengths and angles, provided we do not go out of the two dimensions of the surface in our measurements. Conversely, likewise, the curvature of the surface does not depend on its conformation in the third dimension

of space, but solely upon its interior proportionalities. Riemann, now, conceived the idea of generalizing the notion of measure of curvature and applying it to spaces of three or more dimensions. Conformably thereto, he assumes that finite unbounded spaces of constant positive curvature are possible, corresponding to the unbounded but finite two-dimensional surface of the sphere, while what we commonly take to be infinite space would correspond to the unlimited plane of curvature zero, and similarly a third species of space would correspond to surfaces of negative curvature. Just as the figures drawn upon a surface of determinate constant curvature can be displaced without distortion upon this surface only (for example, a spherical figure on the surface of its sphere only, or a plane figure in its plane only), so should analogous conditions necessarily hold for spatial figures and rigid bodies. The latter are capable of free motion only in spaces of constant curvature, as Helmholtz[12] has shown at length. Just as the shortest lines of a plane are infinite, but on the surface of a sphere occur as great circles of definite finite length, closed and reverting into themselves, so Riemann conceived in the three-dimensional space of positive curvature of the analogues straight line and the plane as finite but unbounded. But there is a difficulty here. If we possessed the notion of a measure of curvature for a four-dimensional space, the transition to the special case of three-dimensional space could be easily and rationally executed; but the passage from the special to the more general case involves a certain arbitrariness, and, as is natural, different enquirers have adopted here different courses[13] (Riemann and Kronecker). The very fact that for a one-dimensional space (a curved line of any sort) a measure of curvature does not exist having the significance of an interior measure, and that such a measure first occurs in connection with two-dimensional figures, forces upon us the question whether and to what extent something analogous has any meaning for three-dimensional figures. Are we not subject here to an illusion, in that we operate with symbols to which perhaps nothing real corresponds, or at least nothing representable to the senses, by means of which we can verify and rectify our ideas?

Thus were reached the highest and most universal notions regarding space and its relations to analogous manifolds which resulted from the conviction of Gauss concerning the empirical foundations of geometry. But the genesis of this conviction has a preliminary history of two thousand years, the chief phenomena of which we can perhaps better survey from the height which we have now gained.

12. The unsophisticated men, who, rule in hand, acquired our first geometric knowledge, held to the simplest bodily objects or figures: the straight line, the plane, the circle, etc., and investigated, by means of forms which could be conceived as combinations of these simple figures, the connection of their measurements. It could not have escaped them that the mobility of a body is restricted when one and then two of its points are fixed, and that finally it is altogether checked by fixing three of its points. Granting that rotation about an axis (two points), or rotation about a point in a plane, as likewise displacement with constant contact of two points with a straight line and of a third point with a fixed plane laid through that straight line – granting that these facts were separately observed, it would be known how to distinguish between pure rotation, pure displacement, and the motion compounded of these two independent motions. The first geometry was of course not based on purely metric notions, but made many considerable concessions to the physiological factors of sense[14]. Thus the appearance is explained of two different fundamental measures: (the straight) length and the angle (circular measure). The straight line was conceived as a rigid mobile body (measuring-rod), and the angle as the rotation of a straight line with respect to another (measured by the arc so described). Doubtless no one ever demanded special proof for the equality of angles at the origin described by the same rotation. Additional propositions concerning angles resulted quite easily. Turning the line *b* about its intersection with *c* so as to describe the angle α (Figure 22), and after coincidence with *c* turning it again about its intersection with *a* till it coincides with *a* and so describes the angle β, we shall have rotated *b* from its initial to its final position *a* through the angle μ in the same sense.[15] Therefore the exterior angle $\mu = \alpha + \beta$, and since $\mu + \gamma = 2R$, also $\alpha + \beta + \gamma = 2R$. Displacing (Figure 23) the rigid system of lines *a*, *b*, *c* which intersect at 1 within their plane to the position 2, the line *a* always remaining within itself, no alteration of angles will be caused by the mere motion. The sum of the interior angles of the triangle 1 2 3 so produced is evidently $2R$. The same consideration also throws into relief the properties of parallel lines. Doubts as to whether successive rotation about several points is equivalent to rotation

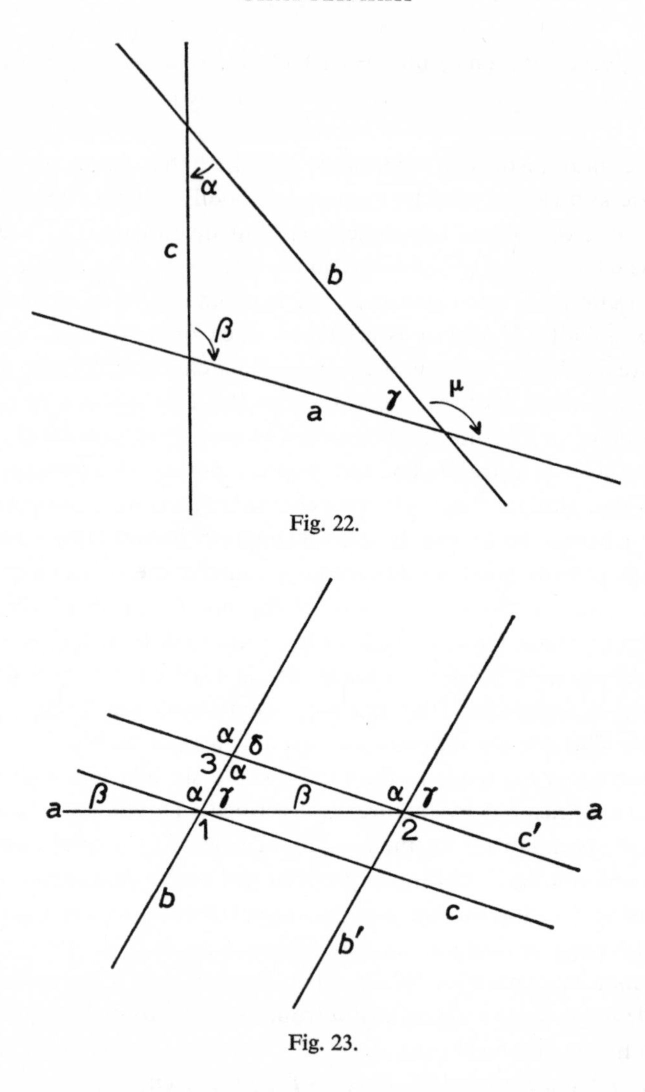

Fig. 22.

Fig. 23.

about one point, whether pure displacement is at all possible – which are justified when a surface of curvature differing from zero is substituted for the Euclidean plane – could never have arisen in the mind of the ingenuous and delighted discoverer of these relations at the period we are considering. The study of the movement of rigid bodies, which Euclid studiously avoids and only covertly introduces in his principle of congruence, is to this day the device best adapted to elementary instruction in geometry. An idea is best made the possession of the learner by the method by which it has been found.

13. This sound and naïve conception of things vanished and the treatment of geometry underwent essential modifications when it became the subject of professional and scholarly contemplation. The object now was to systematize the knowledge of this province for purposes of individual survey, to separate what was directly cognizable from what was deducible and deduced, and to throw into distinct relief the thread of the deduction. For the purpose of instruction the simplest principles, those most easily gained and apparently free from doubt and contradiction, are placed at the beginning, and the remainder based upon them. Efforts were made to reduce these initial principles to the utmost, as may be observed in the system of Euclid. Through this endeavour to support every notion by another, and to leave to direct knowledge the least possible scope, geometry was gradually detached from the empirical soil out of which it had sprung. People accustomed themselves to regard the derived truths more highly than the directly perceived truths, and ultimately came to demand proofs for propositions which no one ever seriously doubted. Thus arose – as tradition would have it, to check the onslaughts of the Sophists – the system of Euclid with its logical perfection and finish. Yet not only were the ways of research concealed by this artificial method of stringing propositions on an arbitrarily chosen thread of deduction, but the varied organic connection between the principles of geometry was quite lost sight of[16]. This system was more fitted to produce narrow-minded and sterile pedants than fruitful, productive investigators. And these conditions were not improved when scholasticism, with its preference for slavish comment on the intellectual products of others, cultivated in thinkers scarcely any sensitiveness for the rationality of their fundamental assumptions and by way of compensation fostered in them an exaggerated respect for the logical form of deductions. The entire period from Euclid to Gauss suffered more or less from this affection of mind.

14. Among the propositions on which Euclid based his system is found the so-called Fifth Postulate (also called the Eleventh Axiom

and by some the Twelfth): "If a straight line meet two straight lines, so as to make two interior angles on the same side of it taken together less than two right angles, these straight lines being continually produced, shall at length meet upon that side on which are the angles which are less than two right angles." Euclid easily proves that if a straight line falling on two other straight lines makes the alternate angles equal to each other, the two straight lines will not meet but are parallel. But for the proof of the converse, that parallels make equal alternate angles with every straight line falling on them, he is obliged to resort to the Fifth Postulate. This converse is equivalent to the proposition that only one parallel to a straight line can be drawn through a point. Further, by the fact that with the aid of this converse it can be proved that the sum of the angles of a triangle is equal to two right angles and that from this last theorem again the first follows, the relationship between the propositions in question is rendered distinct and the fundamental significance of the Fifth Postulate for Euclidean geometry is made plain.

15. The intersection of slowly converging lines lies outside the province of construction and observation. It is therefore intelligible that in view of the great importance of the assertion contained in the Fifth Postulate the successors of Euclid, habituated by him to rigour, should, even in ancient times, have strained every nerve to demonstrate this postulate, or to replace it by some immediately obvious proposition. Numberless futile efforts were made from Euclid to Gauss, to deduce this Fifth Postulate from the other Euclidean assumptions. It is a sublime spectacle which these men offer: labouring for centuries, from a sheer thirst for scientific elucidation, in quest of the hidden sources of a truth which no person of theory or of practice ever really doubted! With eager curiosity we follow the pertinacious utterances of the ethical power resident in this human search for knowledge, and with gratification we note how the enquirers gradually are led by their failures to the perception that the true basis of geometry is experience. We shall content ourselves with a few examples.

16. Among the enquirers notable for their contributions to the theory of parallels are the Italian Saccheri and the German mathematician Lambert. In order to render their mode of attack intelligible, we will remark first that the existence of rectangles and squares, which we

fancy we constantly observe, cannot be demonstrated without the aid of the Fifth Postulate. Let us consider, for example, two congruent isosceles triangles *ABC*, *DBC*, having right angles at *A* and *D* (Figure 24), and let them be laid together at their hypotenuses *BC* so as to form the equilateral quadrilateral *ABCD*; the first twenty-seven propositions of Euclid do not suffice to determine the character and magnitude of the two equal (right) angles at *B* and *C*. For measure of length and measure of angle

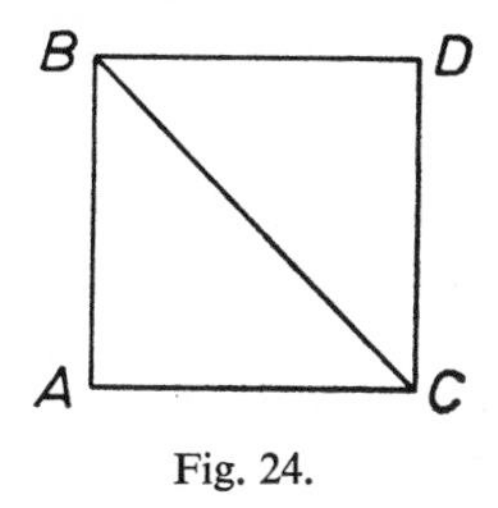

Fig. 24.

are fundamentally different and directly not comparable; hence the first propositions regarding the connection of sides and angles are qualitative only, and hence the imperative necessity of a quantitative theorem regarding angles, like that of the angle-sum. Be it further remarked that theorems analogous to the twenty-seven planimetric propositions of Euclid may be set up for the surface of a sphere and for surfaces of constant negative curvature, and that in these cases the analogous construction gives respectively obtuse and acute angles at *B* and *C*.

17. Saccheri's cardinal achievement was his form of stating the problem[17]. If the Fifth Postulate is involved in the remaining assumptions of Euclid, then it will be possible to prove without its aid that in the quadrilateral *ABCD* (Figure 25) having right angles at *A* and *B* and $AC = BD$, the angles at *C* and *D* likewise are right angles. And, on the other hand, in this event, the assumption that *C* and *D* are either obtuse or acute will lead to contradictions. Saccheri, in other words, seeks to draw conclusions from the hypothesis of the right, the obtuse, or the acute angle. He shows that each of these hypotheses will hold in all cases if it be proved to hold in one. It is necessary to have only one triangle with its angles $\lesseqgtr 2R$ in order to demonstrate the universal validity of the hypothesis of the acute,

the right, or the obtuse angle. Notable is the fact that Saccheri also adverts to physico-geometrical experiments which support the hypothesis of the right angle. If a line *CD* (Figure 25) join the two extremities of the equal perpendiculars erected on a straight line *AB*, and the perpendicular dropped on *AB* from any point *N* of the first line, viz., *NM*, be equal to $CA = DB$, then is the hypothesis of the right angle demonstrated to be correct. Saccheri rightly does not regard it as self-evident that the line which is equidistant from another straight line is itself a straight line.

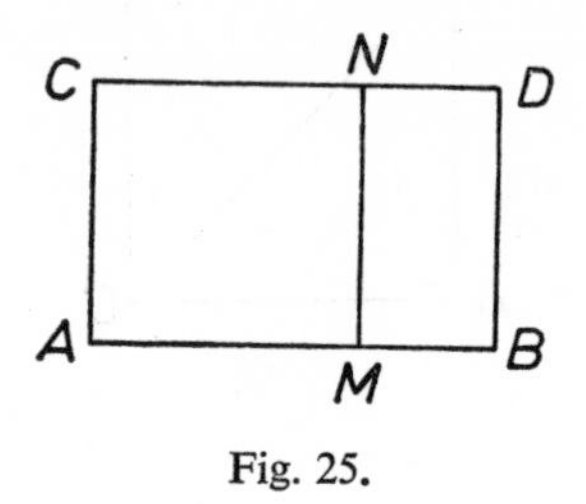

Fig. 25.

Think only of a circle parallel to a great circle on a sphere which does not represent a shortest line on a sphere and the two faces of which cannot be made congruent.

Other experimental proofs of the correctness of the hypothesis of the right angle are the following. If the angle in a semicircle (Figure 26) is shown to be a right angle, $\alpha + \beta = R$, then $2\alpha + 2\beta = 2R$ is the sum of the angles of the triangle *ABC*. If the radius be subtended thrice in a semicircle and the line joining the first and the fourth extremity pass through the centre, we shall have at *C* (Figure 27) $3\alpha = 2R$, and consequently each of the three triangles will have the angle-sum $2R$. The existence of equiangular triangles of different sizes (similar triangles) is likewise subject to experimental

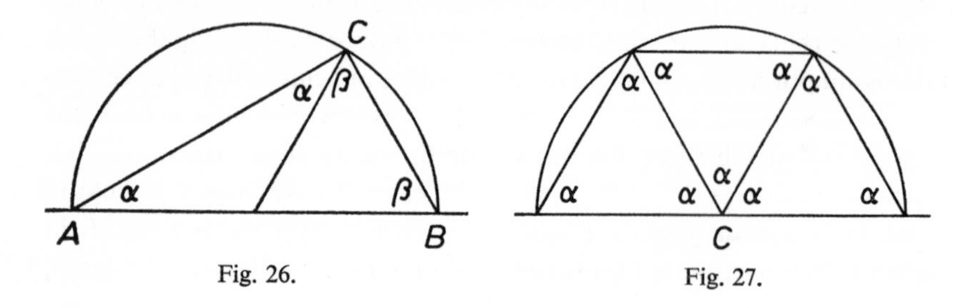

Fig. 26.

Fig. 27.

proof. For (Figure 28) if the angles at B and C give $\beta+\delta+\gamma+\varepsilon=4R$, so also is $4R$ the angle-sum of the quadrilateral $BCB'C'$. Even Wallis[18] (1663) based his proof of the Fifth Postulate on the assumption of the existence of similar triangles, and a modern geometer, Delbœuf, deduced from the assumption of similitude the whole of Euclidean geometry.

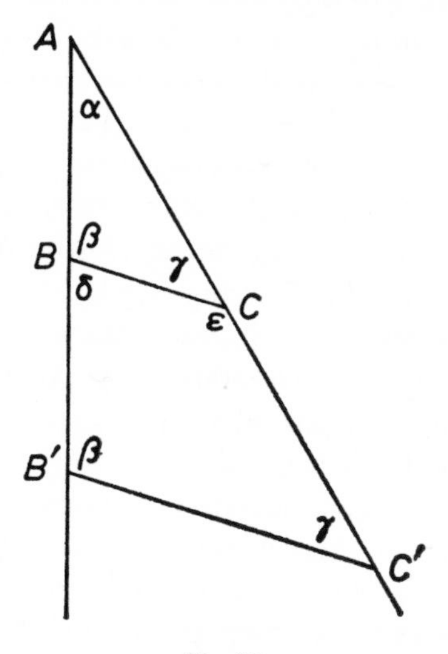

Fig. 28.

The hypothesis of the obtuse angle, Saccheri fancied he could easily refute. But the hypothesis of the acute angle presented to him difficulties, and in his quest for the expected contradictions he was carried to the most far-reaching conclusions, which Lobachevsky and Bolyai subsequently rediscovered by methods of their own. Ultimately he felt compelled to reject the last-named hypothesis as incompatible with the nature of the straight line; for it led to the assumption of different kinds of straight lines, which met at infinity, that is, had there a common perpendicular. Saccheri did much in anticipation and promotion of the labours that were subsequently to elucidate these matters, but exhibited nonetheless a certain bias towards traditional views.

18. Lambert's treatise (1766)[19] is allied in method to that of Saccheri, but it proceeds further in its conclusions, and gives evidence of a less con-

strained vision. Lambert starts from the consideration of a quadrilateral with three right angles, and examines the consequences that would follow from the assumption that the fourth angle was right, obtuse, or acute. The similarity of figures he finds to be incompatible with the second and third assumptions. The case of the obtuse angle, which requires the sum of the angles of a triangle to exceed $2R$, he discovers to be realized in the geometry of spherical surfaces, in which the difficulty of parallel lines entirely vanishes. This leads him to the conjecture that the case of the acute angle, where the sum of the angles of a triangle is less than $2R$, might be realized on the surface of a sphere of imaginary radius. The amount of the departure of the angle-sum from $2R$ is in both cases proportional to the area of the triangle, as may be demonstrated by appropriately dividing large triangles into small triangles, which on diminution may be made to approach as near as we please to the angle-sum $2R$. Lambert advanced very closely in this conception to the point of view of modern geometers. Admittedly a sphere of imaginary radius, $r\sqrt{-1}$ is not a visualizable geometric construct, but analytically it is a surface having a negative constant Gaussian measure of curvature. It is evident again from this example how experimenting with symbols also may direct enquiry to the right path, in periods where other points of support are entirely lacking and where helpful device must be esteemed at its worth.[20] Even Gauss appears to have thought of a sphere of imaginary radius, as is obvious from his formula for the circumference of a circle (*Letter to Schumacher*, July 12, 1831). Yet in spite of everything, Lambert actually fancied he had approached so near to the proof of the Fifth Postulate that what was lacking could be easily supplied.

19. We may turn now to the investigators whose views possess a most radical significance for our conception of geometry, but who announced their opinion only briefly, by word of mouth or letter. "Gauss regarded geometry merely as a logically consistent system of constructs, with the theory of parallels placed at the pinnacle as an axiom; yet he had reached the conviction that this proposition could not be proved, though it was known from experience – for example, from the angles of the triangle joining the Brocken, Hohenhagen, and Inselsberg – that it was approximately correct. But if this axiom be not conceded, then, he contends, there results from its non-acceptance a different and entirely independent geo-

metry, which he had once investigated and called by the name of the Anti-Eclidean geometry." Such, according to Sartorius von Waltershausen[21], was the view of Gauss.

Starting at this point, O. Stolz, in a small but very instructive pamphlet[22], sought to deduce the principal propositions of the Euclidean geometry from the purely observable facts of experience. We shall reproduce here the most important point of Stolz's brochure. Let there be given (Figure 29) one large triangle ABC having the angle-sum $2R$. We draw the perpendicular AD on BC, complete the figure by $BAE \simeq ABD$ and $CAF \simeq ACD$, and add to the figure $BCFAE$ the congruent figure $CBHA'G$. We obtain thus a single rectangle, for the angles at E, F, G, H are right angles and those at A, C, A', B are straight angles (equal to $2R$), the boundary lines therefore straight lines and the opposite angles equal. A rectangle can be divided into two congruent rectangles by a perpendicular erected at the middle point of one of its sides, and by continuing this procedure the line of division may be brought to any point we please in the divided

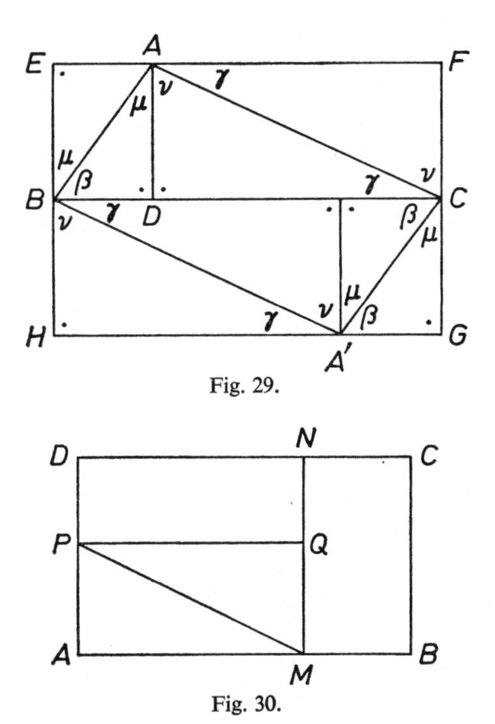

Fig. 29.

Fig. 30.

side. And the same holds true of the other two opposite sides. It is possible, therefore, from a given rectangle *ABCD* (Figure 30) to cut out a smaller *AMQP* having sides bearing any proportion to one another. The diagonal of this last divides it into two congruent right-angled triangles, of which each, independently of the ratio of the sides, has the angle-sum $2R$. Every oblique-angled triangle can by the drawing of a perpendicular be decomposed into right-angled triangles, each of which can again be decomposed into right-angled triangles having still smaller sides – so that $2R$, therefore, results for the angle-sum of every triangle if it holds true exactly of one. By the aid of these propositions which repose on observation we conclude easily that the two opposite sides of a rectangle (or of any so-called parallelogram) are everywhere, no matter how far prolonged, the same distance apart, that is, never intersect. They have the properties of the Euclidean parallels, and may be called and defined as such. It likewise follows, now, from the properties of triangles and rectangles, that two straight lines which are cut by a third straight line so as to make the sum of the interior angles on the same side of them less than two right angles will meet on that side, but in either direction from their point of intersection will move indefinitely far away from each other. The straight line therefore is infinite. What was a groundless assertion stated as an axiom or an initial principle may as inference have a sound meaning.

20. Geometry, accordingly, consists of the application of mathematics to experiences concerning space. Like mathematical physics, it can become an exact deductive science only on the condition of its representing the objects of experience by means of schematizing and idealizing concepts. Just as mechanics can assert the constancy of masses or reduce the interactions between bodies to simple accelerations only within the limits of errors of observation, so likewise the existence of straight lines, planes, the amount of the angle-sum, etc., can be maintained only on a similar restriction. But just as physics sometimes finds itself constrained to replace its ideal assumptions by other more general ones, viz., to put in the place of a constant acceleration of falling bodies one dependent on the distance, instead of a constant quantity of heat a variable quantity – so a similar procedure is permissible in geometry, when it is demanded by the facts or is necessary temporarily for scientific elucidation[23]. And now the endeavours of Legendre, Lobachevsky, and the two Bolyais, the

younger of whom was probably indirectly inspired by Gauss, will appear in their right light.

21. Of the labours of Schweickart and Taurinus, also contemporaties of Gauss, we will not speak. Lobachevsky's works were the first to become known to the world of thinkers and so productive of results (1829). Very soon afterward the publication of the younger Bolyai appeared (1833), which agreed in all essential points with Lobachevsky's, departing from it only in the form of its developments. According to the originals (published 1899)[24], it is permissible to assume that Lobachevsky too undertook his investigations in the hope of becoming involved in contradictions by the rejection of the Euclidean axiom. But after he found himself mistaken in this expectation, he had the intellectual courage to draw all the consequences from this fact. Lobachevsky gives his conclusions in synthetic form. But we can fairly well imagine the general analysing considerations that paved the way for the construction of his geometry.

From a point lying outside a straight line *g* (Figure 31) a perpendicular *p* is dropped and through the same point in the plane *pg* a straight line

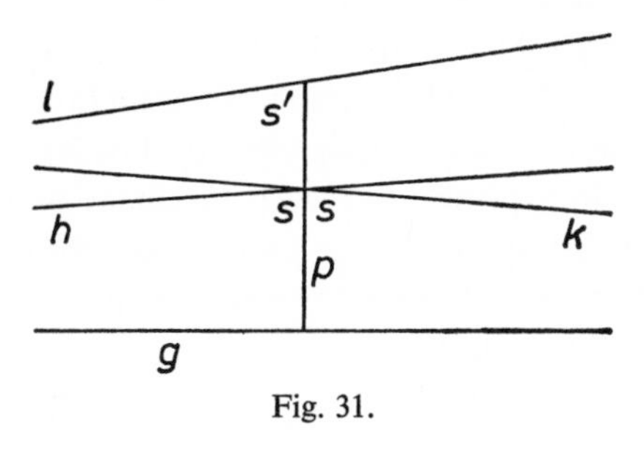

Fig. 31.

h is drawn, making with the perpendicular an acute angle *s*. Making tentatively the assumption that *g* and *h* do not meet but that on the slightest diminution of the angle *s* they would meet, we are at once forced by the homogeneity of space to the conclusion that a second line *k* having the same angle *s* similarly deports itself on the other side of the perpendicular. Hence all non-intersecting lines drawn through the same point are situate between *h* and *k*. The latter form the boundaries between the intersecting and non-intersecting lines and are called by Lobachevsky parallels.

In the Introduction to his *New Principles of Geometry* (1835) Lobachevsky proves himself a thorough natural enquirer. No one would think of attributing even to an ordinary man of sense the crude view that the 'parallel-angle' was very much less than a right angle, when on slight prolongation it could be distinctly seen that they would intersect. The relations here considered admit of representation only in drawings that distort the true proportions, and we have on the contrary to picture to ourselves that with the dimensions of the cut the variation of s from a right angle is so small that h and k are to the eye undistinguishably coincident. Prolonging, now, the perpendicular p to a point beyond its intersection with h, and drawing through its extremity a new line l parallel to h and therefore parallel also to g, it follows that the parallel-angle s' must necessarily be less than s, if h and l are not again to fulfil the conditions of the Euclidean case. Continuing in the same manner, the prolongation of the perpendicular and the drawing of parallels, we obtain a parallel-angle that constantly decreases. Considering, now, parallels which are more remote and consequently converge more rapidly on the side of convergence, we shall logically be compelled to assume, not to be at variance with the preceding supposition, that on approach or on the decrease of the length of the perpendicular the parallel-angle will again increase. The angle of parallelism, therefore, is an inverse function of the perpendicular p, and has been designated by Lobachevsky by $\Pi(p)$. A group of parallels in a plane has the arrangement shown schematically in Figure 32. They all approach one another asymptotically toward the side of their convergence. The homogeneity of space requires that every

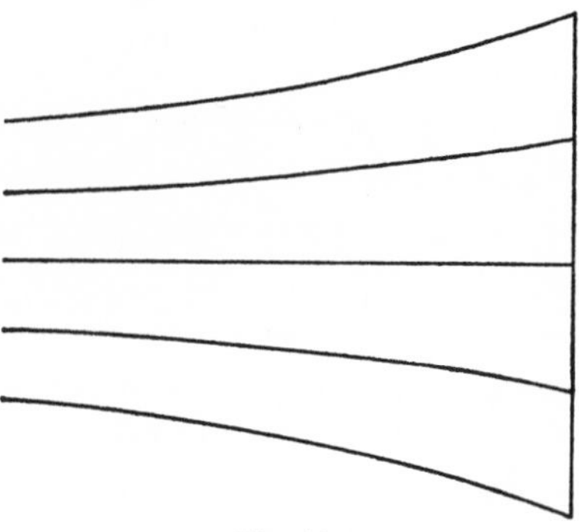

Fig. 32.

'strip' between two parallels can be made to coincide with every other strip provided it be displaced the requisite distance in a longitudinal direction.

22. If a circle be imagined to increase indefinitely, its radii will cease to intersect the moment the increasing arcs reach the point where their convergence corresponds to parallelism. The circle then passes over into the so called 'boundary-line'. Similarly the surface of a sphere, if it indefinitely increase, will pass into what Lobachevsky calls a 'boundary-surface'. The boundary-lines bear a relation to the boundary-surface analogous to that which a great circle bears to the surface of a sphere. The geometry of the surface of a sphere is independent of the axiom of parallels. But since it can be demonstrated that triangles formed from boundary lines on a boundary-surface no more exhibit an excess of angle-sum than do finite triangles on a sphere of infinite radius, therefore the rules of the Euclidean geometry also hold good for these boundary-triangles. To find points of the boundary-line, we determine in a bundle of parallels $a\alpha$, $b\beta$, $c\gamma$, $d\delta$... lying in a plane points a, b, c, d in each of these parallels so situated with respect to the point a in $a\alpha$ that $\angle \alpha ab = \angle \beta ba$, $\angle \gamma ca$, $\angle \alpha ad = \angle \delta da$... (see Figure 33). Owing to the sameness of the entire construction, each of the parallels may be regarded as the 'axis' of the boundary line, which will generate, when revolved about this axis, the boundary-surface. Likewise each of the parallels may be regarded as the axis of the boundary-surface. For the same reason all boundary-lines and all boundary-surfaces are congruent. The intersection of every plane with the boundary-surface

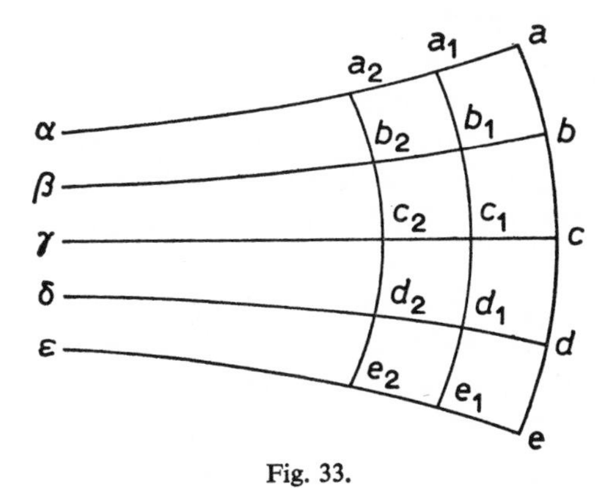

Fig. 33.

is a circle; it is a boundary-line only when the cutting plane contains the axis. In Euclidean geometry there is no boundary-line, nor boundary-surface. The analogues of them are here the straight line and the plane. If no boundary-line exists, then necessarily any three points not in a straight line must lie on a circle. Hence the younger Bolyai was able to replace the Euclidean axiom by this last postulate.

23. Let $a\alpha$, $b\beta$, $c\gamma$ be a system of parallels, and ae, a_1e_1, $a_2e_2 \ldots$ a system of boundary-lines, each of which systems divides the other into equal parts (Figure 33). The ratio to each other of any two boundary-arcs between the same parallels, e.g., the arcs $ae=u$ and $a_2e_2=u'$, is dependent therefore solely on their distance apart $aa_2=x$. We may put generally $u/u'=e^{x/k}$, where k is so chosen that e shall be the base of the Napierian system of logarithms. In this manner exponentials and by means of these hyperbolic functions are introduced. For the angle of parallelism we obtain $s=\cot\frac{1}{2}\Pi(p)=e^{p/k}$. If $p=0$, $s=\pi/2$; if $p=\infty$, $s=0$.

An example will illustrate the relation of the Lobachevskian to the Euclidean and spherical geometries. For a rectilinear Lobachevskian triangle having the sides a, b, c, and the angles A, B, C, we obtain, when C is a right angle,

$$\sinh\frac{a}{k} = \sinh\frac{c}{k}A.$$

Here sinh stands for the hyperbolic sine, $\sinh x=\frac{1}{2}(e^{x}-e^{-x})$ whereas $\sin x=(1/2i)(e^{ix}-e^{-ix})$, or $\sinh x=x/1!+x^3/3!+x^5/5!+x^7/7!$ and sin $x=x/1!-x^3/3!+x^5/5!-x^7/7!+\cdots$.

Considering the relations $\sin(xi)=i(\sinh x)$, or $\sinh(xi)=i\sin x$, involved in the foregoing formulae, it will be seen that the above-given formula for the Lobachevskian triangle passes over into the formula holding for the spherical triangle, viz., $\sin(a/k)=\sin(c/k)\sin A$, when ki is put in the place of k in the former and k is considered as he radius of the sphere, which in the usual formulae assumes the value unity. The re-transformation of the spherical formula into the Lobachevskian by the same method is obvious. If k be very great in comparison with a and c, we may restrict ourselves to the first member of the series for sinh or sin, obtaining in both cases, $a/k=(c/k)\sin A$ or $a=c\sin A$, the formulae of plane Euclidean geometry, which we may regard as a limiting case of both the Lobachevskian and spherical geometries for very large values of k, or for $k=\infty$. It is likewise permissible to say that all three geometries coincide in the domain of the infinitely small.

24. As we see, it is possible to construct a self-consistent, non-contradictory system of geometry solely on the assumption of the convergence of parallel lines. True, there is not a single observation of the geometrical facts accessible to us that speaks in favour of this assumption, and admittedly the hypothesis is at so great variance with our geometrical instinct as easily to explain the attitude toward it of the earlier enquirers such as Saccheri and Lambert. Our imagination, dominated as it is by our modes of visualizing and by the familiar Euclidean concepts, is competent to grasp Lobachevsky's views only piecemeal and gradually. We must suffer ourselves to be led here rather by mathematical concepts than by sensory images derived from a single narrow portion of space. But we must grant, nevertheless, that the quantitative mathematical concepts by which we through our own initiative and within a certain arbitrary scope represent the facts of geometrical experience, do not reproduce the latter with absolute exactitude. Different ideas can express the facts with the same exactness in the domain accessible to observation. The facts must hence be carefully distinguished from the intellectual constructs the formation of which they suggested. The latter – the concepts – must be consistent with observation, and must in addition be logically in accord with one another. Now these two requirements can be fulfilled in more than one manner, and hence the different systems of geometry.

25. Manifestly the labours of Lobachevsky were the outcome of protracted and intense mental effort, and it may be surmised that he first gained a clear conception of his system from general considerations and by analytic (algebraic) methods before he was able to present it synthetically. Expositions in this cumbersome Euclidean form are by no means alluring, and it is possibly due mainly to this fact that the significance of Lobachévski's and Bolyai's labours received such tardy recognition.

26. Lobachevsky developed only the consequences of the modification of Euclid's Fifth Postulate. But if we abandon the Euclidean assertion that "two straight lines cannot enclose a space," we shall obtain a com-

panion-piece to the Lobachevskian geometry[25]. Restricted to a surface, it is the geometry of the surface of a sphere. In place of the Euclidean straight lines we have great circles, all of which intersect twice and of which each pair encloses two spherical lunes. There are therefore no parallels. Riemann first intimated the possibility of an analogous geometry for three-dimensional space (of positive curvature) – a conception that does not appear to have occurred even to Gauss, possibly owing to his predilection for infinity. And Helmholtz[26], who continued the researches of Riemann physically, neglected in his turn, in his first publication, the development of the Lobachevskian case of a space of negative curvature (with an imaginary parameter k). The consideration of this case is in point of fact more obvious to the mathematician than it is to the physicist. Helmholtz treats in the publication mentioned only the Euclidean case of the curvature zero and Riemann's space of positive curvature.

27. We are able, therefore, to represent the facts of spatial observation with all possible precision by both Euclidean geometry and the geometries of Lobachevsky and Riemann, provided in the two latter cases we take the parameter k large enough. Physicists have as yet found no reason for departing from the assumption $k=\infty$ of Euclidean geometry. It has been their practice, the result of long and tried experience, to adhere steadfastly to the simplest assumptions until the facts forced their complication or modification. This accords likewise with the attitude of all great mathematicians toward applied geometry. The behaviour of physicists and mathematicians toward these questions is in the main different, but this is explained by the circumstance that for the former class of enquirers the physical facts are of most significance, geometry being for them merely a convenient implement of investigation, while for the latter class these very questions are the main material of research, and of the greatest technical and particularly epistemological interest. Supposing a mathematician to have modified tentatively the simplest and most immediate assumptions of our geometrical experience, and supposing his attempt to have been productive of fresh insight, certainly nothing is more natural than that these researches should be further prosecuted, from a purely mathematical interest. Analogues of the geometry we are familiar with, are constructed on broader and more general assumptions for any number of dimensions, with no pretension to being regarded as

more than intellectual scientific experiments and with no idea of being applied to reality. In support of my remark it will be sufficient to advert to the advances made in mathematics by Clifford, Klein, Lie, and others. Seldom have thinkers become so steeped in reverie, or so far estranged from reality, as to imagine for our space a number of dimensions exceeding the three of the given space of sense, or to conceive of representing that space by any geometry that appreciably departs from the Euclidean. Gauss, Lobachevsky, Bolyai, and Riemann were perfectly clear on this point, and certainly cannot be held responsible for the grotesque fictions which subsequently arose in this field.

28. It little accords with the principles of a physicist to make suppositions regarding the behaviour of geometrical constructs in infinity and non-accessible places, then subsequently to compare them with our immediate experience and adapt them to it. He prefers, like Stolz, to regard what is directly given as the source of his ideas, which he considers applicable also to what is inaccessible until obliged to change them. But he too may be extremely grateful for the discovery that there exist several adequate geometries, that we can manage also with a finite space, etc., – grateful, in short, for the abolition of certain conventional barriers of thought.

If we lived on the surface of a planet with a turbid, opaque atmosphere, and, on the supposition that the surface of the earth was a plane and our only instruments were square and chain, we undertook measurements, the increase in the excess of the angle-sum of large triangles would soon compel us to substitute a spherometry for our planimetry. The possibility of analogous experiences in three-dimensional space the physicist cannot as a matter of principle reject, although the phenomena that would compel the acceptance of a Lobachevskian or a Riemannian geometry would present so odd a contrast with that to which we have been hitherto accustomed, that no one will regard their actual occurrence as probable.

29. The question whether a given physical object is a straight line or the arc of a circle is not properly formulated. A stretched cord or a ray of light is certainly neither the one nor the other. The question is simply whether the object so reacts spatially that it conforms better to the one concept than to the other, and whether with the exactitude which is suffi-

cient for us and obtainable by us it conforms at all to any geometric concept. Excluding the latter case, the question arises, whether we can practically remove, or at least determine in thought and make allowance for, the deviation from the straight line or circle, in other words, correct the result of the measurement. But we are dependent always, in practical measurements, on the comparison of physical objects. If on direct investigation these coincided with the geometric concepts to the highest attainable point of accuracy, but the indirect results of the measurement deviated more from the theory than the consideration of all possible errors permitted, then certainly we should be obliged to change our physico-metric notions. The physicist will do well to await the occurrence of such a situation, while the mathematician will always have free scope for his speculations.

30. Of all the concepts which the natural enquirer employs, the simplest are the concepts of space and time. Spatial and temporal objects conforming to his conceptual constructs can be framed with great exactness. Nearly every observable deviation can be eliminated. We can imagine any spatial or temporal construct realized without doing violence to a fact. The remaining physical properties of bodies are so intimately connected that here arbitrary fictions are subjected to narrow restrictions by the facts. A perfect gas, a perfect fluid, a perfectly elastic body does not exist; the physicist knows that his fictions conform only approximately and by arbitrary simplifications to the facts; he is perfectly aware of the deviation, which cannot be removed. We can conceive a sphere, a plane, etc., constructed with unlimited exactness, without running counter to any fact. Hence, if any physical fact happens to render a modification of our concepts necessary, the physicist will prefer to sacrifice the less perfect concepts of physics rather than to give up the simpler, more perfect, and more lasting concepts of geometry, which forms the solidest foundation of all his theories.

31. But from another direction the physicist can derive substantial assistance from the labours of geometers. Our geometry refers always to objects of sensory experience. But the moment we begin to operate with mere things of thought like atoms and molecules, which from their very nature *can never be made the objects of sensory contemplation*, we are

under no obligation whatever to think of them as standing in spatial relationships which are peculiar to the Euclidean three-dimensional space of our sensuous experience. This may be recommended to the special attention of thinkers who deem atomistic speculations indispensable.[27]

32. Let us go back in thought to the origin of geometry in the practical needs of life. The recognition of the spatial substantiality and spatial invariability of spatial objects in spite of their movements is a biological necessity for human beings, for spatial quantity is related directly to the quantitative satisfaction of our needs. When knowledge of this sort is not sufficiently provided for by our physiological organization, we employ our hands and feet for comparison with the spatial object. When we begin to compare bodies with one another, we enter the domain of physics, whether we employ our hands or an artificial measure. All physical determinations are relative. Therefore, all geometrical determinations likewise possess validity relatively to the measure. The concept of measurement is a concept of relation, which contains nothing not contained in the measure. In geometry we simply assume that the measure will always and everywhere coincide with that with which it has at some other time and some other place coincided. But this assumption is determinative of nothing concerning the measure. In place of spatial physiological equality is substituted an altogether differently defined physical equality, which must not be confounded with the former, no more than the indications of a thermometer are to be identified with the sensation of heat. The practical geometer, it is true, determines the dilatation of a heated measure, by means of a measure kept at a constant temperature, and takes account of the fact that the relation of congruence in question is disturbed by this non-spatial physical circumstance. But to the pure theory of space all assumptions regarding the measure are foreign. Simply the physiologically created habit of regarding the measure as invariable is tacitly but unjustifiably retained. It would be quite superfluous and meaningless to assume that the measure, and therefore bodies generally, suffered alterations on displacement in space or that they remained unchanged on such displacement – a fact which in its turn could only be determined by the use of a new measure. The relativity of all spatial relations is made manifest by these considerations.

33. If the criterion of spatial equality is substantially modified by the introduction of measure, it is subjected to a still further modification, or intensification, by the introduction of the notion of number into geometry. There is nicety of distinction gained by this introduction which the idea of congruence alone could never have attained. The application of arithmetic to geometry leads to the notion of incommensurability and irrationality. Our geometric concepts therefore contain adscititious elements not intrinsic to space; they represent space with a certain latitude, and arbitrarily also with greater precision than spatial observation could possibly realize. This imperfect contact between fact and concept explains the possibility of different systems of geometry.[28] Exactly the same can be said of physics.[29]

34. The entire movement which led to the transformation of our ideas of geometry must be characterized as a sound and healthful one. This movement, which began centuries ago but is now greatly intensified, is not to be looked upon as having terminated. On the contrary, we are quite justified in the expectation that it will redound not only to the great advancement of mathematics and geometry, especially in an epistemological regard, but also to that of the other sciences. This movement was, it is true, powerfully stimulated by a few eminent men, but it sprang, nevertheless, not from an individual, but from a general need. This will be seen from the difference in the professions of the men who have taken part in it. Not only the mathematician, but also the philosopher and the educationist, have made large contributions to it. So, too, the methods pursued by the different enquirers are not unrelated. Ideas which Leibniz[30] uttered recur in slightly altered form in Fourier[31], Lobachevsky, Bolyai, and H. Erb[32]. The philosopher Ueberweg[33], closely approaching in his opposition to Kant the views of the psychologist Beneke[34], and in his geometrical ideas starting from H. Erb (who mentions K. A. Erb[35] as his predecessor) anticipates a goodly portion of Helmholtz's labours.

35. The results to which the preceding discussion has led, may be summarized as follows:

1. The source of our geometric concepts has been found to be experience.
2. The multiplicity of the concepts satisfying the same geometrical facts has been revealed.

3. By the comparison of space with other manifolds, more general concepts were reached, of which the geometric represented a special case. Geometric thought was thus freed from conventional limitations, hitherto imagined insuperable.

4. By the demonstration of the existence of manifolds allied to but different from space, entirely new questions were suggested. What is space physiologically, physically, geometrically? To what are its specific properties to be attributed, since others are also conceivable? Why is space three-dimensional, etc.?

36. With questions such as these, though we must not expect the answer today or tomorrow, we stand before the entire profundity of the domain to be investigated. We shall say nothing of the inept strictures of the 'Boeotians', whose coming Gauss predicted, and whose attitude determined him to reserve. But what shall we say to the acrid and captious criticisms to which Gauss, Riemann and their associates have been subjected by men of high standing in the scientific world. Have they never experienced in their own persons the truth that enquirers at the outermost boundaries of knowledge discover many things that do not slip smoothly into all heads, but which are not on that account nonsense? True, such enquirers are liable to error, but even the errors of some men are often more fruitful in their consequences than the discoveries of others.

NOTES

* Translated from Professor Mach's manuscript by T. J. McCormack for *The Monist*, 1903.

[1] In conversation and by letter, Professor F. Brentano has raised objections to the present account. These have given me cause for reflection, but pressure of work prevents me from considering them adequately at this moment.

[2] Riemann, *Über die Hypothesen, welche der Geometrie zu Grunde liegen*, Göttingen 1867.

[3] Gauss, Letter to Bessel, 27 January 1829.

[4] Gauss, Letter to Bessel, 9 April 1830. The phrase, 'Number is a product or creation of the mind' has since been repeatedly used by mathematicians. Unbiassed psychological observation informs us, however, that the concept of number is just as much initiated by experience as the formation of geometric concepts. We must at least know that virtually equivalent objects exist in multiple and unalterable form before concepts of number can originate. Experiments in counting also play an important part in the development of arithmetic....

[5] When acoustic pitch, intensity, and timbre, when chromatic tone, saturation and luminous intensity are proposed as analogies of the three dimensions of space, few persons will be satisfied. Timbre, like chromatic tone, is dependent on several variables.

Hence, if the analogy has any meaning whatsoever, several dimensions will be found to correspond to timbre and chromatic tone. Cf. Benno Erdmann, *Die Axiome der Geometrie*, Leipzig 1877.

[6] My attention was drawn to this analogy in 1863 by my study of the organ of hearing, and I have since then further developed the subject: *A* 4, pp. 222f.

[7] Cf. Ch. XX, sn. 10.

[8] I confess that as a young student I was always incensed with symbolic deductions of which the meaning was not perfectly clear and palpable. But historical studies are well adapted to eradicating the tendency to mysticism which is so easily fostered and bred by the dreamlike employment of these methods, in that they clearly show the heuristic function of them and at the same time elucidate epistemologically the points wherein they furnish their essential assistance. A symbolical representation of a method of calculation has the same significance for a mathematician as a model or a visualizable working hypothesis has for the physicist. The symbol, the model, the hypothesis runs parallel with the thing to be represented. But the parallelism may extend farther, or be extended farther, than was originally intended on the adoption of the symbol. Since the thing represented and the device representing are after all different, what would be concealed in the one is apparent in the other. It is scarcely possible to light directly on an operation like $a^{2/3}$. But operating with such symbols leads us to attribute to them an intelligible meaning. Mathematicians calculated for many decades with expressions like $\cos x + \sqrt{-1} \sin x$ and with exponentials having imaginary exponents until in the struggle for adapting concept and symbol to each other the idea that had been germinating for a century finally found expression in 1806 in Argand, viz., that a relationship could be conceived between magnitude and direction by which $\sqrt{-1}$ was represented as a mean direction-proportional between $+1$ and -1.

[9] If the six fundamental colour-sensations were totally independent of one another, the system of colour-sensations would represent a five-fold manifold. Since they are contrasted in pairs, the system corresponds to a three-fold manifold.

[10] Cf. Ch. XXI, sns. 32, 33.

[11] Gauss, *Disquisitiones generales circa superficies curvas*, 1827.

[12] Helmholtz, 'Über die Tatsachen, welche der Geometrie zu Grunde liegen', *Göttinger Nachrichten*, 3 June, 1868.

[13] Compare for example Kronecker, 'Über Systeme von Funktionen mehrerer Variablen' *Berlin. Berichte* 1869.

[14] Cf. Ch. XX, sn. 17; Ch. XXI, sn. 35.

[15] C. R. Kosack, *Beiträge zu einer systematischen Entwicklung der Geometrie aus der Anschauung*, Nordhausen 1852. I was able to see this programme through the kindness of Professor F. Pietzker of Norhausen. Similar simple deductions are found in Bernhard Becker's *Leitfaden für den ersten Unterricht in der Geometrie*, Frankfurt a.M. 1845, and in the same author's treatise *Über die Methoden des geometrischen Unterrichts*, Frankfurt 1845. I gained access to the first-named book through the kindness of Dr M. Schuster of Oldenburg.

[16] Euclid's system fascinated thinkers by its logical excellences and its drawbacks were overlooked amid this admiration. Great inquirers even in recent times, have been misled into following Euclid's example in the presentation of the results of their inquiries and so into actually concealing their methods of investigation, to the great detriment of science. But science is not a feat of legal casuistry. Scientific presentation aims so to expound all the grounds of an idea that is can at any time be thoroughly examined as to its tenability and power. The learner is not to be led half blindfolded. There therefore

arose in Germany among philosophers and educationists a healthy reaction, which proceeded mainly from Herbart, Schopenhauer, and Trendelenburg. The effort was made to introduce greater perspicuity, more genetic methods, and logically more lucid demonstrations into geometry. Cf. M. Pasch, *Vorlesungen über neuere Geometrie*, Leipzig 1882; D. Hilbert, *Grundlagen der Geometrie*, Leipzig 1899.

[17] Saccheri, *Euclides ab omni naevo vindicatus*, Milan 1733.

[18] Wallis apud Engel and Staeckel, *Die Theorie der Parallellinien*, Leipzig 1895, pp. 21ff.

[19] *Ibid.*, pp. 152ff.

[20] Cf. note 8 above.

[21] Sartorius von Waltershausen, *Gauss zum Gedächtnis*, Leipzig 1856.

[22] O. Stolz, 'Das letzte Axiom der Geometrie', *Berichte des naturw.-medicin. Vereins zu Innsbruck*, 1886, pp. 25–34.

[23] The difference between geometry and physics, taken by Duhem (*La Théorie physique*, p. 290) as basic and qualitative, I regard as only a difference of degree.

[24] F. Engel, *N. I. Lobatschefskij, Zwei geometrische Abhandlungen*, Leipzig 1899.

[25] Cf. the essay of Tilly mentioned in Ch. XXI, note 32.

[26] Helmholtz, 'Über die tatsachlichen Grundlagen der Geometrie' 1866, *Wissensch. Abhandl.* II, pp. 610ff.

[27] While still an upholder of the atomic theory, I sought to explain the line-spectra of gases by the vibrations of the atomic constituents of a gas-molecule with respect to another. The difficulties which I here encountered suggested to me (1863) the idea that non-sensory things did not necessarily have to be pictured in our sensory space of three dimensions. In this way I also lighted upon analogues of spaces of different numbers of dimensions. The collateral study of various physiological manifolds (see fn. 6 to this chapter) led me to the problems discussed in the conclusion of this paper. The notion of finite spaces, converging parallels, etc., which can come only from a historical study of geometry, was at that time remote from me. I believe that my critics would have done well had they not overlooked the italicized phrase in the text. For details see the notes to my *Erhaltung der Arbeit,* Prague 1872. Cf. also the explanations in Vaihinger, *Die Philosophie des Als-Ob,* Berlin 1911.

[28] It would be too much to expect of matter that it should realize all the atomistic fantasies of the physicist. So, too, space, as an object of experience, can hardly be expected to satisfy all the ideas of the mathematician, though there be no doubt whatever as to the general value of their investigations.

[29] Cf. note 23 above.

[30] Cf. Ch. XXI, sns. 22, 24.

[31] Fourier, *Séances de l'École Normale. Débats* I (1800) 28.

[32] H. Erb (Grossherzoglich Badischer Finanzrat), *Die Probleme der geraden Linie, des Winkels, und der ebenen Fläche*, Heidelberg 1846. He there gives the completion of elementary geometry demanded in Gauss's letter to Bessel. Similar views in J. Schram, 'Leibnizens Definitionen der Ebene und der Geraden', printed ms., Obersteig (North Tyrol) 1903.

[33] Ueberweg, 'Die Principien der Geometrie wissenschaftlich dargestellt', *Archiv für Phil. und Pädag.* 1851. Reprinted in Brasch, *Welt- und Lebensanschauung F. Überwegs*, Leipzig 1889, pp. 263–317.

[34] Beneke, *Logik als Kunstlehre des Denkens*, Berlin 1842, II. pp. 51–5.

[35] K. A. Erb, *Zur Mathematik und Logik*, Heidelberg 1821. I was unable to examine this work. Readers whose interests are principally philosophical are referred to the article by C. Siegel cited in Ch. XXI, note 37.

CHAPTER XXIII

PHYSIOLOGICAL TIME IN CONTRAST WITH METRICAL TIME

1. If in an environment that is as uniform and constant as possible with the least possible changes in ideas, we doze along as though waking from sleep and we hear the uniform striking of a clock, we clearly distinguish the second stroke from the first, the third from the second and the first, in short the later from the earlier, although all have the same strength pitch and timbre. Nor are we in the least doubt about the equality of interval between strokes, noticing at once without artificial aids if this state of affairs is disturbed. We sense time and position in time as immediately as space and position in space. Without this temporal sensation there would be no chronometry, just as without spatial sensations there would be no geometry.

2. The existence of peculiar physiological processes that lie at the root of temporal sensations becomes very likely given the circumstance that we recognize equality of rhythm or temporal shape in temporal structures of the most varied kind, for example in melodies that have no similarity beyond rhythm[1]. We sense the rhythm of a process unhindered by its quality. Conspicuous physiological facts weigh in favour of the view that the elementary organs themselves contribute to the foundation of temporal sensations. Such facts are for example the after-image of the motion of a twisted spring or of flowing water (Plateau and Oppel)[2] or Dvořák's after-image of brightening or darkening of a long drawn-out change in brightness[3]. The rate of change of position and brightness, within the limits of immediate perception (that is, leaving out the extreme cases of a clock hand or a projectile) is not only a concept of mathematical and physical measure, but also a physiological object.

3. Between our physiological intuition of time and metrical time obtained by temporal comparison of physical processes amongst each other, there are differences analogous to those between physiological space and metrical space. Both seem indeed continuous; to a steady displacement

in physical time there corresponds another such in physiological time, and both run only in one direction. This, however, exhausts the agreements. Physical time runs now faster now more slowly than physiological time; that is, not all processes of equal physical duration seem to be so to immediate observation. The physical discrimination of points in time is very much more refined than the physiological. To our temporal intuition the present appears not as a point of time, which would always have to be empty of content, but as a segment of time of considerable duration with variable limits that are blurred and indeed difficult to determine and displaceable from case to case. Temporal intuition is properly confined to this, and is completed quite imperceptibly through memory of the past and the future imagined by phantasy, both appearing in very foreshortened temporal perspective. This makes it unintelligible why the limits of temporal intuition are imprecise. For physics one periodically repeating individual rhythm is only one individual structure; for our temporal intuition the form of this structure changes at the point where attention sets in[4]. In the same way the form of one geometrical structure changes for spatial intuition according to the orientation and the fixed point, which for one-dimensional time contract into a single determining factor.

4. Today we can hardly doubt that temporal, like spatial, intuition is conditioned by our inherited bodily organization. We should labour in vain to rid ourselves of these intuitions, but in thus adopting the innate theory we are not asserting that at the moment of birth they are completely developed into full clarity; nor do we renounce the account of how they are linked with the biological need or how the latter influences their phylogenetic and ontogenetic development. Finally, none of this as yet excludes an investigation of how spatial and temporal intuitions are connected with geometrical and chronometric concepts. The intuitions are necessary but not sufficient for the concepts; to form metrical concepts we need complementary experience on the mutual spatial behaviour of bodies and on the temporal behaviour of physical processes.

5. Let us first try to clarify the biological significance of temporal sensation. In Spencer we find the apt remark, that the development of a temporal sense is tied to that of a spatial sense and depends on it. An animal that needs to maintain or adapt itself only against stimuli of contact

whether mechanical or chemical, will manage with corresponding simultaneous reactions. These last may be connected with an organically conditioned temporal course of processes unaffected by the environment; such automatic processes will not create a need for conscious notions of time. However, when the spatial radius of action of the senses becomes bigger, so that the approaching prey will announce itself by smell, sound or signs visible at a distance before it comes within reach, there is then a need for consciously reproducing such sequences of approach in their natural temporal order. For without this mental reproduction the reactions could not enter into play with the temporally ordered and measured stages that are required, for example for capture. However, once the prey has been swallowed, the process of digestion is independent of consciousness and therefore no longer enters it. The sensation and idea of time develop in the course of adaptation to temporal and spatial environment. Man, whose interests span the most extensive stretches of space and time, has indeed the most developed temporal sensations and ideas.[5]

6. It is in fact a basic trait of our mental reproductions of experience that they resemble the original not only as to quality of the sensible elements and their combination but also as to spatial and temporal relation and size. What determines accuracy is practice and the degree of attention, but even an unattentive person does not see houses in memory with their roofs pointing downwards and large buildings do not appear to him with Lilliputian dimensions or with disproportionately tall chimneys. In remembering a piece of music we do not invert the temporal sequence of notes or rhythm, an adagio is not reproduced as an allegro or viceversa. All this points to the fact that over and above those experiential elements we call sensation proper, there are other elements that form a relatively firm basis (like a photographic plate or phonographic cylinder) that are always reproduced as well and prevent excessive spatio-temporal distortion of memory images.

7. Various considerations have been tried to obtain a grasp of our views of time. At the outset it is clear that a temporal course of mental elements, whether sensations or ideas, does not as such include consciousness of such a course. If our mental field of vision were always temporally confined to a sufficiently narrow present, the fact of change could not be

perceived at all. Therefore consciousness must always encompass a finite stretch of time in which there are fading sensations and ideas at the same time as well as newly emerging ones, for the first to be viewed as earlier and the second as later. If in addition we conceive of the relatively stable ego-complex, characterized by common feelings and so on, it constitutes a kind of rock with the temporally ordered stream of change flowing past it. This seems a fairly tolerable picture and seems to correspond to the way in which we range the separate members into the chain of experiences. The sensible experience of the present is easily distinguished from the fainter and more fugitive memories of the immediate past and the still more faded ones of the remoter past. The thread of association leads us from the oldest memories to the most recent, and to the present and through it to the expectations that phantasy mirrors in front of us.[6] However, merely to put things into numerical order as we might call this process, does not seem to me fully to correspond to a view of the temporal course of things. It is a procedure that we may perhaps be using when a remote past is remembered in heavily foreshortened perspective. This will hardly create a real temporal view for example of a piece of music according to beat and rhythm, both in the sensible present and in vivid memory. The above-mentioned rigid and undistorted background on to which experience is projected is absent as it were.

8. In order better to grasp this last circumstance, consider a physical case. Disturbances are to enter along various paths into a homogeneous physical body; for example currents, first through electrodes at *a* and *b*, and secondly at *c* and *d*. The equipotential surfaces, the surfaces of equal current density and equal generation of heat and so on will be quite different in the two cases. Next we introduce unsynchronized shock waves at points *m* and *n*, the leading wave going through *m* on the first occasion and through *n* on the second. In the first case the surface of interference lies much closer to *n* and in the second closer to *m*[7]. Such phenomena are more marked still in organized animal bodies. Stimuli that enter along different paths will generally determine different reactions that influence the environment along different paths. The temporal order in which the same organs are affected by given stimuli is not indifferent, but generally will, if changed, produce different reactions. Just as it is not indifferent to the resulting reaction, whether we stimulate the frog's skin on the left or right

of the back, so it is not indifferent in what temporal state the same organ is hit by the same stimulus, for example whether a stimulus of taste or smell occurs when the animal is hungry or not.

9. To facilitate a grasp of an organ's view of space, we have assumed that every stimulated organ supplies not only a sensation proper partly determined by the quality of the stimulus, but also a sensation permanently tied to the individual organ. If we regard this last sensation as consisting of a constant part and one that varies temporally with the organ's activity, then there is some prospect of making a temporal view intelligible in terms of this last, variable part. These are of couse not theories or explanations of physiological space and time, but only possibly useful paraphrases and analyses of the facts in which spatial and temporal views express themselves. How, then, must we take the temporal variation of the part that depends on the organ's activity, so as best to meet the facts of observation?

10. In man, or higher vertebrates near to him, the body has an almost invariable temperature which is necessary for preserving life, and usually maintains a constant temperature difference from the environment for considerable periods. Physically considered, this presupposes a very uniform course of vital functions which suffers only slight disturbances from discontinuous temporal reactions on the environment. Only the smallest and simplest organisms are in conditions that allow a steady supply of food and consequently steady restitution corresponding to steady consumption. In larger and more developed organisms, perioaic processes are unavoidable for the preservation of an imperfect but adequate uniformity of the vital functions. The organism changes between states of sleep and waking, hunger and satiety. The amount of air required to maintain life can be conveyed to the blood only by a periodically acting bellows while the blood must be conveyed to the organs by the periodic action of the heart's pump. Adaptation to environment and the obtaining of food require locomotion, which is carried out through measured periodic movements of the extremities and rhythmic contractions of the muscles[8], which themselves exhibit rhythmic phenomena in a single contraction. Even optical after-images and visual impressions produced by dazzling have a periodic course. The organism indeed possesses a great

many periods of the most varied duration[9]. If we view life in Hering's sense as a state of dynamic equilibrium between consumption and restitution, then this plethora of periodic processes surprises us as little as the great variety of physical vibrations. Indeed, vibrations must arise wherever stable equilibrium is disturbed and damping is insufficient to make the adjustment aperiodic. The tendency of organic functions to periodicity further shows itself in the fact that they easily adapt to an externally forced repeated period of arbitrary duration, which they assume and continue spontaneously. An obvious example is the adapting of one's pace to the rhythm of an accidentally encountered marching tune. If I clench my fist several times in beat and then cease attending to this movement, it often requires a special decision to stop.

11. In lower or younger animals, biologically important stimuli release reflexes of adaptation. If a sequence of sensations attracts the attention of a more highly developed animal, the sensations are accompanied by an activity that consists of reflexes modified by experience (memory). Action is inseparable from sensation. Even observation is a mild form of co-operation[10], for man and animal alike. Animals no doubt are awakened from a state of mental indifference only for the short duration of a voluntary action, and that only by sensations proper, while human attention is often excited by ideas of memory as well. However, in this case we do not just passively let pictures run past us, but we are gently active, as indeed we notice as soon as we think for example of an experienced exchange of words, or even a probable or possible one. If the mental life is strong, attention may be maintained for longer periods, although even then it is not constant but changes in alternations of sudden tightenings and relaxations, as any student and teacher can observe. Thinking about the solution of problems consists in different run-ups to the same goal. Often we believe that we can see what we are seeking, but if we fail to grasp it completely it will elude us again. For this time, then, that is the end, and a new run-up must be tried after a while.

12. Attention, too, is subject to fluctuations that can last perhaps several seconds, roughly covering the physical time that physiologically we call and view as the present. If man in his reactions has adapted himself to the sense experience of his environment, whether it consists of intense

bodily behaviour or only in alert observation, then as soon as attention sets in, to each physical moment there corresponds one phase of the attention. If we think of these phases from the onset to the vanishing or deflection of the attention as roughly equal in length, but of the sensations of these phases as associated with sensations proper, then the reproduction in physical fact and in ideas will be nearly equal in time as well, whatever function of physical time the phase of attention may be. Such coincidence responds to biological need. If an experience is to be met by a conscious voluntary action (think of the behaviour of the hunter), then one must somehow or other have sensation of the phase of attention. Should this view turn out viable, it would supply the rigid undistortable temporal backdrop of memory or the uniformly running phonograph cylinder. Naturally this view helps us to understand only the reproduction of the state of affairs in small periods. For the order of experiences that span long periods, the thread of association is enough, and only a few more important scenes are viewed in microscopic detail. Otherwise our memories would take up as much time as the experiences themselves took in the first place, and there would be no time left for new experiences.[11]

13. After the acts of attention have encompassed the most varied experiences, we become acquainted with temporal sensation as permanent and independent of the remaining experiential content, and as constantly recurring. The sequence of temporal sensations becomes a register, into which we range the other qualities of sense experience. To this is added that there are such processes as pulse beats, paces, pendulum oscillations and the like, whose duration remains the same and which thus exhibit physiological temporal constancy. Although the same events seem to take different times depending on one's various bodily states, both normal and morbid, sleep, fever, being drugged with hashish and so on, nevertheless, we notice that the oscillations of the same pendulum are markedly constant in duration whenever we attend to them with normal wakeful attention.

14. On the lowest level of life we are concerned only with processes affecting our bodies. However, as soon as needs can no longer be satisfied immediately, but only by detours through temporal processes in our environment, these detours must gain an indirect interest that is often much

stronger than the interest in momentary sensations. To judge the temporal course of processes in the environment, physiological sensation of time is too imprecise and unreliable. That is when we begin to compare physical processes with each other, for example pendulum oscillations with falling motions through determinate distances, or with the rotation of the Earth during the oscillations. Here we find that a pair of precisely defined physical processes that coincide in time at both ends and are thus temporally congruent retain this property at all times. Such a precisely defined process can now be used as a time scale, and that is the basis of chronometry. We are indeed accustomed instinctively to transfer the idea of time as a substance to the standard of time measurement, but we must notice that this idea loses all sense in the physical field. Measurement indicates the ratio to the standard, about which latter the definition says nothing. Between the immediate sensation of duration and a numerical measure we must distinguish as sharply as between heat sensation and temperature.[12] Everybody has his own view of time which cannot be transferred; but chronometric concepts are common to all educated men and are transferable. We were able to be brief on this because mutatis mutandis we can repeat everything that was to be said in relation to space.

NOTES

[1] For the inadequacy of earlier theories of space and time and attempts at improvement cf. my short article 'Bemerkungen zur Lehre vom räumlichen Sehen', *Fichtes Zeitschr. f. Philos.* 1865, printed in *P* 3; 'Über den Zeitsinn des Ohres', *Ber. d. Wiener Akademie*, Jan. 1865; *A* 4.

[2] Plateau, *Poggendorffs Annalen* **80**, p. 287. Oppel, *ibid.* Vol. 99, p. 543 [actually 540].

[3] Dvořák, 'Über Nachbilder von Reizveränderungen', *Ber. d. Wiener Akademie* **61**. Mach, *Lehre von den Bewegungsempfindungen*, Leipzig 1875, pp. 59–64.

[4] *A* 4, p. 29.

[5] Spencer, *The Principles of Psychology*, 2nd ed. 1870, **I**, pp. 320–328; **II**, pp. 207–215.

[6] Cf. as regards these general considerations the accounts of psychology, especially the original book by Höffding (*Psychologie in Umrissen*, Leipzig 1893, pp. 250–260) and the fascinating account by W. James (*The Principles of Psychology* **I**, pp. 505–542), and finally the careful work of Ebbinghaus (*Grundzüge der Psychologie*, Leipzig 1902, **I**, pp. 457–466).

[7] Cf. *A* 4, pp. 192–193.

[8] That there are no continuous rotations in an animal body, as used with advantage in machines, is of course due to the fact that this would cause a break in organic connection.

[9] If all these periodic processes of such variable duration were conscious, as is usual for the movement of our legs while less so for breathing and only exceptionally for

heartbeats, they would afford an excellent means for estimating time. The beginning of physical chronometry doubtless lies in the use of these means. Incidentally, there are no perfectly periodic processes, either in physics or in physiology. Each period produces an irreversible remainder. Every moment of life leaves its uneffaceable traces. Their sum is age and death. Cf. W. Pauli, *Ergebnisse d. Physiologie* 1904, Vol. III, pt.1, p. 159, and *A* 4, p. 184.

[10] A man who has once been actively involved therefore observes quite differently than if he had not been. A musician observes and enjoys music differently from an unmusical person and so on.

[11] The conception of attention here involved was developed from the physiological idea in my article 'Zur Theorie des Gehörorgans' (*Ber. d. Wiener Akademie*, July 1863, pp. 15–16 of the off-print). It is from this that I developed my first ideas of physiological time ('Über den Zeitsinn des Ohres', *ibid.*, Jan. 1865, pp. 14–15 of the off-print). Then followed the account in *A*, 1886. Cognate views have been held by Riehl, *Der philosophische Kritizismus*, Vol. II, pt. 1, p. 117; Münsterberg, *Beiträge zur experimentellen Psychologie*, 2nd fasc. 1899; and Jerusalem, *Laura Bridgman* 1891, pp. 39, 40.

[12] Cf. *W*, pp. 39f., and Ch. XXII, sn. 32 of the present volume.

CHAPTER XXIV

SPACE AND TIME PHYSICALLY CONSIDERED

1. As regards physiology, time and space are systems of sensations of orientation that determine the release of sensations proper and of biologically appropriate reactions of adaptation. As regards physics, they are special dependences of physical elements on each other. This comes out in the fact that numerical measures of time and space occur in all equations of physics, and that chronometric and geometrical concepts are gained by the comparison respectively of physical processes and of physical bodies with each other. Consider first physical time.

2. To allow temporal dependence in pure form, consider the fictitious example of a process in which space is eliminated as it were by the fact that we consider only bodies that are completely identical as to spatial relations. Imagine three equal masses of infinite internal thermal conductivity and equal specific heat, each touching the other two in areas of equal size and equal external thermal conductivity (Figure 34). Let the masses have unequal temperatures u_1, u_2, u_3 and examine their temporal change. Given our assumptions, the mean temperature is constant, and therefore $u_1 + u_2 + u_3 = c$. For the change of u_1 with the time t we obtain from Newton's law of conduction $\mathrm{d}u/\mathrm{d}t = k(c - 3u_1)$ and similarly for the other two temperatures. Integrating, $c - 3u_1 =$

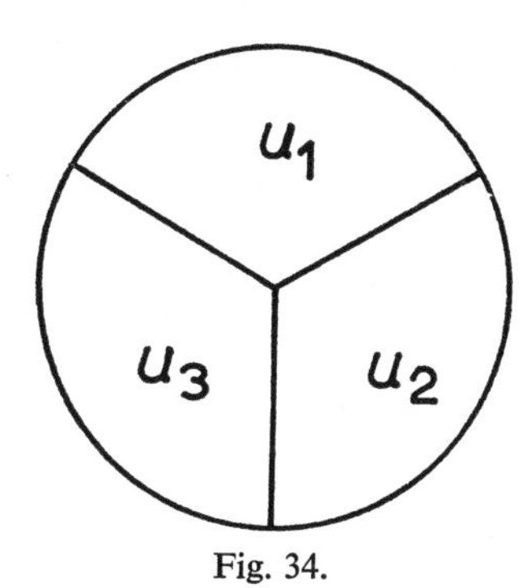

Fig. 34.

$=K.e^{-3kt}$ and replacing the constant of integration K by the initial value U_1 of u_1 and dividing by 3, $(c/3-u_1)=(c/3-U_1)e^{-3kt}$. Thus each temperature tends to the mean $c/3$ which it reaches after an infinite lapse of time. If we denote the first body's variable deviation from the mean by v_1 and its initial value by V_1, we obtain

$$v_1 = V_1 e^{-3kt} \tag{1}$$

and similar relations for v_2 and v_3. Using the first to determine e^{-3kt} and inserting this value in the other two we obtain $v_2 = V_2 \cdot v_1/V_1$, $v_3 = V_3 \cdot v_1/V_1$ or, combining them,

$$v_1/V_1 = v_2/V_2 = v_3/V_3 \tag{2}$$

3. Considering Equation (1), we see that according to the usual measure of time, where t is proportional to the Earth's angle of rotation with regard to the fixed stars, the deviation from the mean temperature decreases exponentially with t. If conversely we express t in terms of V_1 and v_1, we obtain $t=(1/3k)$ (log (V_1/v_1). Since it is entirely a matter of convention, which process we use as standard of comparison for the measurement or counting of time, we could chose log (V_1/v_1) or V_1/v_1 itself instead of t. It would be merely that in the first case we obtain a different unit of time and in the second a different (also infinite) time scale and a different origin.

4. Pursuing this last idea and measuring the temperature changes in terms of each other, the case represented by Equation (2) shows what is typical for the temporal dependence. The differences can only diminish, not increase; the course of time is undirectional. The deviations from mean temperature undergo simultaneous interdependent changes which in the case of immediate mutual relation are proportional to each other. These characteristic features of temporal dependence are fairly intelligible. Every process which is to be investigable at all must be thought of as being determined by some differences or other. Where there are no accessible differences, we cannot find any determining factors. If we imagine, for a moment, that the differences were to become greater, we should recognize that this idea does not agree with the most ordinary traits of our world picture, which never shows changes without bound

but everywhere reveals a striving towards a determined state. It may happen that certain differences become bigger if other more influential ones diminish, but an uncompensated increase of differences never occurs. There are other processes in which a difference may equally well grow as diminish, so that they seem to be able to run in opposite directions, and indeed sometimes actually seem to run periodically in this manner. However, these are never cases of uncompensated differences. Indeed, if we look at these processes carefully and not just schematically, then like all kinds of oscillations they are not strictly periodic but contain irreversible components. The second characteristic of temporal dependence, the mutual measurability of simultaneous changes, is readily intelligible in the case of bodies that are immediately related to each other. The determination of change by means of differences between the bodies is mutual, since no body is privileged above the rest, and what is gained by one is lost by another, as in our example. In cases of mediate dependence we cannot expect that simultaneous changes can thus simply be measured in terms of each other as in our previous example, but here too each change will run parallel to the other provided nature is homogeneous and no unexpected disturbances interfere with the normal course. Consider for example the orbital period of one of Jupiter's satellites and use it as a clock. Although nobody is likely to imagine that this motion has any noticeable influence on terrestrial processes, a cooling process on Earth will be equally well represented by the formula Ke^{-kt}, with different coefficients of course, whether t is derived from the satellite motion or from the axial rotation of the Earth. Only if in the course of our observation the satellite were to change its velocity because of collision with a meteorite, would the formula cease to hold, and it would become apparent that the heat process was not immediately dependent on the satellite motion.[1]

5. Let us modify our previous example in such a way that the influence of different spatial relations manifests itself in the simplest form alongside the temporal ones. Consider four equal masses in immediate contact in pairs forming a ring (Figure 35). Here there are only two different spatial relations: contact between adjacent masses and non-contact of opposite ones. For the rest we retain the assumptions of the previous case. Again we have an equation $u_1+u_2+u_3+u_4=c$. For the change of u_1, we find

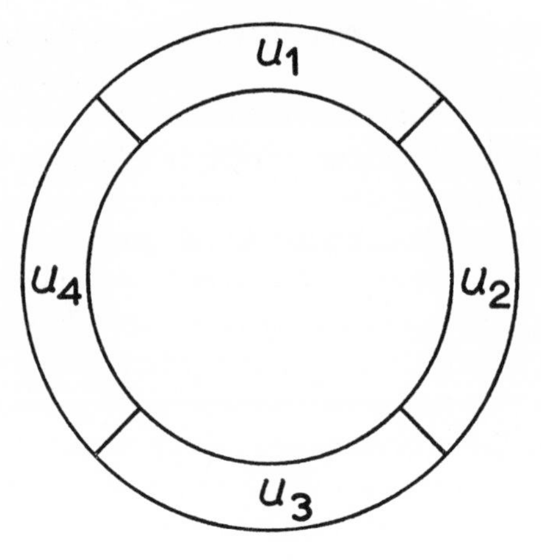

Fig. 35.

$\mathrm{d}u_1/\mathrm{d}t = k(c - u_3 - 3u_1)$. By cyclic interchange we obtain three further similar equations. Combining those for u_1 and u_3 gives

$$\mathrm{d}(u_1 + u_3)/\mathrm{d}t = k[2c - 4(u_1 + u_3)]$$

whose integral is

$$2c - 4(u_1 + u_3) = [2c - 4(U_1 + U_3)]\, e^{-4kt}. \tag{a}$$

The meaning of the symbols is to be taken as in the previous example. Next, we form the equations for $\mathrm{d}(u_1 + u_2)/\mathrm{d}t$ and $\mathrm{d}(u_2 + u_3)/\mathrm{d}t$. Subtract the first from the second and integrate, obtaining

$$2(u_3 - u_1) = 2(U_3 - U_1) \cdot e^{-2kt}. \tag{b}$$

Adding to Equation (a) the double of (b) we obtain an expression for u_1 which is readily transformed into

$$u_1 = \tfrac{1}{4}[c + (U_1 + U_3 - U_2 - U_4)\, e^{-4kt} + \\ + 2(U_1 - U_3)\, e^{-2kt}].$$

For $t = \infty$ we have $u_1 = c/4$; and, of course, fort $t = 0$, $u_1 = U_1$. During the temperature equalization the temperatures of spatially dissimilarly placed masses contribute unequally to changes in u_1. By cyclic interchange we obtain similar expressions for u_2, u_3, u_4.

6. Returning to the first example for some further points, we observe that instead of the equal spatial relations of three masses we could also have produced one of four, if we had been considering the four equal compartments formed in a regular tetrahedron by joining the four corners to the centre of gravity. An analogous division of a regular hexahedron would no longer be usable for our purpose, because in that case each mass would touch four others but not the fifth, so that we should have an instance corresponding to the schema of our second example. Still, we can maintain the physical fiction of an arbitrary number of masses in the same mutual thermal relation by imagining them to be isolated except for infinitely conducting wires from each to each. The number of such masses does not alter the result of our considerations. A single isolated body cannot determine any change in itself, but two bodies are enough to determine a mutual change. The need for an unambiguous determination impels us to notice experiences that decide the two possible (thinkable) directions of the change. If this has been done and the direction is that of decreasing differences, we try to ascertain the part each body plays in the process of equalization. For example, the simultaneous temperature changes may be inversely proportional to the heat capacities, so that both bodies reach the joint mean temperature at the same time. In other cases we find analogous rules. What manifests itself in pure temporal dependence are the simplest immediate physical relations.

7. A closer look at the influence of the spatial arrangement in the second example shows that the regular disposition of four masses in a ring corresponds to the simplest, finite unbounded linear Riemann space of four elements. The annular shape has the advantage that cyclic interchange affords greater clarity. Without essentially altering the result, we might have used a hundred masses instead of four, or even, like Fourier, considered a homogeneous ring with a continuous initial temperature distribution. A two-dimensional Riemann space is obtained by filling a thin spherical shell with such an arrangement of masses. By means of the fiction of suitable conducting links we can simulate further spatial arrangements as regards their physical consequences. The result of our considerations remains always the same. The influence of mediate physical relations shows itself later, and is masked and effaced by relations that are immediate or conveyed through a small number of

intermediate links. What manifests itself in spatial relations is mediate physical dependence.

8. How does this result, perhaps a small step towards clarifying the problem of space though not solving it, agree with current views of space? To appreciate how difficult it was to form the abstraction of 'space', one had best study Book 4 of Aristotle's *Physics*[2]. He is much exercised by the questions whether space (location) is or is not, and how and what it is. He cannot regard space as a body, for how could one body be inside another. Yet neither can he separate space from corporeality, for he takes the place of a body to be that which surrounds or encloses it. He emphasizes that if motion did not exist we should not ask about space. All the difficulties of his spatial views recur in his account of motion[3]. The merging of the idea of space with that of body makes it obvious that a vacuum is unthinkable, as Aristotle and many other thinkers of antiquity held[4]. Those who like Leucippus, Democritus, Epicurus and others assumed a vacuum, thus had spatial views nearer our own. Space to them was a container that might be filled or empty. Indeed geometry, which neglects all bodily properties except rigidity of boundaries, must lead in that direction, and this development is reinforced by the naïve sense observation of the motion of bodies in a thin transparent medium like air, which may occasionally be viewed as nothing or emptiness. This notion is confirmed by a passage in Guericke.[5]

9. That empty space cannot be conceived continues to be held into modern times. Descartes[6] is still so imbued with this view that he assumes the walls of a vessel that could be completely emptied would have to touch at once. We know of the labours of Guericke[7], Boyle[8] and Pascal[9] to give their contemporaries convincing proof that the forbidden vacuum did in fact exist. Still, theirs was not a vacuum in the sense of modern physics. After discussing ancient and modern views of place, time and vacuum, Guericke in his *Experimenta Magdeburgica* (1672) says that he will demonstrate by experiment that there are true vacua in nature. In Book 3, chapters 35 and 36, he refutes at length the objections to the existence of the vacuum and the doubts raised against his experiments. It was through philosophic studies that he had been led to these tests. In considering the enormous celestial spaces he often found the question

obtruding itself, whether these were not the vacuum that had always been denied.[10]

10. The proof that the vacuum existed has doubtless contributed much to render spatial ideas independent. However, other important circumstances intervened too. Galileo had found his dynamic laws by observing motions on Earth. As main representative of the Copernican system he often had occasion to discuss objections to it in terms of his own dynamics. This almost automatically and unobtrusively led to the attempt to relate this dynamics not to the Earth but to the sphere of fixed stars[11] considered as rigid. In this way he found for example his theory of tides, thus giving the Copernican theory a putative prop that he regarded as correct only because he as yet lacked the means to recognize it as defective. Newton's completion of celestial mechanics on the foundations of Galileo and Huygens made the new and successful reference frame positively indispensable. Newton perceived the assumption of gravitational forces depending on distance as the fruitful basic idea. Even if he might have preferred to regard this space as filled and the forces as transmitted through the medium, he nevertheless provisionally had to abide by a view that emphasized space as such and held the field almost alone until after the middle of the 19th century. Considering that Newton's gravitational mechanics could no longer regard the fixed stars as an absolutely unchanging, stationary and rigid system, his daring attempt to relate the whole of dynamics to an absolute space, and correspondingly to absolute time[12], appears in some measure intelligible. In practice, this seemingly senseless assumption did not alter the use of the fixed stars as space-time coordinates, so that it remained harmless and long escaped serious criticism. We may fairly say that mainly since Newton's assertions space and time have been the sort of independent but incorporeal entities as which they are considered today.

11. Newton's idea of forces at a distance was a great intellectual feat that within a century enabled enquirers to complete a homogeneous mathematical physics[13]. This feat rests on breadth of intellectual vision. He saw the actual accelerations at a distance and recognized them as important; how they were transmitted was less clear and he provisionally ignored this question. However, even the merest details must be in-

vestigated, and for this a sharp-eyed myopia is more productive. Looking at large questions of wide scope must alternate with scrutiny of what is near, small and particular, if progress is to continue steadily. The greatest enquirers, Newton above all, were masters of both. The questions that Newton left behind as to contiguous action transmitting action at a distance were taken up with great success by Faraday in the 19th century. However, until Maxwell translated these ideas into the more familiar language of mathematics, enquirers imbued with the physics of action at a distance could not understand them.

12. A naïve observer begins by noticing the strong and close connection between sensible elements at one position in space and time, whether taken physiologically or physically. This connection we call body. Insofar as observation allows us to subdivide a spatio-temporal place into smaller parts, we find that the connection becomes even tighter inside the subdivisions. The parts of bodies are again bodies. Changes usually do not occur in the whole body at once but they engulf one part after another, for example in the process of dissolving or heating and so on. Nothing is more natural than to regard exceptions as merely apparent, and to hope that we may be able to reduce sudden changes of a whole body (for example electrification) and influences at a distance (illumination, gravitational acceleration) to a gradual change transmitted from part to part. This is a naïve view that commended itself to thinkers in ancient times too, and Faraday has revived it through his great successes: from his standpoint we readily understand the proposition that temporal dependence is immediate while spatial dependence is mediate.

13. From this point of view we now gain the prospect of a physical understanding of space and time by grasping it in terms of the most elementary physical facts. For Newton space and time are something hyper-physical, not immediately accessible or at least not precisely determinable independent arch-variables that rule the whole world which runs according to them. Just as space regulates the motion of the remotest planets round the sun, so time keeps the most distant celestial motions in agreement with terrestrial processes. On this view the world becomes an organism or, if the expression be preferred, a machine, whose parts run in full agreement according to the motion of a single one of them, as

though ruled by a uniform will, except that the goal of this motion remains unknown to us[14]. As an after-effect of Newton, this view is still at the basis of contemporary physics, even if we feel perhaps disinclined to admit it openly. However, it will have to be modified according to Faraday's position. The world remains a whole so long as no element is isolated, but all parts are connected, if not immediately then at least mediately through others. The concordant behaviour of members not immediately connected (the unity of space and time) then arises only apparently by failure to notice the mediating links. The goal of the cosmic motion remains unknown only because the segment that we can look at has narrow boundaries beyond which enquiry does not reach. This view is less poetic and grandiose, but more naïve and sober for that.

14. The physical view of space is favoured by progress in the recognition of the 'vacuum'. For Guericke, the vacuum had only negative properties. Even air at first offers only negative qualities to the naïve observer: it is invisible, it becomes tangible only through intense motion and then reveals its temperature as well. By trapping it in a hose or vessel we come to know that it is impenetrable and has weight. Later still, visibility is added, until in the end all characteristics of a body have been demonstrated. Similarly for the vacuum. At first it has no physical qualities. Boyle shows that a burning glass and a magnet act across it. According to Young and Fresnel, in a vacuum traversed by light we must suppose the same physical states as existing simultaneously at very short intervals and imagine that these states are displaced at immense speed in the light. The work of Faraday, Maxwell, Hertz and others has shown the existence of electric and magnetic forces in empty space, connected in such a way that every change of the one will condition the appearance of the other at the same place. In general we cannot immediately perceive anything of these forces, except for rapid periodic changes in which they manifest themselves as light. However, a physical detour readily demonstrates these forces, and it is the rarest exception if they are completely absent. Thus, the vacuum is by no means nothing, but has very important physical properties. Whether one wants to call it a body (aether) is of no importance, but one cannot deny that variable and mutually dependent properties attach to it.[15]

15. Lobachevsky[16] as a natural scientist of geometry, observes that in each measurement we use bodies, so that in setting up geometrical concepts we must start from bodies. He regards contact as the distinguishing mark of bodies in virtue of which we call them geometrical[17]. This seems to point to the fact that bodies are rigid and impenetrable, which manifests itself when they touch each other, which is the basis of all measurement. However, things have moved on since the beginning of the 19th century. We still need rigid bodies to build our equipment, but we can use light interference to mark points and measure stretches by wave length in seemingly empty space much more accurately than would be possible by means of rigid bodies that abut and touch each other. It is even likely that light waves in vacuo will furnish future physical standards of length and time, in terms of wave length and period of oscillation respectively and that these basic standards will be more appropriate and generally comparable than any others. Through such changes space and time increasingly lose their hyperphysical character.[18]

16. We ascribe three dimensions to space, and our geometry regards these as indifferently equivalent, so that space is isotropic. Indeed, if we consider only the fact that bodies are impenetrable, there is no difference. However, if we regard geometry as a physical science, it is questionable whether it is always appropriate to uphold this view. Indeed, already in vector algebra we must take notice that directions are not equivalent. An amorphous or tesseral body, a dilute solution of sulphuric acid in which zinc powder dissolves and so on, show no difference as regards direction; but in a triclinic body, or in a bodily element in which a current is being induced so that magnetic lines of force circle round it in a determinate sense, the three directions are not equivalent. If we could order and suitably direct the random currents produced by the dissolving zinc powder, the three dimensions would likewise cease to be equivalent. Thus equivalence here appears to rest on the effacement of non-equivalence in special, frequent, simpler cases. Physiologically, too, the dimensions are non-equivalent. Perhaps this anisotropy lies in the elementary organs from which the body is made up[19]. If we can use our bodies to gain our bearings in physical processes, witness Ampère's left-hand rule and other similar electro-dynamic rules that are applied with unvarying success, this points to a deep-seated connection between the physical

environment and our physiological constitution which share a common anisotropy.[20]

17. Our intuitions of space and time form the most important foundations of our sensory view of the world and as such cannot be eliminated. However, this does not prevent us from trying to reduce the manifold of qualities of place sensations to a physiologico-chemical manifold. We might think of a system of mixtures in all proportions of a number of chemical qualities (processes)[21]. If such an attempt were one day to succeed, it would lead also to the question whether we might not give a physical sense to the speculations that Herbart, following Leibniz, conducted as regards the construction of intelligible space, so that we might reduce physical space to concepts of quality and magnitude. There is of course much that can be objected to in Herbart's metaphysics. His tracking down of contradictions that are in part contrived artificially and his eleatic tendencies are not too attractive, but he will hardly have produced nothing but errors. His stopping of the construction of space at the third dimension is quite unfounded and it is precisely here that the heart of the matter lies[22]. After a century, such questions can show quite a new complexion.

18. That physiologically space and time represent only an apparent continuum and are probably composed of discontinuous though imprecisely discriminable elements may be mentioned in passing. How far in physics we can uphold the assumption of spatial and temporal continuity is merely a question of what is appropriate and what agrees with experience. These are mere beginnings of thoughts; whether they are capable of development I am unable to decide.

NOTES

[1] Cf. *A* 4, p. 272. I must mention that in these reflections I have been much advanced by objections from Petzoldt ('Das Gesetz der Eindeutigkeit', *Vierteljahrschr. f. wiss. Philosophie* **19**, 146f.).

[2] Especially Chs. 1–9.

[3] Cf. Lange, *Die geschichtliche Entwicklung des Bewegungsbegriffes*, Leipzig 1886.

[4] *Physics*, **IV**, Chs. 6–9.

[5] Guericke, *Experimenta Magdeburgica*, 1672, **III**, Ch. 4, p. 59 "When we look at the distance or interval between two towers or mountains, it is easy to think that this is not

produced by the intervening mass of air, but simply exists by itself, so that if all the air is removed as well, the mountains and towers would not become mutually contiguous."

[6] Descartes, *Principia* **II**, 18. "If one ask, what happens if God takes away all bodies contained in any vessel and allows nothing to take the place of what was removed, we must reply that the sides of the vessel will by this very fact become mutually contiguous." The learned world was much amazed when the experiment that was hardly entrusted to a divine agency was carried out with quite the opposite result by a simple and skilful burgomaster.

[7] Guericke, *l.c.*

[8] Boyle, *New experiments, physico-mechanical*, Oxford 1660.

[9] Pascal, *Nouvelles expériences touchant le vide*, Paris 1647.

[10] Guericke, *l.c.* L. I, Ch. I, p. 55. Amongst the various views as to what fills cosmic space, he comes to the question whether space without matter, namely the vacuum, has always been denied.

[11] This theory too is mentioned in Galileo's *Dialogue on the two world systems*. For a brief report on this, see *M* 5, pp. 227–229.

[12] Cf. the detailed account of the positions of contemporaries towards Newton's view, in Lange, *Die geschichtliche Entwicklung des Bewegungsbegriffes*, 1886.

[13] In Ch. XIV, sn. 20 we pointed to the grave disadvantage that would have accrued, had Newton dropped action at a distance because he could not 'explain' it.

[14] Cf. *Erhaltung der Arbeit*, Prague 1872, pp. 35–37.

[15] Spontaneously, these forces occur no more in an undifferentiated vacuum than in any other body, in which they would have to be caused by a second body or by differences between the parts of the one body.

[16] F. Engel, *N. I. Lobatschefskij, Zwei geometrische Abhandlungen*, Leipzig Teubner 1899, pp. 80–81. Lobachevsky here thinks like Leibniz.

[17] *Ibid.*, p. 83.

[18] The considerations of this chapter show that space and time cannot well be severed during investigation. Cf. the ingenious philosophical joke of Fechner in *Vier Paradoxen* on space having four dimensions. A seriously intended explanation of this kind is given by M. Palágyi, *Neue Theorie des Raumes und der Zeit*, Leipzig 1901. For a view related to Fechner's, see *A* 1, 1886, p. 156. That space and time are inseparable I emphasized in a brief note in *Fichtes Zeitschr. f. Philosophie* 1866. While the present volume was being printed I received K. C. Schneider, 'Das Wesen der Zeit', *Wiener klinische Rundschau* 1905, Nos. 11 and 12; it contains some views reminiscent of Fechner and Palágyi, but bare mention of it must here suffice.

[19] Concerning anisotropy in plant organs cf. Sachs, *Vorlesungen über Pflanzenphysiologie*, Leipzig 1887, pp. 742–762. Analogous questions about anisotropy in the elementary organs of animals are treated by O. zur Strassen, 'Über die Mechanik der Epithelbildung', *Verh. d. D. Zoolog. Gesellsch.* 1903.

[20] Cf. *A*, pp. 264, 265.

[21] Cf. also *W* 1, 1896, pp. 360–361.

[22] Leibniz too thought he could prove four-dimensional space to be impossible from the fact that in (three-dimensional!) space there can be only three mutual perpendiculars!

CHAPTER XXV

SENSE AND VALUE OF THE LAWS OF NATURE

1. One often speaks of laws of nature. What does this expression mean? The usual opinion will be that the laws of nature are rules, which processes in nature must obey, similarly to civil laws, which the actions of citizens ought to obey. A difference is usually seen in that civil laws can be broken while deviations from natural processes are regarded as impossible. However, this view of the laws of nature is shaken by the reflection that we read off and abstract these laws from those processes themselves, and that in doing this we are by no means immune to error. Of course in that case any breaking of the laws of nature may be explained by our mistaken view, and the idea that these laws are unbreakable loses all sense and value. If once we emphasize the subjective side of our view of nature, we easily reach the extreme opinion that our intuition and our concepts alone prescribe laws to nature. However, an unprejudiced consideration of the development of natural science makes us see its origin in the fact that we begin by noticing in processes those aspects that are of immediate biological importance to us, and only later progressively extend our interest to aspects, that are mediately important. In the light of this reflection, the following obvious formulation may perhaps be acceptable: in origin, the 'laws of nature' are restrictions that under the guidance of our experience we prescribe to our expectations.

2. K. Pearson[1], whose views are rather close to mine, expresses himself on these questions in the following manner:

The civil law involves a command and a duty; the scientific law is a description, not a prescription. The civil law is valid only for a *special* community at a *special* time; the scientific law is valid for *all* normal human beings, and is unchangeable so long as their perceptive faculties remain at the same stage of development. For Austin[2], however, and for many other philosophers too, the law of nature was not the mental formula, but the repeated sequence of perceptions. This repeated sequence of perceptions they projected out of themselves, and considered as a part of an external world unconditioned by and independent of man. In this sense of the word, a sense unfortunately far too common to-day, natural law could exist before it was recognised by man.

The term 'description' which appears already in the discussion between J. S. Mill and Whewell has been generally adopted since Kirchhoff; in contrast, let me suggest the expression 'restriction on expectations' as pointing to the biological importance of the laws of nature.

3. A law always consists in a restriction of possibilities,whether as a bar on action, as an invariable course of natural events, or as a road sign for our thoughts and ideas that anticipate events by running ahead of them in a complementary manner. Galileo and Kepler imagine the various possibilities of free fall and planetary motion, trying to guess those that correspond to observation and making them more precise. The law of inertia which assigns uniform rectilinear motion to a body once all forces disappear, selects one of infinitely many possible thoughts as the decisive one for our ideas. Lange's view[3] of inertial motion of a system of free masses represents it as a selection of one mode of motion from infinitely many kinematic possibilities. The mere circumstance that a factual field can be classified, and that concepts can be set up to correspond to the classes, constitutes a limitation of possibilities. A law need not necessarily be expressed as a theorem. The fact that the concept of mass can be applied contains the limitation that the sum of the masses in a closed system, measured in terms of any one body as unit, is invariable. Two bodies that are equal to a third are also equal to each other.[4]

4. It is a need of all living beings endowed with memory, that under given circumstances their expectations are regulated in the direction of preservation. As regards the immediate and simplest biological needs, our mental organization responds to them instinctively, by producing through the mechanism of association the readiness to function appropriate to the great majority of cases. If the conditions of existence are complicated, so that the satisfaction of needs can be achieved only by long detours, only a well-furnished mental life can meet those needs. The individual steps of the detour with the attendant circumstances as such will then acquire mediate interest. Every scientific interest may be viewed as a mediate biological interest in a step of the detour. Whether a case be close to or far from the immediate biological interest, the only thing that responds to our needs is always the correct and appropriate expectation under the circumstances. However, in different circumstances we make

very different demands as regards correctness of expectation. If we are hungry and we find food where we expect it under the circumstances, that alone will satisfy our expectation. However, if given the aim of a gun and the weight of shell and gunpowder, we expect a certain range, then even a slight deviation from expectation can constitute a disturbing deception. If a goal is to be reached by a fairly long route involving several or many steps, a small error in the size and direction of each step will be enough to make us miss the goal. Thus small errors in several numbers that enter a calculation can greatly falsify the final result[5]. Since in science we are dealing with precisely such intermediate steps, which are used in theory or in technical practice, it will be important to gain an especially accurate determination of expectation by means of the given circumstances.

5. With the progress of natural science there arises in fact an increasing restriction of expectations, which assume a gradually more determined shape. The first restrictions are qualitative in kind. It is not important whether the factors $A, B, C, \ldots$ that determine an expectation M can be scientifically formulated together in a single proposition, or whether science provides instructions for adducing these factors one after the other as for example in a table in botany or in chemical analysis. If in qualitatively like cases one can distinguish the various qualities quantitatively as well, so that a quantitatively determined set $A_1, B_1, C_1, \ldots$ determines a correspondingly quantitative expectation M_1, then we have a further restriction, whose sharpness is limited only by the attainable accuracy of observation and measurement. Here too the restriction may occur all at once or in successive stages. The latter happens if a restriction is further narrowed to within smaller limits by a second and complementary determination. In a plane convex rectilinear polyhedron of n sides the sum of the internal angels in Euclidean space is $(n-2)\cdot 2$ right angles. For a triangle ($n=3$) this becomes 2 right angles, so that each angle is determined by the other two. These most narrow restrictions thus rest on a whole sequence of conditions that complete each other or of which some are basic in that they are needed to give the others any sense in the first place. Likewise in physics. The equation $pv/T=$constant holds for a gaseous body of invariable mass for which p, v, T are the same in all its parts and provided the conditions are distant enough from liquefaction.

The limitation contained in the law of refraction $\sin\alpha/\sin\beta = n$ is further narrowed by being related to a definite pair of homogeneous substances at a definite temperature and density or pressure as well as absence of internal differences of electric and magnetic potential. If a physical law is related to a definite substance, this means that the law is valid for a space in which the known reactions of this substance are also found. These additional conditions are usually covered and concealed by the mere name of the substance. Physical laws that hold for empty space (vacuum, aether) always and only relate to definite values of the electric and magnetic constants and so on. By applying a proposition to a given substance we introduce further determinations or equations expressing conditions, just as when we say, or tacitly assume, of a geometrical theorem that it applies to a triangle, parallelogram or a rhombus. If one finds a case where a law stops holding under circumstances where it has hitherto always been found to hold, this makes us look for so far unknown complementary conditions of the law. Finding these is always an important discovery. Thus electricity and magnetism were discovered by the mutual attraction and repulsion of bodies that had been assumed to be mutually indifferent. Not only an explicit hypothesis, but also tacitly assumed attendant conditions form the basis of a geometrical thesis or even a physical one. It is always well to remember that there might be as yet unknown determinant conditions (whose observable variations had so far escaped us).

6. In our view, laws of nature are a product of our mental need to find our way about in nature, so that we do not stand estranged and baffled in front of natural processes. This manifests itself clearly in the motive behind these laws, which always correspond to this need as well as to the current state of culture. The first rough attempts at orientation are mythological, demonological, poetic. At the time of the scientific renaissance of natural science in the period of Copernicus and Galileo, which strove after a mainly qualitative provisional orientation, ease, simplicity and beauty were the leading motives in the search for rules for intellectually reconstructing the facts. More accurate quantitative enquiry aims at determining facts as completely as possible, looking for unambiguous determination as is found already in the older history of the development of mechanics. If individual findings later accumulate, there arises a powerful urge to minimize mental effort, to attain economy, continuity,

constancy, and as general a scope as possible for profitably applying the rules set up. We need only point to the later history of mechanics or of any fairly advanced part of physics.

7. It is very natural that in times of fairly lax epistemological criticism psychological motives were projected into and attributed to nature herself. God or nature strive after simplicity and beauty, then after rigid lawful connection and determinacy, and finally after economy in all processes, that is after achieving the greatest result with the least effort. Even in recent times, Fresnel[6] in emphasizing the universal applicability of wave theory as against the older emission theory, ascribes to nature a tendency to attain much by slender means.

The first hypothesis has the advantage of leading to consequences that are more evident, because mechanical analysis is more readily applied to it; the second, on the contrary, occasions great difficulties here. However, in choosing a system one must consider only the simplicity of the hypotheses, that of the calculations can have no weight in the balance of probabilities. Nature is not burdened by difficulties of analysis, it has avoided only the complexity of means. It seems to have decided to do much with little: it is a principle which the perfection of physical science constantly supports with new proofs.

8. The progressive refinement of the laws of nature and the increasing restriction of expectations corresponds to a more precise adaptation of thought to fact. It is of course not possible to achieve perfect adaptation to every individual and incalculable future fact. It requires abstraction, simplification, schematizing and idealization of the facts, if the laws of nature are to become applicable, repeatedly and as far as possible generally, to actual concrete cases: we must decompose the facts mentally, into such simple elements that from them we can mentally reconstruct and reassemble the facts with sufficient accuracy. Such simple idealized factual elements, which never exactly occur in reality, are uniform and uniformly accelerated motions of masses, stationary (steady) thermal and electric currents, as well as uniformly increasing or decreasing currents and so on. Every arbitrarily variable motion and current may be regarded as made up to any degree of accuracy from such elements, so that the laws of nature can be applied to it, which occurs through the differential equations of physics. Our laws of nature thus consist of a series of theorems, appropriately chosen for this use and lying ready for application. Natural science may be viewed as a kind of collection of instruments for the intel-

lectual completion of any partially given facts or for the restriction, as far as may be, of expectations in future cases.[7]

9. The facts are not compelled to conform to our thoughts, but our thoughts and expectations conform to other thoughts, namely to concepts that we have formed of the facts. The instinctive expectation that attaches to a fact always has a fair amount of play, but if we assume that a fact corresponds exactly to our simple ideal concept, then our expectation will agree with them and thus will be precisely determined. A proposition of natural science always has a merely hypothetical sense: if a fact *A* corresponds precisely to conepts *M*, then the consequence *B* corresponds precisely to concepts *N*; the two correspondences have the same degree of accuracy. Absolutely exact and perfectly precise and unambiguous determination of consequences from a presupposition in natural science, as in geometry, does not exist in sensible reality, but only in theory. All progress aims at making theory more conformable to reality. When we have observed and measured many cases of refraction in a pair of media, our expectation as to a refracted ray for a given incident ray is still subject to the range of inaccuracy in observation and measurement. Only after the law has been fixed and a value chosen for the index is there one unique refracted ray for a given incident ray.

10. How important it is to distinguish sharply between concept and law on one side and fact on the other has been underlined several times. Oerstedt's case (current and needle in one plane) is absolutely symmetrical according to the concept valid before his time, while the facts revealed themselves as unsymmetrical. Circularly polarized light behaves in several respects with the same indifference as unpolarized light, and it requires closer study to reveal its double helicoidal dissymmetry thus forcing us to represent the facts by new and more completely descriptive concepts. If our ideas of nature are ruled by concepts that we regard as adequate and we have correspondingly become used to expectations of unambiguous precision, we are easily led to using the notion of unambiguous determination negatively as well. Where a certain effect, say in motion, is not unambiguously determined, as with three equal forces in a plane acting at a point at 120° of each other, we are going to expect no effect to occur at all. If we are not to be misled by applying the principle of sufficient

reason in this form (see the above example), we must be sure that all operative conditions are known.

11. Only a theory that represents facts more simply and precisely than can really be guaranteed by observation (because of the influence of always numerous and complicated subsidiary circumstances) corresponds to the ideal of unambiguous determinacy[8]. This precision of theory enables us to deduce, by series of equal or unequal steps, far-reaching consequences that will agree with that theory. However, agreement or disagreement of the consequences with experience is usually, because of cumulative deviations, a better test of correctness of theory or of its need for improvement, than comparison of principle with observation. Think of Newton's principles of mechanics and the astronomic consequences drawn from it.

12. The general and often repeated forms of the propositions of theory become understandable if they are considered from the point of view of our need for determinacy, particularly of the unambiguous kind. This makes everything clearer and more perspicuous. For the physicist a few comments will suffice. Physical differences determine everything that happens, and the decrease of differences is preponderant in the segment of reality that we are contemplating. Where many differences of the same kind similarly determine events at a point, the determining factor is the mean of these differences. The equations of Laplace, and Poisson which are applicable to so many areas of statics, dynamics, heat, electricity and so on, respectively state[9] the value of this mean as zero or whatever else it happens to be. Symmetrical differences with regard to a point determine symmetrical events there; or, in special cases of multiple symmetry, an absence of events. When applied, the conjugate functions representing families of orthogonal potential surfaces and lines of force or current, and so on, determine a symmetry of events in the infinitesimal elements. A maximum or minimum in a set of multiple adjacent possibilities can always be viewed as being subject to some symmetry conditions. If the differences always change in the same sense for an arbitrary small change in the arrangement, then the given arrangement is always in some respect maximal or minimal. Cases of equilibrium, and not only those in statics and dynamics, are as a rule of this kind. We have explained elsewhere that in dynamical laws such as the principle of least action and others that

are stated in the form of maxima or minima, it is not these last that are decisive, but the idea of unambiguous determinacy.[10]

13. Are the laws of nature then useless, because they are merely subjective prescriptions for an observer's expectations to which reality need not conform? Not at all: for all that expectations are met by sensible reality only within certain limits, they nevertheless have often proved right and continue to do so daily. Thus we have not been mistaken in postulating the uniformity of nature, even if, because our experience is inexhaustible, we shall never be able to prove that the postulate is applicable with absolute precision everywhere in space and time: like any tool of science, it will remain an ideal. Besides, the postulate relates only to uniformities, without specifying of what kind. If expectation is disappointed we are therefore always free to seek new uniformities instead of the previously expected ones.

14. If, as an enquirer into nature, one regards the human individual and his psyche not as an isolated and alien element opposed to nature, but views the events of the physical senses and of the ideas as an inseparable whole, one will not be amazed that the whole cannot be exhausted by the part. Yet the rules that manifest themselves in the part will suggest to him that there are rules for the whole. He will hope that just as he succeeds in explaining one fact by another in a small field, so the two fields of the physical and mental will gradually clarify each other. It is only a matter of bringing the results of detailed physical and psychological observation into closer agreement than has so far been achieved; nobody now doubts the general connection between the two. We can no longer think of two independent or only loosely related worlds. To connect them by means of an unknown third is senseless: such explanations have, one hopes, lost all credit for ever.

15. It is quite understandable that the views mentioned should arise. When man discovered by analogy that there were other living beings that resembled him and behaved similarly, other men and animals, and thus had to become clearly aware that he must judge their behaviour in regard to circumstances that he could not immediately perceive by his senses, though the analogues were familiar to him from his own experience,

he could not but divide processes into two classes: those that are perceivable by all and those perceivable only by one. This to him was the simplest and practically most helpful solution. In this way he formed the clear thought both of his own ego and that of others. Somebody who by accident were to grow up without living companions would hardly oppose his scant ideas to sensations nor attain the thought of an ego, much less oppose it to the world. Everything that happens would for him be one. However, once the thought of an ego has been grasped, we readily form the abstractions of physical and mental and of our own and others' sensations and ideas. Both modes of looking at things are beneficial for a comprehensive orientation: one leads to regard for detail, the other ensures that we do not lose the whole from sight.[11]

16. If the world is sawn and cut to bits by abstractions, the parts look so airy and insubstantial that doubts arise whether we can glue them together again. One might occasionally ask by way of joke and irony, whether such a sensation or idea that belongs to an ego, could go a-roaming by itself in the world. Thus, mathematicians after breaking the world into differentials were a bit apprehensive whether out of such nothings they could integrate the world together again without damage. My answer is this: a sensation will indeed always occur in a complex, but that this latter should always be a complete and wakeful human ego is doubtful; after all, there is consciousness in dreams, hypnosis, ecstasy, and there is animal consciousness, all in different degrees. Even a body, a lump of lead, the crudest item known to us, always belongs to a complex and so to the world; nothing exists in isolation[12]. Just as the physicist must be free to analyse the material world for the purpose of scientific investigation, and to dismantle it into parts without therefore forgetting the general connected nature of the world, so the psychologist too must be equally free, if he is to obtain any results. Using the phrase of Demonax the Cynic, we might say that sensation exists alone as little as anything else.

Introspectively I find that my ego is exhausted by the concrete contents of consciousness. If at times one imagines that one perceives something in addition, the reason seems to be the following. The abstract idea of one's own ego is closely linked with that of the ego of others and with the difference between them, as well as with the thought that the ego's behaviour towards its content is not indifferent. However, one must ask oneself whether these abstract thoughts do not themselves have and coincide with concrete contents of consciousness alone, and whether one could have discovered them through mere introspection. Still, almost everything certainly remains to be investigated where the physical and physiological foundations of the ego are concerned. This is certainly something alongside the momentary vivid content of consciousness, which in any case only ever represents a minute part of its total wealth.

17. The traditional opinion that there are unsurpassable barriers between ego and the world, just as between ego and ego, is equally understandable psychologically. If I have a sensation or idea of something, this does not seem to influence the world or any other ego, but only apparently so. A slight accompanying movement of my muscles belongs already to the world and to any alert observer, and this holds even more if my ideas break out into speech and action. If one person sees the colour blue and another the shape of a sphere, this indeed cannot result in the judgement "the sphere is blue". What is lacking is the "synthetic unity of apperception", as this trivial fact has been grandiosely described[13]. The two ideas must come close enough to interact, rather like bodies in physics. Such expressions however solve no problem, but rather are apt to coincide with it or conceal it. The ego is not a bowl into which blue and sphere need merely to be dropped for a judgement to result. The ego is more than a mere unity and certainly no Herbartian simple. The same spatial elements that make up the sphere must be blue, and the colour blue must be recognized as different and separable from their location if a judgement is to be possible. The ego is a mental organism to which corresponds a physical organism. It is hard to believe that this must forever remain a problem about which psychology and physiology together could provide no further clarifications. Introspection alone, without the help of physics, would not even have led to the analysis of sensation. Philosophers one-sidedly overrate introspective analysis, while psychiatrists often do likewise as regards physiological analysis, whereas for adequate results we must combine the two. In both groups of enquirers there seems to lurk a vestige of the as yet not quite extinguished prejudice derived from primitive culture, that the mental and the physical are in principle incommensurable. How far the investigation here hinted at will lead, is not predictable at present.

If the ego is not a monad isolated from the world but a part of it, in the midst of the cosmic stream from which it has emerged and into which it is ready to dissolve back again, then we shall no longer be inclined to regard the world as an unknowable something. We are then close enough to ourselves and in sufficient affinity to other parts of the world to hope for real knowledge.

18. Science apparently grew out of biological and cultural development as its most superfluous offshoot. However, today we can hardly doubt that it has developed into the factor that is biologically and culturally the most beneficial. Science has taken over the task of replacing tentative and unconscious adaptation by a faster variety that is fully conscious and methodical. The late physicist E. Reitlinger used to reply to fits of pessimism that man appeared in nature when the conditions were adequate for existing but not for well-being. Indeed these last he must create for himself, and I believe that he has done so. At least today this holds of material well-being, even if so far unfortunately only for some, but we may hope for better things in future[14]. Sir John Lubbock[15] expresses the hope that the fruits of civilization will be extended not only to other countries and peoples, but also to all sections of the developed countries, so that we should not find amongst our own fellow citizens those who lead lives more terrible than savages without the latter's real if simple advantages. Let us remember what miseries our forebears had to endure under the brutality of their social institutions, their laws and courts, their superstition and fanaticism; let us consider how much of these things remains as our own heritage and imagine how much of it we shall still experience in our descendants: this should be sufficient motive for us to start collaborating eagerly in realizing the ideal of a moral world order, with the help of our psychological and sociological insights. Once such an order is established, nobody will be able to say that it is not in the world, nor will anybody need to seek it in the heights or depths.

NOTES

1 K. Pearson, *The Grammar of Science*, 2nd ed. London 1900, p. 87.
2 The English jurist.
3 *M* 5, p. 259.
4 *Ibid.*, pp. 233f.

[5] J. R. Mayer, on the basis of fairly inaccurate figures, found for the mechanical equivalent of heat the value 365 instead of 425.
[6] Fresnel, 'Mémoire couronné sur la diffraction', *Oeuvres*, Paris 1886, Vol. 1, p. 248.
[7] *W*, pp. 461f. Kleinpeter, *Erkenntnistheorie*, Leipzig 1905, pp. 11–13.
[8] Cf. the account of Duhem, *La Théorie physique*, pp. 220f., 320f.
[9] *W*, pp. 117f.
[10] *M* 5, pp. 419–421. Petzoldt, 'Das Gesetz der Eindeutigkeit', *Vierteljahrschr. f. wissensch. Philosophie* **XIX**, pp. 146f.
[11] Cf. W. Jerusalem, *Einleitung in die Philosophie*, 2nd ed. 1903, pp. 118f.
[12] Cf. the controversy between Ziehen (*Zeitschr. f. Psychologie u. Physiologie der Sinnesorgane* **33**, p. 91) and Schuppe (*ibid.* **35**, p. 454). *A* 4, p. 281.
[13] How finally the immutability of the ego is supposed to follow from this I cannot fathom.
[14] E. Metschnikoff, *Studium über die Natur des Menschen. Eine optimistische Philosophie*, Leipzig 1904.
[15] J. Lubbock, *Die Entstehung der Zivilisation*, Jena 1875, p. 399.

ERNST MACH BIBLIOGRAPHY

Compiled by O. Blüh and W. F. Merzkirch

(Adapted from Joachim Thiele, 'Ernst Mach-Bibliographie', *Centaurus* **8** (1963), 189–237 and now reprinted from *Ernst Mach Physicist and Philosopher* (Boston Studies in the Philosophy of Science, Volume VI) with some additions by the editor of *Knowledge and Error.*)

I. WORKS OF ERNST MACH IN ORIGINAL EDITIONS

Prepared by W. F. Merzkirch

This short bibliography contains only the original German titles of Mach's publications. Not listed are: All translations into foreign languages; preliminary communications which have been published later in final form; some short notes to *EJ*, *MU* and *ZphU*; some preliminary lectures which have later been published in *Populär-wissenschaftliche Vorlesungen*; and the text-books written for high-schools. †

The titles marked with* are also printed in *Populär-wissenschaftliche Vorlesungen.* The publications appearing in *SW* and marked with** have later been reprinted in *AP.*

For a more detailed information the reader is referred to the Ernst-Mach-Bibliography by Joachim Thiele (in *Centaurus* **8**, 1963, 189–237).

Abbreviations:

AP *Annalen der Physik*, Leipzig
CR *Carls Repertorium*, München
EJ *Jahrbuch für Photographie und Reproductionstechnik* (ed. by J. M. Eder), Halle
MU *Untersuchungen zur Naturlehre des Menschen und der Tiere* (ed. by J. Moleschott), Giessen
SW *Sitzungsbericht der Kaiserlichen Akademie der Wissenschaften, Mathematisch-Naturwissenschaftliche Classe*, Wien
ZMP *Zeitschrift für Mathematik und Physik*, Leipzig
ZphU *Zeitschrift für den physikalischen und chemischen Unterricht*, Berlin

1859

Blaserna-Mach-Peterin, 'Über elektrische Entladung und Induction', *SW* **37**, 477–524.

1860

'Über die Änderung des Tones und der Farbe durch Bewegung', *SW* **41, 543–560.

† Cf. 'List of Mach's Educational Publications' by Otto Blüh, *Boston Studies*, Vol. VI, pp. 20–21.

1861

'Über das Sehen von Lagen und Winkeln durch die Bewegung des Auges', *SW* **43,** 215–224.
'Über die Controverse zwischen Doppler und Petzval bezüglich der Änderung des Tones und der Farbe durch Bewegung', *ZMP* **6**, 120–126.

1862

'Über die Molecularwirkung der Flüssigkeiten', *SW* **46**, 125–134.
'Zur Theorie der Pulswellenzeichner', *SW* **46**, 157–174.
'Über die Änderung von Ton und Farbe durch Bewegung', *AP* **116**, 333–338.

1863

Compendium der Physik für Mediciner, Wien 1863.
'Über die Gesetze des Mitschwingens', *SW* **47**, 33–48.
'Über eine neue Einrichtung des Pulswellenzeichners', *SW* **47**, 53–56.
'Zur Theorie des Gehörorgans', *SW* **48**, 283–300.
'Vorlesungen über Psychophysik', *Oesterreichische Zeitschrift für praktische Heilkunde* **9**.

1864

'Über einige der physiologischen Akustik angehörige Erscheinungen', *SW* **50**, 342–362.

1865

'Untersuchungen über den Zeitsinn des Ohres', *SW* **51**, 133–150.
'Bemerkungen über die Accomodation des Ohres', *SW* **51**, 343–346.
'Über die Wirkung der räumlichen Vertheilung des Lichtreizes auf die Netzhaut, part I', *SW* **52**, 303–322.
'Über Flüssigkeiten, welche suspendirte Körperchen enthalten', *AP* **126**, 324–330.
'Bemerkungen über den Raumsinn des Ohres', *AP* **126**, 331–333.
'Über die anschauliche Darstellung einiger Lehren der musikalischen Akustik', *ZMP* **10**, 425–428.
'Bemerkungen über intermittirende Lichtreize', *Archiv für Anatomie, Physiologie und wissenschaftliche Medicin.*

1866

Einleitung in die Helmholtz'sche Musiktheorie. Populär für Musiker dargestellt, Graz 1866.
'Über wissenschaftliche Anwendungen der Photographie und Stereoskopie', *SW* **54**, 123–126.
'Über den physiologischen Effect räumlich vertheilter Lichtreize, part II', *SW* **54**, 131–144.
'Über die physiologische Wirkung räumlich vertheilter Lichtreize, part III', *SW* **54**, 393–408.
'Über eine Vorrichtung zur mechanisch-graphischen Darstellung der Schwingungscurven', *AP* **129**, 464–466.
'Über die Wellen mit Flüssigkeit gefüllter elastischer Röhren', *MU* **10**, 71–74.

'Bemerkungen über die Entwicklung der Raumvorstellungen', *Fichtes Zeitschrift für Philosophie* **49**, 227–232.

1867

**Zwei populäre Vorträge über Optik*, Graz 1867.
'Über eine Longitudinalwellenmaschine', *AP* **132**, 174–176.

1868

'Über die physiologische Wirkung räumlich vertheilter Lichtreize, part IV', *SW* **57**, 11–19.
'Beobachtungen über monoculare Stereoskopie', *SW* **58**, 731–736.
'Einfache Demonstration des Huyghens'schen Princips', *AP* **134**, 310–311.
'Einfache Demonstration der Schwingungsgesetze gestrichener Saiten', *AP* **134**, 311–312.
'Über die Definition der Masse', *CR* **4**, 355–359.
'Über die Versinnlichung einiger Sätze der Mechanik', *CR* **4**, 359–361.
'Über die Versinnlichung der Poinsot'schen Drehungstheorie', *CR* **4**, 361.
'Über die Abhängigkeit der Netzhautstellen voneinander', *Vierteljahresschrift für Psychiatrie*, Neuwied **2**.

1870

'Mittheilungen über einfache Vorlesungsversuche', *CR* **6**, 8–12.

1871

'Optische Vorlesungsversuche', *CR* **7**, 261–264.
'Die mechanische Nachahmung des Fermat'schen Brechungsgesetzes', *CR* **7**, 375–377.
'Eine Bemerkung über den zweiten Hauptsatz der mechanischen Wärmetheorie', *Lotos* **21**, 17–18.
'Über die physikalische Bedeutung der Gesetze der Symmetrie', *Lotos* **21**, 139–147.

1872

Die Geschichte und die Wurzel des Satzes von der Erhaltung der Arbeit, Prag 1872; 2nd edition: 1909.
**Die Gestalten der Flüssigkeit. Die Symmetrie*, Prag 1872.
'Über die stroboskopische Bestimmung der Tonhöhe', *SW* **66, 267–274.
Mach-Kessel, 'Die Funktion der Trommelhöhle und der Tuba Eustachii', *SW* **66**, 329–336.
Mach-Kessel, 'Versuche über die Accomodation des Ohres', *SW* **66**, 337–343.
'Über die temporäre Doppelbrechung der Körper durch einseitigen Druck', *AP* **146**, 313–316.
'Spectrale Untersuchung eines longitudinal tönenden Glasstabes', *AP* **146**, 316–317.

1873

Optisch-akustische Versuche. Die spectrale und stroboskopische Untersuchung tönender Körper, Prag 1873.

Mach-Fischer, 'Die Reflexion und Brechung des Schalles', *SW* **67, 81–88.
'Über die Stefan'schen Nebenringe am Newton'schen Farbenglas und einige verwandte Interferenzerscheinungen', *SW* **67**, 371–381.
'Physikalische Versuche über den Gleichgewichtssinn des Menschen, part I', *SW* **68**, 124–140.
'Zur Geschichte des Arbeitsbegriffs', *SW* **68**, 479–488.
'Resultate einer Untersuchung zur Geschichte der Physik', *Lotos* **23**, 189–191.

1874

'Versuche über den Gleichgewichtssinn, part II', *SW* **69**, 121–135.
'Über den Gleichgewichtssinn, part III', *SW* **69**, 44–51.
Mach-Kessel, 'Beiträge zur Topographie und Mechanik des Mittelohres', *SW* **69**, 221–243.

1875

Grundlinien der Lehre von den Bewegungsempfindungen, Leipzig 1875.
Mach-Wosyka, 'Über einige mechanische Wirkungen des elektrischen Funkens', *SW* **72, 44–52.
Mach-Rosicky, 'Über eine neue Form der Fresnel-Arago'schen Interferenzversuche', *SW* **72**, 197–212.
Mach-Merten, 'Bemerkungen über die Änderung der Lichtgeschwindigkeit im Quarz durch Druck', *SW* **72**, 315–328.
'Über einen Polarisationsapparat mit rotirendem Zerleger', *AP* **156**, 169–172.
'Bemerkungen über die Funktion der Ohrmuschel', *Archiv für Ohrenheilkunde* **9**, 72–76.

1877

Mach-Sommer, 'Über die Fortpflanzungsgeschwindigkeit von Explosionsschallwellen', *SW* **75**, 101–130.

1878

Mach-Tumlirz-Kögler, 'Über die Fortpflanzungsgeschwindigkeit der Funkenwellen', *SW* **77**, 7–32.
'Neue Versuche zur Prüfung der Doppler'schen Theorie der Ton- und Farbenänderung durch Bewegung', *SW* **77**, 299–310.
'Über den Verlauf der Funkenwellen in der Ebene und im Raume', *SW* **77**, 819–838.
Mach-Gruss, 'Optische Untersuchung der Funkenwellen', *SW* **78**, 476–480.
Mach-Weltrubsky, 'Über die Formen der Funkenwellen', *SW* **78**, 551–560.
Mach-Doubrava, 'Über die elektrische Durchbrechung des Glases', *SW* **78, 729–732.

1879

Mach-Doubrava, 'Beobachtungen über die Unterschiede der beiden elektrischen Zustände', *SW* **80, 331–345.
Mach-Simonides, 'Weitere Untersuchung der Funkenwellen', *SW* **80**, 476–486.

1882

**Die ökonomische Natur der physikalischen Forschung*, Wien 1882.
'Über Herrn A. Guébhard's Darstellung der Aequipotentialcurven', *SW* **86, 8–14.

1883

Die Mechanik in ihrer Entwickelung historisch-kritisch dargestellt, Leipzig 1883; further German editions: 1888, 1897, 1901, 1904, 1908, 1912, 1921, 1933.
'Versuche und Bemerkungen über das Blitzableitersystem des Herrn Melsens', *SW* **87**, 632–639.

1884

**Über Umbildung und Anpassung im naturwissenschaftlichen Denken*, Wien 1884.
'Über die Grundbegriffe der Elektrostatik', *Lotos* **33, 90–112.

1885

March-Arbes, 'Einige Versuche über totale Reflexion und anomale Dispersion', *SW* **92, 416–426.
Mach-Wentzel, 'Ein Beitrag zur Mechanik der Explosionen', *SW* **92, 625–638.
'Zur Analyse der Tonempfindungen', *SW* **92**, 1283–1289.

1886

Beiträge zur Analyse der Empfindungen, Jena 1886; further German editions: 1900, 1902, 1903, 1906, 1911, 1918, 1919, 1922.
**Der relative Bildungswert der philologischen und der mathematischen-naturwissenschaftlichen Unterrichtsfächer der höheren Schulen*, Prag 1886.

1887

Mach-Salcher, 'Photographische Fixirung der durch Projectile in der Luft eingeleiteten Vorgänge', *SW* **95, 764–778.
'Über den Unterricht in der Wärmelehre', *ZphU* **1**, 3–7.

1888

'Über die Fortpflanzungsgeschwindigkeit des durch scharfe Schüsse erregten Schalles', *SW* **97**, 1045–1052.
'Über eine Lichtquelle zum Photographiren nach der Schlierenmethode', *EJ* **2**, 284.
'Ergebnisse der Momentphotographie', *EJ* **2**, 287–290.
'Über die Anordnung von quantitativen Schulversuchen', *ZphU* **1**, 197–199.
'Die experimentelle Darstellung der Linsenabweichungen', *ZphU* **2**, 52–55.

1889

Mach-Salcher, 'Über die in Pola und Meppen angestellten ballistischphotographischen Versuche', *SW* **98**, 41–50.

'Über die Schallgeschwindigkeit beim scharfen Schuss nach von dem Krupp'schen Etablissement angestellten Versuchen', *SW* **98**, 1257–1276.

Mach-Salcher, 'Optische Untersuchung der Luftstrahlen', *SW* **98, 1303–1309.

Mach-Mach (Ludwig), 'Weitere ballistisch-photographische Versuche', *SW* **98**, 1310–1326.

Mach-Mach (Ludwig), 'Über longitudinale fortschreitende Wellen im Glase', *SW* **98**, 1327–1332.

Mach-Mach (Ludwig), 'Über die Interferenz der Schallwellen von grosser Excursion', *SW* **98**, 1333–1336.

1890

'Sphärische Concavspiegel zur Photographie mittels des Schlierenapparates', *EJ* **4**, 108–109.

'Über das psychologische und logische Moment im naturwissenschaftlichen Unterricht', *ZphU* **4**, 1–5.

1891

Leitfaden der Physik für Studierende, Prag & Wien; Leipzig 1891; 2nd edition: Leipzig 1891.

'Über weitere Fortschritte in der Momentphotographie', *EJ* **5**, 166–167.

1892

'Ergänzungen zu den Mitteilungen über Projectile', *SW* **101**, 977–983.

'Zur Geschichte und Kritik des Carnot'schen Wärmegesetzes', *SW* **101**, 1589–1612.

'Über eine elementare Darstellung der Fraunhofer'schen Beugungserscheinungen', *ZphU* **5**, 225–229.

1893

Mach-Doss, 'Bemerkungen zu den Theorien der Schallphänomene bei Meteoritenfällen', *SW* **102**, 248–252.

1894

'Über ein Princip der Verstärkung unterexponierter Bilder', *EJ* **8**, 152–153.

'Über die rasche Ermittlung der richtigen Expositionszeit', *EJ* **8**, 154–156.

*'Über das Princip der Vergleichung in der Physik', *Verhandlungen der Gesellschaft Deutscher Naturforscher und Ärzte.*

'Einfache Versuche über strahlende Wärme', *ZphU* **7**, 113–116.

1896

Die Principien der Wärmelehre. Historisch-kritisch entwickelt, Leipzig 1896; further German editions: 1900, 1919.

Populär-wissenschaftliche Vorlesungen, Leipzig 1896 (15 lectures); further German editions: 1897 (15 lectures), 1903 (19), 1910 (26), 1923 (33).

'Durchsicht-Stereoskopbilder mit Roentgenstrahlen', *Zeitschrift für Elektrotechnik*, Wien **14**, 259–261.

'Über Gedankenexperimente', *ZphU* **10**, 1–5.

1897

'Bemerkungen über die historische Entwicklung der Optik', *ZphU* **11**, 3–8.

1902

'Die Ähnlichkeit und die Analogie als Leitmotiv der Forschung', *Annalen der Naturphilosophie*, Leipzig **1**, 5–14.

1904

'Objektive Darstellung der Interferenz des polarisierten Lichtes', in *Festschrift Ludwig Boltzmann*, Leipzig 1904, pp. 441–447.
Mach-Mach (Ludwig), 'Versuche über Totalreflexion und deren Anwendung', *SW* **113**, 1219–1230.

1905

Erkenntnis und Irrtum. Skizzen zur Psychologie der Forschung, Leipzig 1905; further German editions: 1906, 1917, 1920, 1926.

1906

'Über den Einfluss räumlich und zeitlich variierender Lichtreize auf die Gesichtswahrnehmung', *SW* **115**, 633–648.
'Über die Phasenänderung des Lichtes durch Reflexion', in *Festschrift Adolf Lieben*, Leipzig 1906, pp. 291–296.

1907

'Die Phasenverschiebung durch Reflexion an den Jamin'schen Platten', *SW* **116**, 997–1000.

1910

'Die Leitgedanken meiner naturwissenschaftlichen Erkenntnislehre', *Physikalische Zeitschrift* **11**, 599–606.
*'Eine Betrachtung über Zeit und Raum', in *Das Wissen für Alle*, Wien 1910.
'Sinnliche Elemente und naturwissenschaftliche Begriffe', *Pflüger's Archiv für Physiologie* **136**, 263–274.
'Die Organisierung der Intelligenz', *Neue Freie Presse*, Wien, July 24.

1911

*'Psychisches und organisches Leben', *Österreichische Rundschau*, Wien **29**, 22–31.
*'Allerlei Erfinder und Denker', *Naturwissenschaftliche Wochenschrift* **26**, 497–501.

1912

*'Das Paradoxe, das Wunderbare und das Gespenstische', *Kosmos* **9**, 17–20.
*'Psychische Tätigkeit, insbesondere Phantasie, bei Mensch und Tier', *Kosmos* **9**, 121–125.

1915

Kultur und Mechanik, Stuttgart 1915.

'Einige Experimente über Interferenz, insbesondere über complementärfarbige Interferenzringe', in *Festschrift Wilhelm Jerusalem*, Wien 1915, pp. 154–173.

1916

*'Einige vergleichende tier- und menschenpsychologische Skizzen', *Naturwissenschaftliche Wochenschrift* **31**, 241–247.

1921

Die Principien der physikalischen Optik. Historisch und erkenntnispsychologisch entwickelt, Leipzig 1921.

II. ENGLISH TRANSLATIONS OF BOOKS BY ERNST MACH

(Other than the Present Volume)

The Science of Mechanics. A critical and historical exposition of its principles (transl. from the 2nd German edition by Thomas J. McCormack), The Open Court Publ. Co., Chicago, 1893; Watts & Co., London 1893.

2nd rev. and enl. ed., The Open Court Publ. Co., Chicago, 1902.

3rd ed., The Open Court Publ. Co., Chicago, 1907.

Supplement to the 3rd English translation, containing the author's addition to the 7th German ed. (transl. by Ph. E. B. Jourdain), The Open Court Publ. Co., Chicago, 1915.

4th ed., The Open Court Publ. Co., Chicago, 1919.

5th ed. La Salle, Ill., The Open Court Publ. Co., London, 1942. (Containing additions and alterations up to the 9th German ed.)

6th edition. New Introduction by Karl Menger, La Salle, Ill., The Open Court Publ. Co. 1960.

Popular Scientific Lectures (Transl. by Thomas J. McCormack), The Open Court Publ. Co., Chicago, 1895 (12 lectures).

2nd ed., rev. and enl. The Open Court Publ. Co., Chicago, 1897 (13 lectures).

3rd ed., rev. and enl., The Open Court Publ. Co., Chicago, K. Paul, Trench, Truebner & Co., London, 1898. Reprint of 3rd ed., 1910.

Contributions to the Analysis of the Sensations (transl. by C. M. Williams), The Open Court Publ. Co., Chicago, 1897.

The Analysis of Sensations and the Relation of the Physical to the Phychical (transl. from the 1st German ed. by C. M. Williams, rev. and suppl. from the 5th German ed. by Sidney Waterlow), The Open Court Publ. Co., Chicago and London, 1914. (With a new introduction by Thomas S. Szasz) Dover Publ. New York, 1959.

History and Root of the Principle of the Conservation of Energy (transl. from the German and annotated by Philip E. B. Jourdain), The Open Court Publ. Co., Chicago, 1911.

Space and Geometry in the light of physiological, psychological and physical inquiry. (transl. from the German by Thomas J. McCormack), The Open Court Publ. Co., Chicago, 1906 (Reprinted from *The Monist*, April 1901, July 1902, October 1903).

The Principles of Physical Optics. An historical and philosophical treatment (transl. by John S. Anderson and A. F. A. Young), Methuen & Co., London, 1926; Dover Publ., New York, 1953.

III. ENGLISH TRANSLATIONS OF ARTICLES BY ERNST MACH

In: *Philosophical Magazine*, Ser. 4. (London)

'On the Visual Sensations Produced by Intermittent Excitations of the Retina', **30** (1865), 319.
'On an Arrangement for the Graphical Representation of Curves of Vibration by Means of Mechanism', **33** (1866), 159.

In: *The Open Court. A Weekly Journal devoted to the Religion of Science.* (Chicago)

'Transformation and Adaptation in Scientific Thought', **2** (1888), 1087, 1115.
'The Forms of Liquids', **8** (1894), 3935.
'Symmetry', **8** (1894), 4015.
'The Fibres of Corti', **8** (1894), 4087.
'On the Causes of Harmony', **8** (1894), 4136.
'The Velocity of Light', **8** (1894), 4167.
'Why has Man two Eyes?' **8** (1894), 4175.
'On the Fundamental Concepts of Electrostatics', **8** (1894) 4247, 4255.
'The Economical Character of Physical Research', **8** (1894), 4263, 4271.
'On the Principle of Comparison in Physics', **8** (1894), 4283, 4288.
'On the Relative Educational Value of the Classics and the Mathematico-Physical Sciences in Colleges and High Schools', **8** (1894), 4295, 4308, 4311.
N.B. All the above reprinted in *Popular Scientific Lectures* (1895).
'Names and Numbers', **14** (1900), 37.
'Language' (Its Origin, Development and Significance for the Development of Thought), **14** (1900), 171.
'The Concept', **14** (1900), 348.
'The Notion of a Continuum' (An Essay in the Theory of Science), **14** (1900), 409.
'The Propensity toward the Marvellous', **14** (1900), 539.
'The Theory of Heat', **16** (1902), 641.
'Sketch of the History of Thermometry', **16** (1902), 733; **17** (1903), 26.
'Critique of the Concept of Temperature', **17** (1903), 95, 154.
'On the Determination of High Temperatures', **17** (1903), 297.
N.B. The last 9 articles are translations of chapters from Mach's *Wärmelehre* (*Principles of Heat*) (1896).

In: *The Monist. A Quarterly Magazine Devoted to the Philosophy of Science.* (Chicago)

'The Analysis of Sensations – Antimetaphysical', **1** (1890), 48.
'Facts and Mental Symbols', **2** (1892), 198.
'On the Principle of Conservation of Energy', **5** (1894), 22.
'On the Part Played by Accident in Invention and Discovery', **6** (1896), 161.
'On the Stereoscopic Application of Röntgen's Rays', **6** (1896), 321.

'On Sensations of Orientation', **8** (1897), 79.
'On Physiological, as Distinguished from Geometrical Space', **11** (1906), 321.
'On the Psychology and Natural Development of Geometry', **12** (1902), 481.
'Space and Geometry from the Point of View of Physical Inquiry', **14** (1903), 1. (The German text of this and the preceding article was published in *Erkenntnis und Irrtum,* 1905).
'Inventors I have met', **22** (1912), 230.
'Psychic and Organic Life', **23** (1913), 1.

In: Floyd Ratliff, *Mach Bands: Quantitative Studies on Neural Networks in the Retina.* (San Francisco 1965)
Translation of six articles (1865–1906) on this topic.

IV. PUBLICATIONS ON ERNST MACH IN ENGLISH

A Symposium on Ernst Mach (ed. by J. Hintikka), *Synthese* **18** (1968) 132–301. This includes the following articles:
Robert S. Cohen, Ernst Mach: Physics, Perception and the Philosophy of Science. [Also in *Boston Studies in the Philosophy of Science,* vol. 6, pp. 126–164.]
Milič Čapek, Ernst Mach's Biological Theory of Knowledge. [Also in *Boston Studies in the Philosophy of Science*, vol. 4, pp. 400–420.]
S. G. Brush, Mach and Atomism.
A. Koslow, Mach's Concept of Mass: Program and Definition.
Wolfgang Yourgrau and Alwyn van der Merwe, Did Ernst Mach 'Miss the Target'?
M. Strauss, Einstein's Theories and the Critics of Newton. An Essay in Logico-Historical Analysis.
Joachim Thiele, Briefe deutscher Philosophen an Ernst Mach.
Blackmore, John T., *Ernst Mach: His Work, Life, and Influence*, University of California Press, Berkeley &c 1972 (contains a substantial bibliography).
Blanshard, Brand, *Reason and Analysis* (*In memory of Paul Carus*, 1852–1919), chap. III: 'The Rise of Positivism', Allen and Unwin, Ltd., London, 1962.
Blüh, Otto, 'Ernst Mach as Teacher and Thinker,' *Physics Today* **20** (1967), 32.
Blüh, Otto, 'Ernst Mach as an Historian of Physics', *Centaurus* **13** (1968), 62.
Bradley, J., *Mach's Philosophy of Science*, The Athlone Press, London, 1971.
Bridgman, P. W., 'Significance of the Mach Principle', *Am. J. Phys.* **29** (1961), 32.
Bunge, Mario, 'Mach's Criticism of Newtonian Mechanics', *Am. J. Phys.* **34** (1966), 585.
Carus, Paul, 'Feelings and the Elements of Feelings', *The Monist* **1** (1891), 401.
Carus, Paul, 'Professor Mach's Term Sensation', Supplementary to his Controversy with the Editor (Carus), *The Monist* **3** (1893), 298.
Carus, Paul, 'Professor Mach's Philosophy', *The Monist* **16** (1906) 627.
Carus, Paul, 'Professor Mach and his work', *The Monist* **21** (1911), 19.
Cohen, M. R., *Reason and Nature. An Essay on the Meaning of Scientific Method.*, chap. II, Sec. 2., Free Press, Glencoe, Ill., 1953.
Cohen, Robert S., and Raymond J. Seeger (eds.), *Ernst Mach: Physicist and Philosopher*, Boston Studies in the Philosophy of Science VI, Dordrecht, 1970.
Dicke, R. H., 'Cosmology, Mach's Principle and Relativity', *Am. J. Phys.* **31** (1963), 500.
Dicke, R. H., and C. Brans, 'Mach's Principle and a Relativistic Theory of Gravitation', *Physical Review* **124** (1961), 925.

Feld, James W., 'Mach's Principle Revisited', *Laval Theologique et Philosophique* **20** (1964), 35.

Frank, Philipp, 'Ernst Mach. The Centenary of his Birth'. *Erkenntnis Journal for Unified Science* **7** (1937/38), 247.

Frank, Philipp, 'Einstein, Mach and Logical Positivism', in: *Albert Einstein – Philosopher – Scientist* (ed. by P. A. Schilpp), The Library of Living Philosophers, Inc., Evanston, Ill.,vol. VII, 1949.

Gardner, Martin, *The Ambidextrous Universe*, chap. 19, p. 189: 'Mach's Shock', Basic Books, Inc. New York, 1964.

Hiebert, Erwin, *The Conception of Thermodynamics in the Scientific Thought of Mach and Planck,* Wissenschaftlicher Bericht Nr. 5/58, 106 p. (Freiburg i.Br., 1968).

Hiebert, Erwin,' Mach's Philosophical Use of the History of Science', pp. 184–203 in Roger H. Stuewer (ed.), *Historical and Philosophical Perspectives of Science* (Minneapolis, 1970).

Hiebert, Erwin, 'The Energetics Controversy and the New Thermodynamics', pp. 67–86 in Duane H. D. Roller (ed.), *Perspectives in the History of Science and Technology* (Norman, Oklahoma, 1971).

Hiebert, Erwin, Article Mach, Ernst, in *Dictionary of Scientific Biography* **8**, New York, 1973.

Hiebert, Erwin, 'Mach's Conception of Thought Experiments in the Natural Sciences', pp. 339–348 in Yehuda Elkana (ed.), *The Interaction Between Science and Philosophy* (New York, 1975).

Hiebert, Erwin, 'An Appraisal of the Work of Ernst Mach: Scientist-Historian-Philosopher', in Peter Machamer & Robert Turnbull (eds.), *Motion and Time, Space and Matter* (Columbus, Ohio, 1975).

Høffding, Harald, *Modern Philosophers* (transl. by Alfred C. Mason), section on Mach, pp. 115–121. Macmillan & Co., London, 1920.

Holton, Gerald, 'Mach, Einstein, and the Search for Reality', *Daedalus* **97** (1968), 636.

Jourdain, Ph. E. B., 'The Principle of Least Action. Remarks on Some Passages of Mach's *Mechanics', The Monist* **22** (1912), 285.

Kleinpeter, Hans, 'On the Monism of Professor Mach', *The Monist* **16** (1906), 161.

Kolakowski, Leszek, *The Alienation of Reason. A History of Positivist Thought.* (Original Polish title: *Positivist Philosophy. From Hume to the Vienna Circle*) (transl. by Norbert Guterman), Doubleday, New York, 1968. On Mach, p. 118ff.

Lenin, Vladimir Il'ich, *Materialism and Empirio-Criticism*, New York, 1927.

Mayerhöfer, Josef, 'Ernst Mach as a Professor of the History of Science', *Proc. 10th Intern. Congr. Hist. Sci.* **I** (Paris 1964), 337.

Mises, Richard von, *Positivism. A Study in Human Understanding* (transl. by J.Bernstein and R. G. Newton), Harvard University Press, Cambridge, 1951. Reprint, unchanged, Dover Publ., New York, 1968.

Moon, P. and D. E. Spenser, 'Mach's Principle', *Philosophy of Science* **26** (1959), 125.

Pittenger, H. W., 'Ernst Mach: Biographical Notes', *Science* **150** (1965), 1120.

Ratliff, Floyd, *Mach Bands: Quantitative Studies on Neural Networks in the Retina*, Holden-Day, Inc., San Francisco, 1965.

Seaman, Francis, 'Mach's Rejection of Atomism', *J. Hist. Ideas* **29** (1968), 381.

Titchener, E. B., 'Mach's Lectures on Psychophysics', *Am. J. Psych.* **33** (1922), 213.

Weinberg, C. B., Mach's Empirio-Pragmatism in Physical Science, thesis, Columbia University, New York, 1937.

V. NON-ENGLISH PUBLICATIONS ON ERNST MACH SINCE 1930

Theodor Ackermann Antiquariat, *Bibliothek Ernst Mach,* Teil I, Katalog 634, 93 p. (München, 1959); Teil II, Katalog 636, 101 p. (München, 1960). A list of some 3700 works from Mach's working library.

Birk, Alfred, 'Zur Hundertjahrfeier zweier grossen Freunde. Ernst Mach – Josef Popper', *Ingenieur-Zeitschrift*, Teplitz Schönau **18** (1938), 12.

Blüh, Otto, 'Ernst Mach', *Brünner Tagesbote*, 15 February 1938.

Chmelka, Fritz, 'Ernst Mach: Physiker und Philosoph', *Universum* **21** (1966), 74.

D'Elia, Alfonsina, *Ernst Mach*, La Nuova Italia, Florence 1971 (contains a substantial bibliography).

Dugas, René, *La Théorie Physique au Sens de Boltzmann.* Part III. Chap. VI. 'Planck et Mach', p. 256. Griffon, Neuchâtel, 1959.

Elek, Tibor, 'Zur Geschichte der Beziehungen zwischen Physik und Philosophie: Die philosophische Diskussion zwischen E. Mach, M. Planck u. A. Einstein zu Beginn des 20. Jh', *Periodica Polytechnica* (Budapest) **1** (1967), 135.

Heller, K. D., *Ernst Mach – Wegbereiter der Modernen Physik*, Springer Verlag, Vienna, 1964.

Herneck, Friedrich, 'Über eine Unveröffentlichte Selbstbiographie Ernst Machs', *Wissenschaftliche Zeitschrift der Humboldt-Universität Berlin*, Mathem. naturwiss. Reihe, **6** (1956/57), 209.

Herneck, Friedrich, 'Ernst Mach, Eine bisher unveröffentlichte Autobiographie', *Phys. Bl.* **14** (1958), 385.

Herneck, Friedrich, 'Zu einem Brief Albert Einsteins an Ernst Mach', *Phys. Bl.* **15** (1959), 563.

Herneck, Friedrich, 'Wiener Physik vor 100 Jahren', *Phys. Bl.* **17** (1961), 455. N.B. Photographs of Mach and Boltzmann on p. 457.

Herneck, Friedrich, 'Zum Briefwechsel Albert Einsteins an Ernst Mach', *Forschungen und Fortschritte* **37** (1963), 239.

Herneck, Friedrich, 'Die Beziehungen zwischen Einstein und Mach dokumentarisch dargestellt', *Naturwiss. Zeitschrift der Friedrich-Schiller-Universität Jena* **15** (1966), 45.

Herrmann, D. B., 'Ernst Mach und seine Stellung zur Doppler-Theorie', *Forschungen und Fortschritte* **40** (1966), 362.

Hönl, H., 'Ein Brief Albert Einsteins an Ernst Mach', *Phys. Bl.* **16** (1960), 571.

Kerkhof, F., 'Ernst-Mach Symposium am 11./12. März 1966 in Freiburg/Br.', *Phys. Bl.* **22** (1966), 223.

Kraft, Viktor, 'Ernst Mach als Philosoph', *Almanach d. Österr. Akad. d. Wissenschaften* **116** (1966), 373.

Löwy, Heinrich, 'Die Erkenntnistheorie von Popper-Lynkeus und ihre Beziehung zur Machschen Philosophie', *Die Naturwissenschaften* **20** (1932), 770.

Lohr, Ernst, 'Ernst Mach als Physiker', *Zeitschr. für die gesammte Naturwissenschaft* **4** (1938/39), 108.

Lübbe, Hermann, 'Positivismus und Phänomenologie', (Mach und Husserl) in: *Beiträge zur Philosophie und Wissenschaft*, Munich 1960, p. 161.

Mayerhöfer, Josef, 'Ernst Machs Berufung an die Wiener Universität 1895', *Clio Medica* **2** (1967), 47.

Mises, Richard von, 'Ernst Mach und die Empiristische Wissenschaftsauffassung. Zu Ernst Machs 100. Geburtstag am 18. Februar 1938', *Einheitswissenschaft* **7** (1938),

35. [Transl. in *Boston Studies in the Philosophy of Science*, Vol. 6, pp. 245–270.]
Nicolle, Jacques, 'Lénine, Mach et Paul Langevin', *La Pensée* **57** (1954), 66.
Oakland, Fridthjof, *Machs Elementlaere og Biologien*, Oslo 1947.
Sommerfeld, Arnold, 'Ernst Mach als Physiker, Psychologe und Philosoph', *Verhandl. Deutschen Physik. Gesellschaft*, Reihe 3, **19** (1938), 51.
Symposium aus Anlass des 50. Todestages von Ernst Mach, veranstaltet vom Ernst-Mach-Institut, Freiburg/Br. am 11/12. März 1966 in Freiburg/Br. With contributions by G. v. Békésy, M. Bunge, F. v. Hayek, F. Herneck, D. B. Herrmann, N. Hiebert, H. Hönl, H. P. Jochim, R. E. Kutterer, E. Lesky, J. Mayerhoefer, W. F. Merzkirch, J. Pachner, F. Ratliff, N. Strauss, J. Thiele.
Thiele, J., 'Bemerkungen zu einer Äusserung im Vorwort der *Optik* von Ernst Mach', *Schriftenreihe für Geschichte der Naturwiss., Technik und Medizin* **2** (1965), 10.
Thiele, J., 'William James und Ernst Mach', *Philosophia Naturalis* **9** (1966), 298.
Thiele, J., 'Briefe von Gustav Theodor Fechner und Ludwig Boltzmann an Ernst Mach', *Centaurus* **11** (1967), 222.
Thiele, J., 'Naturphilosophie und Monismus um 1900' (Briefe von W. Ostwald, Ernst Mach, Ernst Haeckel und Hans Driesch), *Philosophia Naturalis* **10** (1968), 295.
Thiele, J., 'Briefe Robert Lowies an Ernst Mach', *Isis* **59** (1968), 84.
Thiele, J., 'Ernst Mach und Heinrich Hertz, Zwei unveröffentlichte Briefe aus dem Jahre 1890', *Schriftenreihe für Geschichte der Naturwiss., Technik und Medizin* **5** (1968), 132.
Thiele, J., 'Ein zeitgenössisches Urteil über die Kontroverse zwischen Max Planck und Ernst Mach', *Centaurus* **13** (1968), 85.
Thiele, Joachim, 'Schulphysik vor 70 Jahren: Hinweis auf Ernst Machs Lehrbücher &c', *Zeits. f. Math. u. naturw. Unterricht* **19** (1966), 15.
Thirring, Hans, 'Ernst Mach als Physiker', *Almanach d. Österr. Akad. d. Wissenschaften* **116** (1966), 361.

BIBLIOGRAPHY OF WORKS CITED BY ERNST MACH IN *KNOWLEDGE AND ERROR*

The following bibliography attempts to give tolerably adequate bibliographical information about the works cited by Mach. In many cases these have now become, if they were not always, obscure. Readers should remember that any departure from, or addition to, the reference given in a footnote is an editorial inference or speculation. In a few cases either a periodical quoted by Mach was not available in England or there was no easily discoverable record there of a book referred to by him: in those cases it has not been possible to go beyond the jejune bibliographical information, no doubt intelligible enough at the time, that he himself provided.

The attempt has been made to discover English translations or originals of works quoted from originals or translations in other languages. Copies of the English and other volumes involved are held in such scattered libraries that it was not in general possible to undertake a comparison of the text known to Mach with the corresponding English.

Abel, Niels Henrik, 'Démonstration de l'impossibilité de la résolution algébrique des équations générales qui dépassent le quatrième degré', *Crelles Journal* **1** (1826.)

Aepinus, Franz Ulrich Theodor, *Tentamen theoriae Electricitatis et Magnetismi*, 1759.

Alembert, Jean Lerond d', *Traité de Dynamique*, 1743.

Ampère, André Marie, *Théorie des Phénomènes électrodynamiques*, Paris 1826.

Apelt, Ernst Friedrich, *Die Theorie der Induktion*, Leipzig 1854.

Apuleius, *Metamorphoses*.

Arago, Dominique François Jean, 'Expériences relatives á l'aimantation du fer et de l'acier par l'action du courant voltaique', *Ann. de chimie et de physique* **15** (1820), 93.

Arago, Dominique François Jean, Report of the session of 7 March, *Ann. de chimie et de physique* **28** (1825).

Archimedes, (G.T. by Nizze, Stralsund 1824; E.T. by T. L. Heath, Cambridge 1897).

Argand, Jean Robert, *Essai sur la manière de représenter les quantités imaginaires*, Paris 1806 (E.T. by A. S. Hardy, N.Y. 1881).

Aristotle, *Physics*.

Autenrieth, Johann Heinrich Ferdinand, *Ansichten über Natur- und Seelenleben* 1836.

Becker, Bernhard, *Leitfaden für den ersten Unterricht in der Geometrie*, Frankfurt 1845.

Becker, Bernhard, *Über die Methoden des geometrischen Unterrichts*, Frankfurt 1845.

Becquerel, Alexandre Edmond, 'Sur la phosphorescence par insolation', *Ann. de chim. et de phys.* **22** (1848).

Beneke, Friedrich Eduard, *System der Logik als Kunstlehre des Denkens*, Berlin 1842.

Bernouilli, Jacques, 'Ad Examen Perpetui Mobilis' *Acta Eruditorum*, 1686.

Bernouilli, Jacques, *Ars conjectandi*, Basle 1713 (E.T. by F. Maseres, 1795).

Biedermann, P. F., *Die wissenschaftliche Bedeutung der Hypothese*, Dresden 1894.

Boltzmann, Ludwig, *Über die Frage nach der objektiven Existenz der Vorgänge in der unbelebten Natur*, Vienna 1897, (E.T. by P. Foulkes as 'The objective existence of processes in inanimate nature' in *Theoretical Physics and Philosophical Problems*, Dordrecht 1974).

Boole, George, *An investigation of the laws of thought*, London 1854.
Borelli, Giovanni Alfonso, *De motu animalium*, 1680 (E.T. *The Flight of Birds*, London 1911).
Bourdeau, Jean, *Les Forces de l'Industrie*, Paris 1884.
Boyle, Robert, *New experiments, physico-mechanical*, Oxford 1660.
Brasch, Moritz, *Die Welt- und Lebensanschauung F. Ueberwegs*, Leipzig 1889.
Bretschneider, Carl Anton, *Die Geometrie and die Geometer vor Euklides*, Leipzig 1870.
Bücher, Karl, *Arbeit und Rhythmus*, Leipzig 1902.
Buttel-Reepen, Hugo Berthold von, *Die stammesgeschichtliche Entstehung des Bienenstaates*, Leipzig 1903 (E.T. *The Evolution of the Communal Life of the Bee*, n.p., 19–).
Campbell, William Taylor, *Observational Geometry*, N.Y. 1899.
Cantor, Georg, *Grundlagen einer allgemeinen Mannigfaltigkeitslehre*, Leipzig 1883.
Cantor, Moritz Benedikt, *Mathematische Beiträge zum Kulturleben der Völker*, Halle 1863.
Cantor, Moritz Benedikt, *Geschichte der Mathematik*, Leipzig 1880–98.
Cantor, Moritz Benedikt, *Die römischen Agrimensoren*, Leipzig 1875.
Cardano,Girolamo (Cardanus), *De subtilirate* 1560 (E.T. of Book I by M. M. Cass, Williamsport 1934).
Cavalieri, Bonaventura, *Geometria indivisibilibus continuorum nova quadam ratione promota*, Bologna 1635.
Colozza, Giovanni Antonio, *L'Immaginazione nella scienza*, Turin 1900.
Copernicus, Nicolaus, *De revolutionibus orbium coelestium*, 1543 (E.T. by C. G. Wallis, Chicago 1955).
Coulomb, Charles Augustin de, 'Sixième Mémoire sur l'Électricité', *Histoire de l'Académie r. des sciences*, 1788.
Couturat, Louis, *De l'infini mathématique*, Paris 1896.
Couturat, Louis, *La logique de Leibniz*, Paris 1901.
Cyrano de Bergerac, Savinien, *Histoire comique des états et empires de la lune*, 1648 (E.T. by T. St. Serf, *Selenarchia*, 1659).
Czuber, Emanuel, 'Zum Zahl- und Grössenbegriff', *Zeitschr. f. d. Realschulwesen* **29** (1904).
Darwin, Charles Robert, *Kleinere Schriften*, trans. E. Krause [Mach appears to cite *The Expression of the Emotions in Men and Animals*, London 1872, p. 142].
Decremps, Henri, *La magie blanche dévoilée*, Paris 1789 (E.T. by T. Denton, *The Conjurer Unmasked*, 1785).
Desargues, Gérard, *Oeuvres*, ed. Poudra, Paris 1864.
Descartes, René, *Dioptrice* (Latin T. of 'Dioptrique' from *Discours de la méthode &c*, 1637 &c).
Descartes, René, *Principia Philosophiae* 1644 (Various E.TT.).
Detto, Carl Albert Eduard, 'Über den Begriff des Gedächtnisses in seiner Beleuchtung für die Biologie', *Naturwiss. Wochenschr.* **42** (1905).
Diderot, Denis, *Entretien entre D'Alembert et Diderot & le rêve de D'Alembert*, 1769.
Diderot, Denis, *Lettre sur les aveugles* 1749.
Diodorus Siculus.
Drobisch, Moritz Wilhelm, *Neue Darstellung der Logik*, Leipzig 1895.
Dufay, Charles François de Cisternay, 'Sur l'Électricité', *Mém. de l'Académie de Paris*, 1733.
Du Hamel du Monceau, Henri Louis, *La Physique des arbres*, Paris 1758.

Duhem, Pierre Maurice Marie, *Les Origines de la statique*, Paris 1905.
Duhem, Pierre Maurice Marie, *La Théorie physique*, Paris, 1906 (E.T. by P. P. Wiener, *The Aim and Structure of Physical Theory*, Princeton 1954).
Dvořák, V., 'Über Nachbilder von Reizveränderungen', *SB. d. Wiener Akademie* **61** (1870).
Ebbinghaus, Hermann, *Grundzüge der Psychologie*, Leipzig 1902.
Eisenlohr, August, *Ein mathematisches Handbuch der alten Aegypter: Papyrus Rhind*, Leipzig 1877.
Engel, Friedrich and Staeckel, Paul Gustav, *Die Theorie der Parallellinien von Euklid bis auf Gauss*, Leipzig 1895.
Ennemoser, Joseph, *Geschichte der Magie*, Leipzig 1844 (E.T. by W. Howitt, *History of Magic*, London 1893).
Erb, Heinrich, *Die Probleme der geraden Linie, des Winkels, und der ebenen Fläche*, Heidelberg 1846.
Erb, K. A., *Zur Mathematik und Logik*, Heidelberg 1821.
Erdmann, Benno, *Die Axiome der Geometrie*, Leipzig 1877.
Erman, Georg Adolf, *Ägypten*, 1885 &c (E.T. by H. M. Tirard, London and N.Y. 1894).
Euclid, *Elements*, (Mach cites the G.T. by J. F. Lorenz, Halle 1798).
Euler, Leonhard, *Lettres à une princesse d'Allemagne*, St. Petersburg, 1768–74 (Various E.TT.).
Fack, Michael, 'Zählen und Rechnen', *Zeitschr. f. Philos. u. Pädagogik* (Flügel & Rein) **2** (1895).
Faraday, Michael, 'Experimental Research in Electricity I, II', *Philos. Transact. of the Royal Society* 1832.
Faraday, 'Electro-magnetic Rotation Apparatus' (*Experimental Researches in Electricity*, II, London 1839–55).
Faraday, 'On the physical character of lines of magnetic force' (*Experimental Researches in Electricity*, III, London 1839–55).
Fechner, Gustav Theodor, *Elemente der Psychophysik*, Leipzig 1860.
Fechner, Gustav Theodor, *Vier Paradoxa*, Leipzig 1846.
Forel, Auguste Henri, Expériences et remarques sur les sensations des insectes', *Rivista di scienze biologiche*, Como, 1900–1901.
Forel, Auguste Henri, 'Geruchsinn bei den Insekten', 'Psychische Fähigkeiten der Ameisen', *Verh. d. 5. Intern. Zoologenkongresses*, Jena 1902.
Forel, Auguste Henri, *Der Hypnotismus*, 6th ed. (E.T. by H. W. Armit, *Hypnotism*, London 1906 etc.).
Foucault, Léon, *Recueil des travaux scientifiques*, Paris 1878.
Fouillée, Alfred Jules Émile, *La psychologie des idées-forces*, Paris 1893.
Fourier, Jean Baptiste Joseph (*baron*), Opening of a discussion on descriptive geometry, *Séances des Écoles Normales Débats* I (1800) 28.
Fraunhofer, Joseph von, *Gesammelte Schriften*, Munich 1888 (E.T. by J. S. Ames of the relevant paper in Fraunhofer, *Prismatic and Diffraction Spectra*, N.Y. and London 1898).
Fresnel, Augustin Jean, 'Mémoire couronné sur la diffraction', *Oeuvres*, vol. 1, Paris 1886, (E.T. in *The Wave Theory of Light*, ed. by H. Crew, N.Y. 1900).
Fries, Jakob Friedrich, *Versuch einer Kritik der Prinzipien der Wahrscheinlichkeitsrechnung*, Brunswick 1842.
Fries, Jakob Friedrich, *System der Logik*, Heidelberg 1819.
Fries, Jakob Friedrich, *Die mathematische Naturphilosophie*, Heidelberg 1822.

Galilei, Galileo, *Dialogue on the two world systems.*

Galilei, Galileo, *Il saggiatore, Opere de Galilei*, Padua 1744 (E.T. by S. Drake, *The Assayer*, in *Discoveries and Opinions of G.*, New York 1957).

Galilei, Galileo, *Sydereus nuncius, ibid.* (E.T. by S. Drake, *ibid.*)

Galton, (Sir) Francis, 'History of Twins', *Inquiries into human faculty*, London, 1883.

Gauss, Karl Friedrich, *Disquisitiones generales circa superficies curvas*, 1827. (E.T. by J. C. Morehead & A. M. Hiltebeitel, *General Investigations of Curved Surfaces*, Princeton 1902).

Gauss, Karl Friedrich, *Briefwechsel zwischen Gauss und Bessel*, Leipzig 1880.

Geiger, Lazarus, *Ursprung und Entwicklung der menschlichen Sprache und Vernunft*, Stuttgart 1868.

Gercken, Albert Wilhelm Ernst, *Die philosophischen Grundlagen der Mathematik*, Perleberg 1887.

Gerhardt, Karl Immanuel, *Die Entdeckung der höheren Analysis*, Halle 1855.

Gilbert, William, *De Magnete*, 1600 (Various E.TT.)

Goltz, Friedrich Leopold, *Die Nervenzentren des Frosches*, Berlin 1869.

Gomperz, Heinrich, *Zur Psychologie der logischen Grundtatsachen*, Vienna 1897.

Gomperz, Theodor, *Griechische Denker*, Leipzig 1896 (E.T. by L. Magnus and G. G. Berry, *Greek Thinkers*, London 1901–12).

Gow, James, *A Short History of Greek Mathematics*, Cambridge 1884.

Graeser, Kurt, *Der Zug der Vögel*, Berlin 1905.

Groos, Karl, *Die Spiele der Tiere*, Jena 1896 (E.T. by E. T. Baldwin, *The Play of Animals*, N.Y. 1898).

Gruithuisen, Franz von Paula, *Beiträge zur Physiognosie und Eautognosie*, Munich 1812.

Gruithuisen, Franz von Paula, *Die Naturgeschichte im Kreise der Ursachen und Wirkungen*, Munich 1810.

Guericke, Otto von, *Experimenta Magdeburgica*, Amsterdam 1672.

Haberlandt, Gottlieb Friedrich Johann, *Physiologische Pflanzenanatomie*, 1904 (E.T. by M. Drummond, *Physiological Plant Anatomy*, London 1914).

Haberlandt, Gottlieb Friedrich Johann, *Über den tropischen Urwald*, Schr. d. Vereins z. Verb. naturw. Kenntnisse, Vienna 1898.

Haddon, Alfred Cort, *Evolution in Art*, London 1895.

Hankel, Hermann, *Zur Geschichte der Mathematik*, Leipzig 1874.

Hecker, Justus Friedrich Karl, *Die grossen Volkskrankheiten des Mittelalters*, Berlin 1865 (E.T. by B. A. Babington, *The Epidemics of the Middle Ages*, London 1859).

Heinrich, J. P., *Die Phosphoreszenz der Körper*, Nuremberg 1820.

Helmholtz, Hermann Ludwig Ferdinand von, *Über die Erhaltung der Kraft*, Berlin 1847 (E.T. by E. Atkinson in *Popular Lectures on Scientific Subjects*, London and N.Y. 1861 &c).

Helmholtz, Hermann Ludwig Ferdinand von, 'Über die Thatsachen, welche der Geometrie zu Grunde liegen'. *Göttinger Nachrichten*, 3 June 1868.

Helmholtz, Hermann Ludwig Ferdinand von, 'Über die tatsachlichen Grundlagen der Geometrie' 1866, *Wissensch. Abhandl.* II, pp. 610ff (E.T. by M. Lowe *et al.* in *Epistemological Writings of H.v.H.*, Dordrecht & Boston 1975).

Helmholtz, Hermann Ludwig Ferdinand von, 'Zählen und Messen', *Philos. Aufsätze. E. Zeller gewidmet* (1887) (E.T. *ibid.*)

Hering, Ewald, *Die Lehre vom binocularem Sehen*, Leipzig 1868.

Hering, Ewald, *Zur Lehre vom Lichtsinn*, Vienna 1878.

Hering, Ewald, 'Der Raumsinn und Bewegungen des Auges', in L. Hermann, *Handb.*

d. Physiol. III, 1 (E.T. by C. A. Radde, *Spatial Sense and Movements of the Eye*, Baltimore 1942).

Hero, Opera (G.T. by W. Schmidt, Leipzig 1869.)

Herodotus.

Herschel, (*Sir*) John Frederick William, *A preliminary discourse on the study of natural philosophy*, London 1831.

Herschel, (*Sir*) John Frederick William, *Untersuchungen über die Ausbreitung der elektrischen Kraft*, Leipzig 1892 (E.T. by D. E. Jones, *Electric Waves &c.*, London & N.Y. 1900).

Hertz, Heinrich Rudolph, *Die Prinzipien der Mechanik*, 1894 (E.T. by D. E. Jones and T. Walley, *The Principles of Mechanics*, London & N.Y. 1899 &c).

Hertz, Heinrich Rudolph, *Werke*, Leipzig 1895.

Heymans, Gerardus, *Einführung in die Metaphysik auf Grundlage der Erfahrung* 1905.

Hilbert, David, *Grundlagen der Geometrie*, Leipzig 1899 (E.T. by E. J. Townsend, *The Foundations of Geometry*, Chicago 1902 &c).

Hillebrand, Franz, 'Theorie der scheinbaren Grösse bei binokularem Sehen', *Denkschr. d. Wiener Academie, math. naturw. Cl.* **72** (1902).

Hillebrand, Franz, 'Zur Lehre von der Hypothesenbildung', *SB. d. Wiener Akademie, philos.-histor. Cl.* **134** (1896).

Hirn, Gustave Adolphe, *Théorie méchanique de la chaleur*, Paris 1865.

Hobbes, Thomas, *Problemata Physica.*

Høffding, Harald, *Moderne Philosophen* 1905 (E.T. from Danish by A. C. Mason, *Modern Philosophers*, London 1915).

Høffding, Harald, *Psychologie in Umrissen*, Leipzig 1893.

Hoffmann, Fridolin, *Geschichte der Inquisition*, Bonn 1878.

Hölder, Otto, *Anschauung und Denken in der Geometrie*, Leipzig 1900.

Homer, *Odyssey.*

Hönigswald, Richard, *Zur Kritik der Machschen Philosophie*, Berlin 1905.

Hooke, Robert, *Micrographia*, 1665.

Hoppe, Janus, *Die Analogie*, Berlin 1873.

Houdin, *see* Robert-Houdin.

Huygens, Christian, *Traité de la lumière*, Leiden 1690 (E.T. by S. P. Thompson, *A Treatise on Light*, London 1912).

Jaeger, Johann, *et al.*, *Epistulae obscurorum virorum*, 1520.

James, William, *The feeling of effort*, Boston 1880.

James, William, *Principles of Psychology*, New York 1890.

Jerusalem, Wilhelm, 'Ein Beispiel von Association durch unbewusste Mittelglieder', *Philosophische Studien* (ed. Wundt) **10** (1894).

Jerusalem, Wilhelm, *Der kritische Idealismus und die reine Logik*, 1905.

Jerusalem, Wilhelm, *Einleitung in die Philosophie*, 2nd ed. 1903 (E.T. by C. F. Sanders, *Introduction to Philosophy*, London & N.Y. 1911).

Jerusalem, Wilhelm, *Laura Bridgman*, Vienna 1890.

Jerusalem, Wilhelm, *Lehrbuch der Psychologie*, 3rd ed. Vienna 1902.

Jevons, William Stanley, *The principles of science*, London 1892.

Jones, Bence, *The Life of Faraday*, London 1870.

Joule, James Prescott, 'On the changes of temperature produced by the rarefaction and condensation of air', *Phil. Mag.* (1845).

Kahlbaum, George Wilhelm August and Schaer, Eduard, *Ch. F. Schönbein, Ein Blatt zur Geschichte des 19. Jahrhunderts*, 1901.

Kant, Immanuel, *Kritik der reinen Vernunft*, [1]1781, [2]1787
Kant, Immanuel, *Prolegomena zu einer jeden künftigen Metaphysik*, 1783 (Various E.TT.).
Kant, Immanuel, *Metaphysische Anfangsgründe der Naturwissenschaft*, 1786.
Kapp, Ernst, *Grundlinien einer Philosophie der Technik*, Brunswick 1877.
Keibel, M., 'Die Abbildtheorie u. ihr Recht in d. Wissenschaftslehre', *Zeitschr. f. immamente Philos.* **3** (1898).
Kekulé von Stradonitz, August, Reply on the occasion of his Jubilee, *Berichte d. Deutschen chem. Gesellschaft*, **23** (1890), pp. 1306f.
Kempelen, Wolfgang *(ritter)* von, *Mechanismus der menschlichen Sprache, nebst Beschreibung einer sprechenden Maschine*, Vienna 1791.
Kepler, Johann, *Ad Vitellionem paralipomena*, 1604.
Kepler, Johann, *Astronomia Nova, De Motibus Stellae Martis*, 1609.
Kepler, Johann, *Dioptrice*, 1611.
Kepler, Johann, *Epitome astronomiae Copernicanae*, 1619 (E.T. of Books IV & V by E. T. Wallis, Annapolis 1939).
Kepler, Johann, *Harmonice Mundi*, 1619.
Kepler, Johann, *Mysterium cosmographicum*, 1596.
Kepler, Johann, *Opera*, ed. Frisch.
Killing, Wilhelm Karl Joseph, *Einführung in die Grundlagen der Geometrie*, Paderborn 1898.
Kircher, Athanasius, *Ars magna lucis et umbrae*, Amsterdam 1671.
Klein, Felix, *Andwendungen der Differential- und Integralrechnung auf Geometrie, eine Revision der Prinzipien*, Leipzig 1902.
Klein, Felix, *Ausgewählte Fragen der Elementargeometrie*, Leipzig 1895.
Kleinpeter, Hans, *Die Erkenntnistheorie der Naturforschung der Gegenwart*, Leipzig 1905.
Knight, Thomas Andrew, 'On the inverted Action of the alburnous Vessels of Trees', *Philos. Transact. of the Royal Society*, 1806.
Kosack, C. R., *Beiträge zu einer systematischen Entwicklung der Geometrie aus der Anschauung*, Nordhausen 1852.
Koster, W., 'Zur Kenntnis der Mikropie und Makropie', *Graefes Archiv für Ophthalmologie*, **42** (1896).
Kreibig, Joseph Klemens, *Die Aufmerksamkeit als Willenserscheinung*, Vienna 1897.
Kroman, Kristian Frederik Vilhelm, *Unsere Naturkenntnis*, Copenhagen 1883.
Kronecker, Leopold, 'Über Systeme von Funktionen mehrerer Variablen', *Berlin Berichte* 1869.
Kronecker, Leopold, 'Über den Zahlbegriff' in *Philosophische Aufsätze E. Zeller, gewidmet*, 1887.
Kulke, E., *Über die Umbildung der Melodie*, Prague 1884.
Kussmaul, Adolf, *Störungen der Sprache*, Leipzig 1885 (E.T. of earlier version in H. Ziemssen, *Cyclopaedia of the Practice of Medicine*, N.Y. 1877).
La Mettrie, Julien Offray de, *Oeuvres philosophiques*, Berlin 1796.
Lange, Ludwig, *Die geschichtliche Entwicklung des Bewegungsbegriffes*, Leipzig 1886.
Lanner, Alois, *Die wissenschaftlichen Grundlagen des ersten Rechenunterrichts*, Vienna & Leipzig 1905.
Laplace, Pierre Simon (*marquis*) de, *Essai philosophique sur les probabilités*, 6th ed. Paris 1840.
Lea, Henry Charles, *A History of the Inquisition of the Middle Ages*, New York 1888.

Leibniz, Gottfried Wilhelm (*freiherr*) von, *Mathematische Schriften*, ed. Gerhardt, Berlin 1849.

Le Sage, George Louis, 'Sur la méthode d'hypothèse', reprinted in P. Prévost, *Essai de philosophie*, Geneva an XIII [1805].

Liebig, Justus (*freiherr*) von, 'Induktion und Deduktion', *Reden und Abhandlungen*, 1874 (E.T. in *Smithsonian Institution Annual Report*, 1870).

Lieh-tzŭ (Licius), *Der Naturalismus bei den alten Chinesen... oder die sämmtlichen Werke des Philosophen Licius*, trsl. Ernst Faber, Elberfeld 1877 (E.T. of some parts by L. Giles, *Taoist Teachings from the Book of Lieh-Tzŭ*, London 1912 &c).

Listing, Johann Benedict, *Vorstudien zur Topologie*, Göttingen 1847.

Lloyd Morgan, Conwy, *Animal Life*, London 1891.

Lloyd Morgan, Conwy, *Comparative Psychology*, London 1894.

Lobachevsky, Nikolai Ivanovich, *Zwei geometrische Abhandlungen*, ed. F. Engel, Leipzig 1899 (Various E.TT. from Russian, *New Principles of Geometry* &c).

Loeb, Jacques, 'Geotropismus der Tiere', *Pflügers Archiv* (1891).

Loeb, Jacques, *Heliotropismus der Tiere*, Würzburg 1890.

Loeb, Jacques, 'Orienterung der Tiere gegen die Schwerkraft' *SB. d. ph.-med. Gesellschaft zu Würzburg* **1** (1888) 5.

Loeb, Jacques, *Vergleichende Gehirnphysiologie*, Leipzig 1899 (E.T. *Comparative Physiology of the Brain*, N.Y. 1900 &c).

Lotze, Hermann, *Medizinische Psychologie*, 1852.

Lotze, Hermann, *Mikrokosmos* 1856 (E.T. by E. Hamilton and E. E. Constance Jones, Edinburgh 1865 &c).

Lotze, Hermann, *Die vorgeschichtliche Zeit*, Jena 1874 (G.T. of *Pre-historic Times*, London & Edinburgh 1865 &c).

Lubbock, (*Sir*) John, (Lord Avebury), *Die Entstehung der Zivilisation*, Jena 1875 (G.T. of *The Origin of Civilization*, London 1870 &c).

Lucian, *Demonax, A True Story, The Lover of Lies.*

Mann, Friedrich, *Abhandlungen aus dem Gebiet der Mathematik*, Würzburg 1882.

Mann, F., *Die logischen Grundoperationen der Mathematik*, Erlangen & Leipzig 1895.

Van Marum, Martin, *Déscription d'une très grande machine électrique*, 1785.

Mason, Otis T., *The Origins of Invention*, London 1895.

Mauthner, Fritz, *Beiträge zu einer Kritik der Sprache*, Stuttgart 1901.

Maxwell, James Clerk, *A Treatise on Electricity and Magnetism*, Oxford 1873.

Maxwell, James Clerk, 'Dynamical Theory of the Electromagnetic Field', *Phil. Trans. of the Royal Society*, 1865.

Maxwell, James Clerk, 'On Faraday's Lines of Force', *Transact. of the Cambridge Philos. Soc.* **10** (1855).

Menger, Anton, *Neue Sittenlehre*, Jena 1905.

Menger, Anton, *Neue Staatslehre*, Jena 1902.

Metschnikoff, E., Il'ya Il'ich (E. Metchnikoff), *Studium über die Natur des Menschen, Eine optimistische Philosophie*, Leipzig 1904 (E.T. from Russian by P. Chalmers Mitchell, *The Nature of Man*, London & N.Y., 1904).

Meynert, Theodor, *Populärwissenschaftliche Vorträge*, Vienna 1892.

Mill, John Stuart, *System der deduktiven und induktiven Logik*, G.T. by Th. Gomperz, Leipzig 1884 of *A System of Logic* 1843 &c.

Möbius, Carl, *Die Bewegungen der Tiere und ihr psychischer Horizont*, Naturwissensch. Verein f. Schleswig-Holstein 1873.

Morgan, *see* Lloyd Morgan.

Müller, Hermann, *Befruchtung der Blumen durch Insekten*, Leipzig 1873 (E.T. by D. A. W. Thompson, *The Fertilisation of Flowers*, London 1883).
Müller, J., *Über die phantastischen Gesichtserscheinungen*, Koblenz 1826.
Müller, Johann, III, *Handbuch der Physiologie*, Koblenz 1840.
Münsterberg, Hugo, *Beiträge zur experimentellen Psychologie*, 1889.
Münsterberg, Hugo, *Die Willenshandlung*, Freiburg 1882.
Natorp, Paul, *Die logischen Grundlagen der exakten Wissenschaften*, Leipzig 1910.
Naville, Ernst, *La logique de l'hypothèse*, 2nd ed. Paris 1895.
Needham, John Turberville, *New Microscopical Discoveries*, London 1745.
Newton, (*Sir*) Isaac, *Arithmetica Universalis*, 1732.
Newton, (*Sir*) Isaac, *Opera*, ed. Horsely, London 1782.
Newton, (*Sir*) Isaac, *Optice*, London 1719 (2nd and enlarged Latin edition).
Newton, (*Sir*) Isaac, *Philosophiae naturalis Principia mathematica*, 1687.
Noiré, Ludwig, *Logos. Ursprung und Wesen der Begriffe*, Leipzig 1885.
Noll, Fritz, 'Über Geotropismus', *Jahrb. f. wissensch. Botanik* **34** (1900).
Oelzelt-Newin, Anton, *Kleinere philosophische Schriften*, Vienna 1901.
Oelzelt-Newin, Anton, *Uber die Phantasie-Vorstellungen*, Graz 1889.
Ofterdinger, Ludwig Feliks, *Beiträge zur Geschichte der griechischen Mathematik*, Ulm 1860.
Oppel, J. J., 'Neue Beobachtungen und Versuche über eine eigenthümliche, noch wenig bekannte Reactionsthätigkeit des menschlichen Auges', *Poggendorffs Annalen* **99** (1856).
Örstedt, H. C., 'Experimenta circa effectum conflictus electrici in acum magneticam', *Gilberts Annalen* **66** (1820).
Ostwald, (Friedrich) Wilhelm, *Vorlesungen über Naturphilosophie*, 1902.
Palágyi, Menyhért, *Neue Theorie des Raumes und der Zeit*, Leipzig 1901.
Pascal, Blaise, *Expériences nouvelles touchant le vide*, Paris 1647.
Pasch, Moritz, *Vorlesungen über neuere Geometrie*, Leipzig 1882.
Pasteur, Louis, 'Mémoire sur les corpuscules organisés qui existent dans l'atmosphère &c', *Ann. de chimie et de physique*, 3rd ser., **64** (1862).
Pearson, Karl, *The Grammar of Science*, 2nd ed. London 1900.
Pauli, Wolfgang, 'Allgemeine Physiko-Chemie der Zelten und Gewebe' in *Ergebnisse der Physiologie* **III.I** (1904).
Peltier, Jean Charles Athanase, 'Électricité' & 'Courans électriques' in *L'Institut* (1834) 26: iv & 16: viii.
Petronius, *Cena Trimalchionis*.
Petzoldt, Joseph, 'Das Gesetz der Eindeutigkeit', *Vierteljahrschr. f. wissensch. Philosophie* **19** (1895).
Petzoldt, Joseph, 'Solipsismus auf praktischem Gebiet', *Vierteljahrsschrift f. wissensch. Philosophie* **25** (1901).
Plateau, J., 'Vierte Notiz über neue, sonderbare Anwendungen des Verweilens der Eindrücke auf die Netzhaut', *Poggendorffs Annalen* **80**.
Plato, *Euthydemus, Gorgias*.
Plautus, *Menaechmi*.
Plutarch, 'On the face in the orb of the moon'.
Poinsot, Louis, *Eléments de Statique*, 10th ed. Paris 1861 (E.T. of earlier edn. by T. Sutton, Cambridge 1847).
Poisson, Simon Denis, 'Mémoire sur la Distribution de l'Électricité à la Surface des Corps Conducteurs', *Mém. de l'Institut* **12** (1811).

Popper-Lynkeus, Joseph, *Fundament eines neuen Staatsrechts*, 1905.

Popper-Lynkeus, Joseph, *Das Recht zu leben und die Pflicht zu sterben*, Leipzig 1878.

Popper-Lynkeus, Josef, *Die technischen Fortschritte nach ihrer ästhetischen und kulturellen Bedeutung*, Leipzig 1888.

Powell, John Wesley, *Truth and Error*, Chicago 1898.

Preyer, William Thierry, *Die Seele des Kindes*, Leipzig 1882 (E.T. by H. W. Brown, *The Mind of the Child*, London 1894).

Priestley, Joseph, *History and present state of discoveries relating to vision, light and colours*, London 1772.

Reimarus, Hermann Samuel, *Triebe der Tiere* 1790.

Reuter, Hermann Ferdinand, *Geschichte der religiösen Aufklärung in Mittelalter*, Berlin 1875–7.

Ribot, Théodule Armand, *L'évolution des Idées générales*, Paris 1897 (E.T. by F. A. Welby, London 1899).

Ribot, Théodule Armand, *Essai sur l'imagination créatrice*, Paris 1900 (E.T. by A. H. N. Baron, London & Chicago 1906).

Ribot, Théodule Armand, *Les maladies de la personnalité*, Paris 1888.

Ribot, Théodule Armand, *Maladies de la mémoire*, Paris, 1888.

Ribot, Théodule Armand, *Maladies de la volonté*, Paris 1888.

Rickert, Heinrich, 'Zur Theorie der naturwissenschaftlichen Begriffsbildung', *Viertelj. f. wiss. Philosoph.* **18** (1894).

Riehl, Alois, *Der philosophische Kritizismus*, vol. II, (E.T. by A. Fairbanks, *The Principles of Critical Philosophy*, n.p. 1894).

Riemann, Georg Friedrich Bernhard, *Über die Hypothesen, welche der Geometrie zu Grunde liegen*, Göttingen 1867.

Robert-Houdin, Jean Eugène, *Comment on devient sorcier*, Paris 1882 (E.T. by 'Professor Hoffmann', *The Secrets of Conjuring and Magic*, London 1878 &c).

Robert-Houdin, Jean Eugène, *Confidences d'un prestidigitateur*, Paris 1881, I (E.T. by (*Sir*) F. C. L. Wraxall, *Memoirs of Robert-Houdin*, London 1859 &c).

Roskoff, Georg Gustav, *Geschichte des Teufels*, Leipzig 1869.

Roux, Wilhelm, *Vorträge und Aufsätze über Entwicklungsmechanik*, 1905.

Rudio, Ferdinand, *Geschichte des Problems der Quadratur des Zirkels*, Leipzig 1892.

Russell, Bertrand Arthur William, *The Principles of Mathematics*, Cambridge 1903.

Saccheri, Giovanni Girolamo, *Euclides ab omni naevo vindicatus*, Milan 1733 (E.T. by G. B. Halsted, London & Chicago 1920).

Sachs, Ferdinand Gustav Julius von, *Vorlesungen über Pflanzen-Physiologie*, Leipzig 1887 (E.T. by H. M. Ward, *Lectures on the Physiology of Plants*, Oxford 1887).

Schaer, *see* Kahlbaum.

Schmidt, F. J., *Grundzüge der konstitutiven Erfahrungsphilosophie*, Berlin 1901.

Schneider, Georg, *Die Zahl im grundlegenden Rechenunterricht*, Berlin 1900.

Schneider, Georg Heinrich, *Der tierische Wille*, Leipzig 1880.

Schneider, K. C., 'Das Wesen der Zeit', *Wiener klinische Rundschau* **19** (1905).

Schönflies, Arthur M., 'Die Entwicklung der Lehre von den Punktmannigfaltigkeiten', *Jahrb. d. Deutschen Mathematiker-Vereinigung* **8** (1900).

Schopenhauer, Arthur, *Über den Willen in der Natur*, Frankfurt 1836 (E.T. by K. Hillebrand, *On the Will in Nature*, 1889).

Schopenhauer, Arthur, *Über die vierfache Wurzel des Satzes vom zureichenden Grunde*, Leipzig 1864 (E.TT. by K. Hillebrand, *On the Fourfold Root &c*, 1889 and by E. F. J. Payne, La Salle, Illinois, 1974).

Schram, J., *Leibnizens Definitionen der Ebene und der Geraden*, Obersteig 1903.
Schröder, Ernst, 'Note über den Operationskreis des Logikkalküls', *Math. Annal.* **12** (1877).
Schuppe, (Ernst Julius) Wilhelm, *Erkenntnistheoretische Logik*, Bonn 1878.
Schuppe, (Ernst Julius) Wilhelm, *Grundriss der Erkenntnistheorie und Logik*, Berlin 1894.
Schuppe, (Ernst Julius) Wilhelm, 'Meine Erkenntnistheorie und das bestrittene Ich', *Zeitschr. f. Psychologie u. Physiologie der Sinnesorgane* **35** (1904).
Schuppe, (Ernst Julius) Wilhelm, 'Der Solipsismus', *Zeitschr. für immanente Philosophie* III (1898) 327.
Scott, (*Sir*) Walter, (*Bart.*), *Letters on Demonology and Witchcraft*, 4th ed., London 1898.
Seebeck, Thomas Johann, 'Uber den Magnetismus der galvanischen Kette', *Berlin Abhandlungen*, 1820, 1821.
Semon, Richard Wolfgang, *Die Mneme*, Leipzig 1904.
Shakespeare, *Comedy of Errors, Taming of the Shrew.*
Siegel, Carl, *Über Raumvorstellung und Raumbegriff*, Leipzig 1905.
Siegel, Carl, 'Versuch einer empiristischen Darstellung der räumlichen Grundgebilde etc...', *Vierteljahrschr. f. wiss. Philosophie* **24** (1900).
Simon, Paul Max, *Le Monde des rêves*, Paris 1888.
Snell, Carl, *Lehrbuch der Geometrie*, Leipzig 1869.
Soldan, Wilhelm Gottlieb, *Geschichte der Hexenprozesse*, Stuttgart 1843.
Spallanzani, Lazzaro, *Opuscules de Physique animale et végétale*, 1777 (E.T. from Italian by J. G. Dalyell, *Traits on the Nature of Animals and Vegetables*, Edinburgh 1799).
Spear, W. W., Advanced Arithmetic, Boston 1899.
Sprenger, Jacob, *Malleus maleficarum*, 1474 &c.
Stallo, Johann Bernhard, *The Concepts and Theories of Modern Physics*, 1862; German ed., *Die Begriffe und Theorien der modernen Physik*, by H. Kleinpeter with preface by E. Mach, Leipzig 1901.
Spencer, Herbert, *The Principles of Psychology*, London 1870.
Staeckel, *see* Engel.
Steinen, Carl von den, *Unter den Naturvölkern Zentral-Brasiliens*, Berlin 1897.
Steiner, Jakob, *Systematische Entwicklung der Abhängigkeit der geometrischen Gestalten von einander*, Berlin 1832.
Stern, L. William, *Die Analogie im volkstümlichen Denken*, Berlin 1893.
Stern, L. William, *Beiträge zur Psychologie der Aussage*, 1903.
Sterneck, Robert von, 'Über die Elemente des Bewusstseins', *Wissenschaftl. Beilage, Jahresbericht der philosophischen Gesellschaft an d. Univ. Wien* **16** (1903) 77.
Sterneck, Robert von, 'Versuch einer Theorie der scheinbaren Entfernungen', *Ber. d. Wiener Akademie, math. naturw. Cl.* **114** (1905).
Stöhr, Adolf, *Algebra der Grammatik*, Vienna 1898.
Stöhr, Adolf, *Leitfaden der Logik in psychologisierender Darstellung*, Vienna 1905.
Stolz, Otto, 'Das letzte Axiom der Geometrie', *Berichte des naturw.-medicin. Vereins zu Innsbruck*, 1886, pp. 25–34.
Stolz, Otto, *Grössen und Zahlen*, Leipzig 1891.
Strabo.
Strassen, O., zur 'Über die Mechanik der Epithelbildung,' *Verh. d. D. Zoolog. Gesellsch.* **13** (1903).
Strümpell, Adolf, 'Beobachtungen über ausgebreitete Anästhesien und deren Folgen

für die willkürlichen Bewegung und das Bewusstsein', *Deutsch. Archiv f. klin. Medic.* **22** (1878).

Stumpf, Carl, *Uber den physiologischen Ursprung der Raumvorstellungen*, Leipzig 1873.

Suárez, Franciscus, *Disputationes metaphysicae, Opera*, Tom. 22, 23, Venice 1751.

Sundara Rāu, T., *Geometric Exercises in Paper-Folding*, Chicago 1901.

Swoboda, Hermann, *Die Perioden des menschlichen Organismus*, 1904.

Thibaut, Bernhard Friedrich, *Grundriss der reinen Mathematik*, Göttingen 1809.

Thomson, William (Lord Kelvin), *On the Dynamical Theory of Heat*, 1852.

Thomson, William (Lord Kelvin), 'On the uniform motion of heat in solid bodies and its connection with the mathematical theory of electricity', *Cambridge Math. Journal* **III** (1842).

The Thousand and One Nights.

Tilly, J. M. de, *Essai sur les principes fondamentaux de la géometrie et de la mécanique, Mémoires de la société des sciences physiques et naturelles de Bordeaux*, 1880.

Tissandier, Gaston, *La physique sans appareils*, Paris 7th ed. (E.T. by H. Frith, *Half Hours of Scientific Amusement*, London 1890).

Tylor, *(Sir)* Edward Burnett, *Einleitung in das Studium der Anthropologie*, Brunswick 1883 (E. Original, *Anthropology*, London 1881).

Tylor, (*Sir*) Edward Burnett, *Anfänge der Kultur* (E. Original, *Primitive Culture*, London 1871).

Tylor, (*Sir*) Edward Burnett, *Urgeschichte der Menschheit*, Leipzig (n.d.) (E. Original, *Researches into the Early History of Mankind*, London 1865).

Ueberweg, Friedrich, 'Die Principien der Geometrie wissenschaftlich dargestellt', *Archiv für Phil. und Pädag.*, 1851 (reprinted in Brasch above).

Vaihinger, Hans, *Die Philosophie des Als-Ob*, Berlin 1911 (E.T. by C. K. Ogden, *The Philosophy of 'As if'*, London 1924).

Vailati, Giovanni, 'Il Pragmatismo in Germania' [review of *Erkenntnis und Irrtum* in] *Leonardo* **3** (1905) 193.

Vaschide, N., and Vurpas, Cl., *Essai sur la Psycho-Physiologie des humains*, Paris (n.d.).

Veraguth, Otto, 'Über Mikropie und Makropie', *Deutsche Zeitschrift für Nervenheilkunde von Strümpell* **24** (1903).

Verworn, Max, *Naturwissenschaft und Weltanschauung*, 1904.

Vitruvius, *De architectura* **V**, cap. III.

Volkmann, Paul, *Einführung in das Studium der theoretischen Physik*, Leipzig 1900.

Volta, Alessandro Giuseppe Antonio Anastaso (*count*), 'On the Electricity excited by mere Contact &c' *Philos. Transact. of the Royal Society*, 1800.

Voltaire, *Dictionnaire philosophique*, 1764 (Various E.TT.).

Wagner, Rudolph, *Handwörterbuch der Physiologie.*

Wallaschek, Richard, *Primitive Music*, London 1893. (Enlarged German edition, Leipzig 1903.)

Wallaschek, Richard, *Psychologie und Pathologie der Vorstellung*, Leipzig 1905.

Wallis, John, *Algebra*, London 1685 (Latin T. Oxford 1693).

Wallis, John, *Arithmetica infinitorum*, Oxford 1655.

Sartorius von Waltershausen, Wolfgang (*baron*), *Gauss zum Gedächtnis*, Leipzig 1856.

Weber, Ernst Heinrich, 'Über den Raumsinn und die Empfindungskreise in der Haut und im Auge', *Ber. d. kgl. sächs. Gesellsch. d. Wissenschaften, math. naturw. Cl.* (1852).

Weissenborn, Hermann, *Principien der höheren Analysis in ihrer Entwickelung*, Halle 1856.

Wernicke, Carl, *Gesammelte Aufsätze, Berlin* 1893.
Whewell, William, *Geschichte der induktiven Wissenschaften*, Stuttgart 1840 (E. Original, *History of the Inductive Sciences*, London 1837).
Whewell, William, *On the Philosophy of Discovery*, London 1860.
Whewell, William, *The Philosophy of the Inductive Sciences*, 2nd edn. London 1847.
Whitney, William Dwight, *Leben und Wachstum der Sprache*, Leipzig 1876 (E. Original, *The Life and Growth of Language*, London 1875).
Wiener, Otto, *Die Erweiterung der Sinne*, Leipzig 1900.
Wilbrand, Hermann, *Die Seelenblindheit als Herderscheinung*, Wiesbaden 1887.
Wlassak, R., 'Die statischen Funktionen des Ohrlabyrinthes II', *Vierteljahrsch. f. wiss. Philosophie* **17** (1893).
Wohlwill, Emil, *Galilei und sein Kampf für die Kopernikanische Lehre*, Hamburg 1909.
Wuttke, Heinrich, *Geschichte der Schrift*, Leipzig 1872.
Zell, Th., (Leopold Bauke), *Das rechnende Pferd*, Berlin.
Zell, Th., (Leopold Bauke), 'Ist das Tier unvernünftig' in *Tierfabeln*, Stuttgart.
Ziehen, Th., 'Erkenntnistheoretische Auseinandersetzungen', *Zeitschr. f. Psychologie u. Physiologie der Sinnesorgane* **33**, (1903).
Zindler, Conrad, 'Beiträge zur Theorie der mathematischen Erkenntniss', *Wiener Sitzungsberichte* (Phil.-hist. Abt.) **118** (1889).
Zola, Émile Édouard Charles Antoine, *Le Roman expérimental*, Paris 1898 (E.T. by B. M. Sherman, *The Experimental Novel*, N.Y. 1893).
Zoth, O., 'Über den Einfluss der Blickrichtung auf die scheinbare Grösse der Gestirne und die scheinbare Form des Himmelgewölbes', *Pflügers Archiv* **78** (1899).

INDEX OF NAMES

VIENNA CIRCLE COLLECTION

1. OTTO NEURATH, *Empiricism and Sociology*. Edited by Marie Neurath and Robert S. Cohen. With a Selection of Biographical and Autobiographical Sketches. Translations by Paul Foulkes and Marie Neurath. 1973, xvi+473 pp., with illustrations. ISBN 90 277 0258 6 (cloth), ISBN 90 277 0259 4 (paper).
2. JOSEF SCHÄCHTER, *Prolegomena to a Critical Grammar*. With a Foreword by J. F. Staal and the Introduction to the original German edition by M. Schlick. Translated by Paul Foulkes. 1973, xxi+161 pp. ISBN 90 277 0296 9 (cloth), ISBN 90 277 0301 9 (paper).
3. ERNST MACH, *Knowledge and Error. Sketches on the Psychology of Enquiry*. Translated by Paul Foulkes. ISBN 90 277 0281 0 (cloth), ISBN 90 277 0282 9 (paper).
5. LUDWIG BOLTZMANN, *Theoretical Physics and Philosophical Problems. Selected Writings*. With a Foreword by S. R. de Groot. Edited by Brian McGuinness. Translated by Paul Foulkes. ISBN 90 277 0249 7 (cloth), ISBN 90 277 0250 0 (paper).
6. KARL MENGER, *Morality, Decision, and Social Organization. Toward a Logic of Ethics*. With a Postscript to the English edition by the Author. Based on a translation by E. van der Schalie. ISBN 90 277 0318 3 (cloth), ISBN 90 277 0319 1 (paper).